ANZAC MEMORIES

ANZAC MEMORIES

MEMORIES

Living with the Legend

– NEW EDITION –

ALISTAIR THOMSON

MONASH University Publishing

Building 4, Monash University
Clayton, Victoria 3800, Australia
www.publishing.monash.edu

Monash University Publishing brings to the world publications which advance the best traditions of humane and enlightened thought.

Monash University Publishing titles pass through a rigorous process of independent peer review.

www.publishing.monash.edu/books/am-9781921867583.html

Series: Monash Classics

Design: Les Thomas

Cover image: Steve Siewart/Fairfax Syndication.

Acknowledgments

The author and publisher thank copyright holders for granting permission to reproduce illustrative material. Every effort has been made to trace the original source of copyright material contained in this book. The publisher would be pleased to hear from copyright holders to rectify any errors or omissions.

National Library of Australia Cataloguing-in-Publication Data:

Author:	Thomson, Alistair.
Title:	Anzac memories: living with the legend / Alistair Thomson
Edition:	New ed.
Notes:	Includes index and bibliography.
ISBN:	9781921867583 (pbk.)
Subjects:	
	Australia. Army. Australian and New Zealand Army Corps.
	World War, 1914–1918 — Campaigns — Turkey — Gallipoli Peninsula
	World War, 1914–1918 — Australia — Historiography;
	World War, 1914–1918 — Personal narratives, Australian;
	Soldiers — Australia — Interviews;
	Veterans — Australia.
	National characteristics, Australian.
	Australia — History — 20th century
Dewey Number:	940.41294

Printed in Australia by Griffin Press an Accredited ISO AS/NZS 14001:2004 Environmental Management System printer.

PRAISE FOR THE FIRST EDITION OF
ANZAC MEMORIES

'As good a picture of the impact of the Great War on individuals and
Australia as we are likely to get in this generation.'

— Michael McKernan, *Sydney Morning Herald*

'An immense achievement of this book is that it so clearly illuminates the
historical processes that left men like my grandfather forever struggling to
fashion myths which they could live by.'

— Michael Roper, *Oral History*

'One of the last oral histories of the generation that fought in World War
I, it is also one of the best, a superb account of the interaction of private
memory with public myth.'

— Richard White, *American Historical Review*

'The intricate assessment of the experience of war and its persistence as
a memory in survivors' lives is well-demonstrated by Alistair Thomson's
quite exceptional recent book *Anzac Memories*.'

— Patrick Wright, *The Guardian*

'A wonderfully subtle and engaging account of the Anzac tradition.'

— Alan Atkinson, *The Commonwealth of Speech*

'Surely one of the decade's most important books in Australian history.'

— Peter Stanley, *Australian Historical Studies*

CONTENTS

CONTENTS

ACKNOWLEDGMENTS

New edition

Two decades on and many more people have helped with this new edition. My colleagues in History at Monash University have provided a wonderfully stimulating intellectual community, and I'm especially grateful to members of our research group who have read drafts. Thanks also to members of our Melbourne Life Writing Group for lively discussion and generous support.

Archivists have helped with access to the sources for the new chapters in part IV, and thanks in particular to Jessica Reid at the National Archives of Australia, Katie Wood at the University of Melbourne Archives and Stephanie Boyle in the Film and Sound Section at the Australian War Memorial. Bill Langham's niece Margaret Paulsen provided invaluable family history background. Thanks also to the team at Monash University Publishing, including Sarah Cannon, Nathan Hollier and Joanne Mullins.

For comments on draft chapters, and support in many other ways, I thank: Bain Attwood, Johnny Bell, Frank Bowden, Ian Britain, Vicki Davies, Graham Dawson, Siân Edwards, Susan Foley, Ruth Ford, Peg Fraser, David Garrioch, Jim Hammerton, Mike Hayler, Carolyn Holbrook, Katie Holmes, Ken Inglis, Marina Larsson, Jim Mitchell, John Rickard, Bruce Scates, Dorothy Sheridan, Chips Sowerwine, Peter Stanley, Campbell Thomson, Judy Thomson, Christina Twomey and Jay Winter.

I dedicate this new edition to my father David Thomson, and to his parents Hector and Nell.

First edition

I would like to thank the following people who have influenced my understanding of Anzac. Most of the Great War veterans whom I interviewed between 1982 and 1987 have died in the intervening years, but I am profoundly grateful to all of them for sharing their life stories with me. In particular, I thank Percy Bird, Fred Farrall, Fred Hocking, Bill Langham and Ern Morton for participating in my second 'popular memory' interviews with so much enthusiasm and care, and Stan D'Altera and James McNair for their early inspiration. Percy Bird's daughter, Kath Hunter, reminded me of the value of oral history for families—and the role which families play in this work—and I thank her and many other family members who were encouraging and supportive. Anna Young also deserves special mention for her painstaking and sensitive transcription of most of the tapes.

The Australian War Memorial gave me a research grant which paid for tapes and transcription, and I particularly thank Matthew Higgins and Peter Stanley of the Research and Publications Section for their support. I also thank the Association of Commonwealth Universities for a postgraduate scholarship, and the Northcote Scholarship Scheme for providing me with a return air fare to Australia to complete the research in 1990. The Trustees of the Liddell-Hart Centre for Military Archives granted permission for the use of a number of quotations. My parents, Judy and David Thomson, offered practical assistance in times of need, and I value their support for a project that touched their own histories and was sometimes painful and disconcerting. Thanks also to my colleagues in the Centre for Continuing Education at the University of Sussex, for encouraging and supporting me to finish a project that had few obvious connections with the needs of adult learners in Sussex.

Many people helped me to develop the ideas in this book by offering advice and editorial or typing assistance, and for their support I thank: Fi Black, Joanna Bornat, Ruth Brown, Stephanie Brown, Bob Bushaway, Angus Calder, Jane Carver, Mike Cathcart, Drew Cottle, Martin Evans, Stephanie Gilpin, Mike Hayler, Chris Healy, Mary Hoar, Katie Holmes, Ursula Howard, Alun Howkins, Rod Kedward, Terry King, John Lack, Marilyn Lake, Ross McMullin, Jim Mitchell, John Murphy, Richard Nile, Lucy Noakes, Nick Osmond, Rob Perks, Jeff Popple, Lloyd Robson, Alan Seymour, Glenda Sluga, Cliff Smyth, Chips Sowerwine, Mary Stuart, David Thomson, Judy Thomson, Denis Winter and Stephen Yeo. Campbell Thomson brought old photos back to life, and Peter Rose and Katherine Steward at Oxford University Press were supportive and constructive editors. Above all, the approach of this book was influenced by discussions in my reading group, and I am deeply grateful to Graham Dawson and Dorothy Sheridan for their unstinting intellectual and personal support.

ABBREVIATIONS

ADCC	Anzac Day Commemoration Committee
AIDS	Acquired Immunodeficiency Syndrome
AIF	Australian Imperial Force
ANZAC	Australian and New Zealand Army Corps
ASIO	Australian Security Intelligence Organisation
AWL	Absent Without Leave
AWM	Australian War Memorial, Canberra
DVA	Department of Veterans' Affairs
MO	Medical Officer
NAA	National Archives of Australia
NLA	National Library of Australia
Repat	Repatriation Department
RGHH	Repatriation General Hospital Heidelberg
RSL	Returned and Services League
RSSILA	Returned Sailors' and Soldiers' Imperial League of Australia
TB	Tuberculosis
TPI	Totally and Permanently Incapacitated
UMA	University of Melbourne Archives
VC	Victoria Cross

NOTE TO READERS

It should be noted that within quotations in the text, ellipses in square brackets ([...]) indicate an omission, and ellipses without square brackets (...) indicate a pause in speech.

Where an entry in the notes consists solely of a name and page number (e.g. Stabb, pp. 6–7.), it refers to a transcript of an interview. A number following the name in such entries (e.g. Langham 2, p. 6.) indicates a particular interview in a series. See the bibliography for details.

FOREWORD:

MEMORY AND SILENCE

Jay Winter

'Memory is not an instrument for exploring the past', wrote Walter Benjamin, 'but its theatre. It is the medium of past experience, as the ground is the medium in which dead cities lie interred'.[1] This claim is particularly true in the case of oral history, which replays the past through a three-way conversation. The first party is the historian or archivist, who poses the questions, or encourages the subject being interviewed to develop a point or turn to another one. The second party is the interviewee, as she is at the moment of the interview. The third, and most quicksilver presence, is the interviewee, as she was at the particular point in time in the past on which the interview dwells. This triangulation is what gives such power to the record we have of these exchanges.

This is not at all an original or surprising observation to the large and growing number of oral historians who have done so much to enrich our understanding of the past. But Alistair Thomson's book adds a fourth party to the conversation and, by doing so, breaks entirely new ground in the retrieval and interpretation of narratives of the Great War.

What is new is that in this book we follow the interviewer at two points in time. The first is the moment when he did his pioneering work between 1982 and 1987 in collecting an archive that, once digitised, will be available at the Australian War Memorial to all those in the field. The second is the moment when, as Edmund Blunden put it, he went back over the ground again, and explored materials, both oral and written, which enabled him to write about matters excluded 20 years ago.

This quadrilateral of remembrance is striking. First we have the men as they were in the 1980s interviews; then we have the self each of these men recalled, living through the Great War and its long aftermath. These two sides of the encounter — parallel lines if you will — meet Alistair Thomson

---

1   Walter Benjamin, 'A Berlin chronicle', in *One-way street and other writings*, translated by Edmund Jephcott and Kingsley Shorter, Verso, London, 1979, p. 314.

at two points in time in his life; and his story, again in two parts, completes the quadrilateral. He is there in the 1980s, tape recorder running, listening to these life histories and interpreting them in his research journal and then in the first edition of the book. But in the new edition, we have a fourth participant, Thomson in 2013 adding significantly to the book by exploring the life of his grandfather in ways he could not put in print a generation ago.

A similar story might emerge were a psychoanalyst and his analysand to take up the challenge of dealing with the past a second time, two decades after their first encounter. When they start again, they have the memory of their earlier exchanges with them, and their second exchanges reflect who they have become since.

It is in part this layering of remembrance over time which gives this new edition much of its power and its originality. But there is a second way in which in this second edition Thomson has shifted around the furniture in the field of memory studies. He has injected into the narrative the way the stories — both his and his subjects' stories — have silence at their core.

All families, I believe, are defined by their silences. We all know about family photograph albums which need commentary to make sense of them. What someone needs to point out in these images is what and who they do *not* show. Pierre Bourdieu once remarked that with such albums in hand, this was the way mothers-in-law welcomed daughters-in-law into their families, by pointing out who does not talk to whom, or who never comes to various celebrations or funerals. Or who did something unmentionable in the past. What the new family member learns is to see who isn't there and to hear why.

What is taboo in any family or in any society is never fixed. And neither is that body of family information which everybody knows but no one talks about. Mental illness is one such subject, and it created a kind of fence around one central element of Thomson's work in the 1980s — his grandfather Hector's story. He has had the courage to take that fence down and use a range of sources to enter the no man's land of suffering and isolation which was a part of his grandfather's life, and perforce, that of his grandmother and the young child who became his father.

When the first edition was in preparation, Alistair Thomson's father objected strenuously to any mention in the book of his father's (Alistair's grandfather's) mental illness; reluctantly Alistair agreed to leave out the subject. We can understand why the author's father, himself a soldier, felt so strongly. The images were too hard to bear for the man who was a young boy in the 1930s, living through very, very hard times with his disturbed father after his mother's death. Now, afflicted with Alzheimer's disease, but still able to read the text, he gave his son permission to tell the story. And it is a compelling and important one.

From that story, we see the price families and in particular wives paid for the multiple wounds men brought home with them from war. What the new edition shows was the sheer force of survival in his grandmother Nell, who had not only the handful of two small boys to raise, but a damaged husband to support. And making her life harder still was that her husband's disability was very hard to define precisely. Was it malaria, contracted in Palestine? Was it an infection arising from a sequel to the Spanish flu of 1918, about which Oliver Sacks has written in *Awakenings*? Was it depression or a personality disorder? We will never know, because Hector Thomson lived at a time when psychiatric ailments were stigmatised. They still are today, but not to the degree that was the case in the 1920s and 1930s.

Demobilised soldiers knew this, and so did the doctors who tried to assess their war-related disability and validate or invalidate claims to war pensions. If a veteran had lost an arm or a leg, there was no problem; but physicians struggled harder with those who had unclear ailments, and who (like Hector) did not report their full medical condition before returning to civilian life.

This is why historians have consistently underestimated the number of soldiers who left the service suffering from various kinds of mental illness. Soldiers were loath to see themselves as 'mental cases', and doctors knew that their patients' chances of getting a disability pension were higher if they stayed away from this grey area, whenever possible. Hence the documentary record itself is silent on a central feature of the story of what Bill Gammage termed 'the broken years'. Broken indeed, and in the case of Hector Thomson, neither fully recognised nor fully treated, and never repaired.

We know that the damage war does to families is generational; it doesn't stop when the shooting stops. It is passed on indirectly from father to son to grandson, and to the women with whom they live. By retelling his family's story, Alistair Thomson has been able to fashion a moving portrait of his family: his grandmother Nell, and after her death, of their sons, Al's dad and his uncle, still children, having cold mutton for Christmas dinner, alone with their father, a soldier of the Great War.

The Argentine troubadour Atahualpa Yupanqui wrote a wonderful song entitled 'Le tengo rabia al silencio' — 'I feel anger towards silence'. That is what this book makes me feel. To be sure, most of us bypass the painful elements in our families' histories. Few have the chance and the courage and the skill to go back to a locked box, find the key, open it, and share with others the truths he finds. Alistair Thomson has done just that. We are in his debt for telling the story, the fuller story, a second time, and for showing us that breaking these silences is not a choice but a necessity.

Jay Winter
Yale University

INTRODUCTION TO THE
NEW EDITION

The past never alters, but memory and history change all the time. It is 30 years since I interviewed Australian First World War veterans, and a quarter century since I drafted this history of their lives and memories. Those men are all long-dead, almost all of them before the first edition of *Anzac Memories* was published in 1994. They had grown old on the cusp of a resurgence of interest in Australians at war; indeed, their ageing contributed to that interest and a concern about if and how the Great War might be remembered after their passing. Were they alive today, they would be astonished at the extent and influence of Australian war remembrance: the crowds which pack Anzac Day Dawn Services across the country and on overseas battlefields; bookshops with heaving shelves of military history and memoir; lavish government-funded memorial and education schemes; politicians invoking the Anzac tradition as they send Australians to war and peacekeeping; and a nation preparing for what will surely be Australia's largest ever commemorative occasion, the centenary in 2015 of the Anzac landing.

A new, concluding part of this book, 'Anzac memories revisited', considers changes over the past quarter century in Australian history and remembrance of the First World War. New sources have become available that enable historians to explore different aspects of the experience of Australians at war. Within the avalanche of Anzac writing there are some fine recent histories which offer new ways of understanding war and society in Australia's twentieth century and war remembrance in the twenty-first century. 'Memory studies' has boomed in the academy in recent decades — both reflective of and reflecting upon our autobiographical age — and academics from multiple disciplines have developed approaches to interpreting both individual and 'social' or 'collective' memory. My own perspective has changed too. I was a novice historian in my twenties (and an expatriate living in England) when I first wrote *Anzac Memories* as a doctoral thesis. Half a lifetime later, and now back in Australia, the new chapters are influenced by my life experience and the rather different vantage point of later life.

The first new chapter in part IV, chapter 10, 'Searching for Hector Thomson', revisits the autobiographical introduction of the first edition and uses Repatriation Department files to uncover the war and postwar story of my grandfather. It's a story about war damage, mental health and family tragedy that I was unable to tell in the first edition, and which I now use to highlight the impact of the war on postwar families, and the importance of family memory and history in Anzac remembrance today. Chapter 11, 'Repat war stories', is based on the Repatriation medical case files of Percy Bird, Bill Langham and Fred Farrall, First World War veterans whose life stories figured in the first edition. I explore what this new information adds to my earlier account of these men's wartime and postwar lives, and indeed test some of my conclusions from 25 years ago, including how veterans' war stories were recreated through their difficult relationship with the Repat. A Postscript offers concluding reflections on 'Anzac postmemory': how has remembrance of Australia's 1914–18 war and of the Anzacs changed since the last witnesses (the generation of men like Percy, Bill and Fred) died; how might we conceptualise Anzac mythology in the twenty-first century; and how can historians best contribute to public understanding of the war and its significance?

The first three parts of this book remain as they were in the first edition, except I have changed the title of part III, written in the early 1990s, from 'Anzac today' to 'Anzac comes of age', and I have added the years '1939–1990' to the title of chapter 8 about 'The Anzac revival'. The new edition is, in part, an artefact of its time and represents its original historical moment. *Anzac Memories* was one of the last oral histories of Australia's Great War and one of the first Australian oral histories to use memory not just to write about the past but also to explore the changing meanings of that past for individuals and in Australian society. As I explain in more detail in the introduction to the first edition, the three original parts of the book follow a chronological order: 'Making a legend' (wartime), 'The politics of Anzac' (the inter-war years), and 'Anzac comes of age' (from the Second World War through to the early 1990s). Each part comprises three chapters. The initial, oral history chapters in each part (1, 4 and 7) draw upon oral history interviews and outline the main features of the men's experiences as Anzacs during and after the war, and into old age. The second chapter in each part (2, 5 and 8) focuses on the making and re-making of the Anzac legend during and after the war, and into the 1990s. The third chapter in each part (3, 6 and 9) comprises what I call a 'memory biography', focusing in turn on three men: Percy Bird, Bill Langham and Fred Farrall. Here I explore the ways in which these veterans composed their war memories during and after the war, and into old age, and how those memories were influenced by the Anzac legend and by their own later life experiences and understandings.

The introduction to the first edition describes how I came to the topic through family war history, outlines the oral history interviews I conducted in the 1980s, explains how I used the interviews, and introduces Percy, Bill and Fred. The more detailed reflections about 'Oral history and popular memory' in the appendix represent the personal and intellectual context of the 1980s when I created and interpreted the interviews, and are also unchanged. The bibliography and index are updated to include new material from part IV.

Like most oral history books, *Anzac Memories* presents only the text of the stories told in interviews. The written transcript is a poor translation of an interview. It barely hints at the rich layers of meaning expressed in the aural exchange — through silence or emphasis, excitement or pathos — or the ways in which we communicate through face and body, with an arched eyebrow or eloquent hand. My audio cassette recordings could not capture non-verbal expression, though some of that I noted in my research diary of the time, and some of it is stamped in my memory of each occasion: James McNair resplendent in a silk dressing gown as he performed wartime songs; and Fred Farrall pointing to the First War discharge certificate which he had only recently hung up on his living room wall, next to pictures of Marx and Lenin. The Australian War Memorial is digitising my interviews and making the audio available online. Thanks to new technologies you can now download and listen to the original recordings and make your own sense of the meaning of sound as well as word. To access transcripts and audio go to www.awm.gov.au/collection/s01311/ for Fred Farrall; s01305/ for Percy Bird; and s01317/ for Bill Langham.

Each was a performer in his own way and keen to have his story on the record, and I imagine that my interviewees would have been amazed and pleased to know that their words and their voices would be so easily heard down the ages. And that through their interviews, and the ways in which I have interpreted and used oral history, they might continue, beyond living memory, to have an impact upon Australian war remembrance and history.

INTRODUCTION TO THE
FIRST EDITION

Growing up with the Anzac legend

I had a military childhood. For the first twelve years of my life, from 1960 until 1972, my father was a senior infantry officer in the Australian army. With my two brothers, I grew up in army barracks in different parts of Australia and around the world. We were surrounded by soldiers and soldiering. My earliest memories are of starched khaki and green-clad men parading across asphalt squares, trooping and wheeling to echoed commands. When I was five, my father took his battalion to Borneo to fight the secret war of confrontation against the Indonesians. While the men were away, the army brats marched up and down in makeshift uniforms, childish imitations of our soldier fathers.

We also relished the warrior culture of Australian boyhood (girls were rarely included in our play). During the day we raced across school quadrangles between concrete trenches; when we came home we fought an hour of war before tea. We felt strong and proud with our wooden guns and tin hats, exhilarated by ambushes in the park and frontal assaults on unarmed hedges. Pocket money was spent on war comics and Airfix toy soldiers. I was especially proud of my collection of 2000 plastic soldiers, and would set them up in intricate battle formations and then pelt them with matchstick-firing guns. Death was count-to-ten and make-believe; war was an adventure.

At an early age we knew that Australians made the best soldiers. The heroes of my crayon war drawings always wore slouch hats, and Airfix Australians were my favourite toy soldiers. When we played war we dressed in the light khaki of Australian army scouts or bushmen, and were especially brave and stealthy. Of course, the Aussies always won. As we grew older this national fantasy gained historical confirmation. We read the story of Gallipoli in a popular cartoon history series about Australian heroes: navigators, explorers, pioneers and soldiers. During the public holiday on Anzac Day — commemorating the ANZAC (Australian and New Zealand Army Corps) landing at Gallipoli on 25 April 1915 — we listened to speeches about that national 'baptism of fire'. In Canberra, we often visited the Australian War Memorial. My favourite exhibits were the dioramas of the landing at Anzac Cove and of the battle of Lone Pine. The sculptured Australian soldiers seemed strong and attractive, inspiring heroes for our games. Clambering over old tanks and submarines generated excitement and interest more than a horror of war.

Family stories were another way in which I learnt about Australians at war. Soldiering men dominated the family mythology of my childhood; two were recalled as heroes but one was a shadowy memory. My great-uncle Boyd Thomson was the son of a Gippsland pastoral family and was a promising architect and poet. When Gallipoli casualty lists reached Victoria in 1915, Boyd decided to enlist. He wrote a poem 'To the Mother School', justifying his decision:

> Would you wish for your sons a happier aim
> Than that a man go forth to die
> For a faith that is more than an empty name
> For a faith that burns like a scorching flame?
> Mother thy blessing! and so — good-bye![1]

Boyd Thomson died on the Somme in 1916, before he could record his experiences of war. His Scotch College school friends made a memorial booklet of his poems. In 1980 I found his grave when I was touring the battlefields of France. The family had selected an inscription for his gravestone:

> He went forth
> To die for a faith
> That is more
> Than an empty name

Death froze the meaning of Boyd Thomson's life. When I was a student at Scotch College in Melbourne I proudly read Boyd Thomson's name on the school Honour Roll. His memory became a romantic tradition in my family, and my father read to us the poems of the uncle he never knew. Boyd represented talent that was never allowed to mature, but his memory also upheld the values of his family, his school and his social class, and, above all, the duty of service to the nation and a just cause.

In contrast, the war of my grandfather, Hector Thomson, Boyd Thomson's cousin and also a Scotch Collegian, was seldom mentioned. As a child I was told that in 1914 he had enthusiastically galloped from an outback station in Queensland to enlist in the Light Horse, and that during the war he had been awarded the Military Medal for bravery. But that was all I heard. Hector contracted malarial encephalitis while serving in the Jordan Valley. As a consequence, when he returned to Australia he was in and out of Caulfield Repatriation Hospital. After his wife died he struggled to cope with his two young sons and a farm ruined in the 1930s' Depression. One of my father's few positive memories of his childhood is of himself and his younger brother wearing Hector's war medals to school on Anzac Day. They had the most medals. Much of the time their lives were hard and unhappy. Partly to escape from the difficulties of his civilian life, Hector lied about his age and re-enlisted for the Second World War. The soldiering that had perhaps ruined his life now served as a sanctuary of security and good mates. After the Second World War he sold his farm and lived in Melbourne. He died before I was born and his sons seldom talked about him; they had few happy memories to relate. The 1914–18 war was one of several factors responsible for Hector Thomson's troubled life, and for his absence from our family mythology.

Figure 1   *A letter sent by the author, aged five, to his father who was fighting in the war of confrontation against Indonesia, with the Australians wearing their distinctive slouch hats.*

My other grandfather, John Rogers, provided a more positive and vivid connection with Australia's soldiering past. The son of a Methodist clergyman, he won scholarships to Geelong College and then, in 1914, to the University of Melbourne. Midway through the academic year he enlisted in the AIF (Australian Imperial Force) as a private. After the landing at Gallipoli he was promoted rapidly in his depleted battalion, and then survived that campaign and the horrors of France virtually unscathed to finish the war as a captain on General Monash's Australian Corps staff. He postponed a successful business career in 1939 to rejoin the army, and concluded his military career in 1945 as a Brigadier and Director of Military Intelligence. We grandchildren never tired of 'Papa's' stories of the humour and adventure of war. He would recall the cunning of the Anzac evacuation from Gallipoli, or the Australian victories in France of 1918. He kept the sickening memories of war to himself, preferring to tell us funny anecdotes about the bold and cheerful 'diggers', as the Australian soldiers were nicknamed, who scorned military rank and etiquette but were the best fighters of the war. On Anzac Day we watched him march, stiffly but with pride, surrounded by his cobbers of the 6th Battalion. I was fascinated by his stories of wartime camaraderie and adventure, just as I admired Boyd Thomson's dutiful sacrifice.

My family and cultural myths reveal the selective nature of war remembrance. In this version war is fascinating and heroic, at worst a hard time shared by good mates. There is little recognition of the horrors of war or the fate of its victims. Public memorials and remembrance rituals transform personal mourning and sadness, and justify death as sacrifice for the causes of freedom and the nation; 'their name liveth for evermore' to remind us of noble qualities and fine deeds. This war mythology also defines a selective national identity. From war stories and memorials I had learnt that Australians, typified by Australian soldiering men, were the courageous and resourceful adventurers of the New World, and that the Anzacs had established Australian nationhood.

Yet there were tensions and contradictions in my own sense of this national identity. For example, Boyd Thomson identified with the Mother Country and Empire as much as with Australia, and in my years at Scotch College the imperial Remembrance Day was still given precedence over the national Anzac Day. Furthermore, the supposed egalitarianism of the AIF didn't always match the social divisions in the Australian army of my experience, in which officers' families were housed apart from the families of private soldiers. The only private soldiers I ever met were my father's batmen.

I stayed with the family of one batman on a number of occasions when my parents went away, and I was fascinated by different accents, food and family culture.

My family war myths show how only some experiences become highlighted in remembering, while others are repressed and silenced. They also reveal how some 'private' memories attain 'public' significance, both within the family and beyond. For example, as a prominent ex-serviceman and civic figure in Melbourne after the Second World War, John Rogers was invited to promote his version of Australians at war in countless Anzac Day speeches. In contrast, my grandmother's war story remained unheard outside the family. To some degree she internalised the pre-eminence of men in the Anzac legend and believed that her own wartime story — in 1914 she was the first woman to study agricultural science in the Southern hemisphere — was of less public significance than her husband's military career. Similarly, while Boyd Thomson's short life was commemorated through readings of his memorial poetry booklet, Hector Thomson's life was seldom mentioned.

The Anzac tradition that I grew up with articulated a selective family history and generalised it as an influential version of the nation's wartime past. But one of the lessons of growing up in a relatively powerful family and class is a recognition that its members do not simply, or conspiratorially, impose their views upon society. Their views are pervasive because of public power, but they are also sincerely believed and propagated. My father and grandfather Rogers believed their version of the Anzac legend — in which Australian soldiers and Australians in general are all good mates and equally able to achieve their full potential — because it made sense of their own experiences of military and social success and corroborated their personal and political beliefs. By emphasising the qualities of Australian soldiers rather than the nature and effects of war, it also helped them to keep painful personal memories at bay, and to compose a military past that they could live with in relative comfort. It is not surprising that my work on the Anzac legend, including this book, has caused pain and anger within my family. Personal identities are interwoven with national identities, individual memories intersect with public legends, and critical analysis of Anzac thus inevitably collides with powerful emotional investments in the past.

The process of subjective identification thus helps to explain the resonance of national myths. Take, for example, my own childhood fascination with war, or at least with what I imagined war to be. Even today, martial music and marching men make my spine shiver, and I can feel within myself some of the entranced enthusiasm that impels young men to war. Patriotic military ritual and rhetoric touches a sensitive human nerve. It fulfils our common need for a sense of purpose and a proud collective identity. One explanation for the success of the Anzac tradition among generations who have not known war is that we gain vicarious satisfaction from the saga of loyalty, courage and self-sacrifice. Many young Australian men would like to think that we, too, are Anzacs.

Subjective identification works by linking personal experiences and emotions with public meanings. While watching a video of the 'Anzacs' television serial in 1987, I was moved to tears during a scene in which the platoon commander inspired his exhausted and mutinous men to fight on for the memory of their dead mates. As an Australian living in Britain the speech made me feel like a deserter from my homeland; as a critic of the Anzac legend it made me feel like a deserter from the values of my family, class and nation. Yet the rhetoric of Australian 'mateship' was resonant and appealing precisely because it addressed my own experiences and emotional needs and my own dislocated Australian identity. For all my rational scepticism, I was inspired and moved by the officer's speech.

My subconscious identification with the Anzac legend is still strong, but in researching and writing this book I have tried to step outside my family and national myths. As an adolescent I believed the myths but felt incoherent, contradictory emotions about my family background and mythology. I was intrigued by lives — such as Hector Thomson's — that did not seem to fit the story. As I grew older I began to investigate my Anzac tradition. In 1980 I visited Gallipoli and the battlefield of the Somme and was mesmerised by the tranquil beauty of each place. Although I was appalled by the number of Australian graves, the Australian war cemeteries confirmed my old understanding of the war. I felt proud to be Australian. On the Somme the old French gardeners who tended the war graves shook my hand with respect when I said, 'Je suis australien'. The inscriptions on many Australian gravestones suggested that the families of the dead had felt the same pride:

> He was just an Australian soldier
> One of God's bravest and best

> He died the helpless to defend
> An Australian soldier's noble end

I concluded that the greatest significance of the war for Australians was the proud discovery of an Australian identity.[2]

After I came home I started to question whether that discovery had concealed other experiences and meanings of the war. At university, fellow activists in the peace movement argued that the Anzac legend deflected criticism of contemporary military alliances, and that soldier heroes were anachronistic and even dangerous in a nuclear world. Feminist friends pointed out that the tradition ignored or marginalised Australian women. Historical research also undermined my belief in the legend. I was concerned to find that my grandfather had used the example of the first Anzacs to promote conscription during the Vietnam War.[3] While researching the history of the Melbourne working-class suburb of Brunswick between the wars, I discovered that not all diggers were treated as heroes after the Great War, and that there were massive ex-servicemen's riots in the post-war years.[4]

I was beginning to think that my family myths of the war and Australian society might only represent the experience of a particular class. The story of Anzac that I grew up with was very similar to the official Anzac legend — both my father and grandfather Rogers were among the powerful public men who spoke on Anzac Day platforms — but I suspected that other groups in Australian society might have different memories of the war, and different relationships with the legend. In 1983, as part of a postgraduate research project to explore the history and politics of the Anzac legend, I set off into the industrial suburbs of Melbourne to record the memories of working-class diggers.

Oral history and Anzac memories

I conducted interviews with twenty-one Great War veterans, most of whom lived in Melbourne's western suburbs and for whom brief details are provided in Appendix 2. My initial contacts were with members of the local Returned and Services League (RSL) sub-branches, although I was also referred to digger friends who were not RSL members. Readers who are interested in the details of the oral history project, and in the methodological issues of oral history, are referred to Appendix 1. To summarise, in the interviews I adopted a guided life-history approach which focused upon the men's pre-war lives and their experiences as soldiers and ex-servicemen.

The interviews did highlight certain contrasts between the experiences of working-class diggers and my perception of the Anzac legend. There was little romance or heroism in the war stories I was told; many men admitted that if they had their time again they would not enlist. They recited familiar anecdotes about the egalitarian Anzacs and AIF, but their emphasis was sometimes different from that of conventional stories. For example, mateship was a sacred memory, but it was the creed of the diggers in the ranks and did not necessarily include officers. Even men who respected capable officers had often detested the authoritarian practices of the army. Most vividly, many of the old men scornfully contrasted their status as national heroes with the way they were ill-treated after the war. Indeed, a number of 'radical diggers' had become disenchanted with the RSL and official Anzac commemoration, and had joined socialist and pacifist movements in the inter-war years. To a certain extent, the memories of working-class veterans thus represented a forgotten and even oppositional history.[5]

However the interviews also suggested that the memories of working-class diggers had become entangled with the legend of their lives, and that veterans had adopted and used the Anzac legend because it was resonant and useful in their own remembering. For sixty years most of these men had been members of the RSL and attended Anzac Day parades. Many of them had read the official history of the war and quoted anecdotes as if they had come from their own experiences. In some interviews I felt like I was listening to the script of the film *Gallipoli*. Memories were also reshaped by present-day situations and emotions. Lonely old men were sometimes eager to recall the camaraderie of the AIF or the adventure of the war, and to reassert a proud Anzac identity.

I was fascinated by this relationship between the Anzac legend and digger memories, and instead of simply or naively challenging the legend I now wanted to understand how and why it worked, or sometimes didn't work, for veterans of the war. This interest was informed by new theoretical work about memory, subjectivity and 'popular memory', including writings by international oral historians such as Luisa Passerini and Alessandro Portelli, and by members of the Popular Memory Group at the Centre for Contemporary Cultural Studies in Birmingham. From these writings I developed the working model of remembering, and of the relationship between public legends and personal memory, that informs this book and is summarised as follows.[6]

We compose our memories to make sense of our past and present lives. 'Composure' is an aptly ambiguous term to describe the process of memory making. In one sense we compose or construct memories using the public languages and meanings of our culture. In another sense we compose memories that help us to feel relatively comfortable with our lives and identities, that give us a feeling of composure. In practice the two processes of composure are inseparable, but for the sake of clarity I will deal with them separately and identify certain features of each process.

The first sense of the construction of memories using public language and meaning requires a cultural approach to remembering. The basis of this approach is that there is no simple equation between experience and memory, but rather a process in which certain experiences become remembered in certain ways. Only a selection of an individual's myriad experiences are recorded in memory, and for each of these there are a range of ways in which the experience might be articulated. Raymond Williams usefully describes how the initial, inarticulate consciousness or 'structure of feelings' regarding an experience is articulated through public forms and metaphors, which shape and bind that consciousness into a more fixed state. According to the Popular Memory Group, public representations of the past — including the very recent past that has only just been experienced — are thus used as an aid in the constant process of making sense of personal experiences:

> [...] provoking reflection and inviting comparison between the more general accounts and the remembered particulars of personal experiences. For if the effect of the public field of representation is to generalise significance, it must be by offering forms and general interpretative categories by means of which people can locate their own experience in terms of wider social patterns. Popular memories work in just this way, struggling to generalise meanings in such a way as to pull together and give a shared form to a multiplicity of individual and particular experiences, and so to reconstruct people's sense of the past.[7]

The available public languages and forms that we use to articulate and remember experience do not necessarily obliterate experiences that make no acceptable public sense. Incoherent, unstructured and indeed unremembered, these unrecognised experiences may linger in memory and find articulation in another time and place, or in less conscious outlets. New experiences constantly stretch the old forms and eventually require and generate new public forms of articulation.

A further aspect of this cultural theory of remembering is the distinction between the 'general' and 'particular' publics within which we articulate and remember experience. The 'general public' includes the various media which provide generally available representations and interpretative categories ('the public field of representations') in an indirect, impersonal relationship. We also make sense of experience by using the meanings available within the active relationships of a 'particular public' group, such as a wartime platoon.

The particular public is significant in a number of ways. Firstly, a member of a particular public may participate in and contribute to the development of those meanings. Secondly, because of the importance of social acceptance and affirmation the particular public is especially influential, and potentially repressive, in the construction of meaning and identity. Thirdly, a particular public may provide an important site for the maintenance of alternative or oppositional meaning, a source of public strength for its members to filter or even reject and contest more general meanings.[8]

How we make sense of experience, and what memories we choose to recall and relate (and thus remember), changes over time. Memory 'hinges around a past–present relation, and involves a constant process of re-working and transforming remembered experience'. Thus our remembering changes in relation to shifts in the particular publics in which we live, and as the general public field of representations alters.[9]

It also changes in relation to our shifting personal identity, which brings me back to the second, more psychological sense of the need to compose a past we can live with. This sense presumes a dialectical relationship between memory and identity. Our identity (or 'identities', a more appropriate term to suggest the multi-faceted and contradictory nature of subjectivity) is the sense of self that we construct by comparisons with other people and with our own life over time. We construct our identities by telling stories, either to ourselves as inner stories or day-dreams, or to other people in social situations. Remembering is one of the vital ways in which we identify ourselves in storytelling. In our storytelling we identify what we think we have been, who we think we are now and what we want to become. The stories that we remember will not be exact representations of our past, but will draw upon aspects of that past and mould them to fit current identities and aspirations. Thus our identities shape remembering; who we think we are now and what we want to become affects what we think we have been. Memories are 'significant pasts' that we compose to make a more comfortable sense of our life over time, and in which past and current identities are brought more into line.

There are many ways in which our remembered experiences — of both the immediate and distant past — may threaten and disturb identities and thus require composure. Traumatic experiences may violate public taboos or personal comprehension (although in some situations people make a conscious attempt not to repress trauma, like the Holocaust survivors who are determined to recall and relate their experience for the sake of those who died, and as a lesson to us all). Dramatic life changes often render old identities irrelevant and require drastic re-evaluation. Everyday psychological life comprises frustrated desires and debilitating losses which we seek to compose into a safer, less painful sense.

Thus our public remembering and private inner stories often seek to compose a safe and necessary personal coherence out of the unresolved, risky and painful pieces of past and present lives. Yet stories rarely provide complete or satisfactory containment of threatening experiences from the past. Our attempts at composure are often not entirely successful and we are left with unresolved tension and fragmented, contradictory identities. Composure, 'based as it is on repression, and exclusion, is never achieved, constantly threatened, undermined, [and] disrupted'. Repressed feelings and impulses are expressed or 'discharged' (sneaking through the barricades of conscious coherence) in particular forms — such as dreams, errors, physical symptoms and jokes — which reveal hidden, painful and fragmented personal meanings. Thus the dreams of soldiers reveal the repressed significances of a soldier's war, the psychological impact of fear, carnage and guilt. Slips or errors of expression may be more than just carelessness or confusion; for Freud, the error or 'parapraxis' was due to the unconscious association of a particular word or phrase with a repressed desire. Physical symptoms — such as the nervous flicker of an eyelid or the edge of the mouth, or a more serious speech impediment — may also be connected by a psychological chain of association with an unresolved trauma. Jokes and laughter are often ways of discharging difficult or painful memories.[10]

Oral historians sometimes listen only to the spoken narrative and neglect these hidden texts. Just as the stories of remembering reveal the particular ways in which a person has composed his or her past, these hidden forms of meaning can reveal experiences and feelings that have been silenced because they could not fit with public norms or with a person's own identity.

This brings me back to the key theoretical link between the two senses of composure, that the apparently private process of composing safe memories is in fact very public. Our memories are risky and painful if they do not fit the public myths, so we try to compose our memories to ensure that they will fit with what is publicly acceptable. Just as we seek the affirmation of our personal identities within the particular publics in which we live, we also seek affirmation of our memories. 'Recognition' is a useful term to describe the process of public affirmation of identities and memories. Recognition is essential for social and emotional survival; the alternative of alienation and exclusion may be psychologically devastating. We may seek recognition in other, more empathetic publics, but our memories need the sustenance of public recognition, and are composed so that they will be recognised and affirmed.

'Identities', 'general' and 'particular publics', 'recognition', 'affirmation' and 'composure'; these are key concepts for the exploration of the processes of remembering. This new theoretical framework informed my investigations into the relationships between Anzac identities, memories and myth. In 1987 I conducted a second set of 'popular memory' Anzac interviews with five of my initial interviewees. In these second interviews I focused on how each man composed and related his memories, and explored four key interactions: between interviewer and interviewee, public legend and individual memory, past and present, and memory and identity. The relationships that I had developed previously with each of these men facilitated this new approach. In long, detailed interviews they were encouraged to go back over their experiences as soldiers and ex-servicemen, and to reflect upon the ways in which they had come to terms with their wartime past.

The popular memory approach suggested a particular structure for this book, which explores how the Anzac legend works for soldiers and veterans by highlighting the interactions between experiences, memories, identities and the legend, and by showing how these interactions have changed over time. The three main parts of the book focus, respectively, on Anzac experiences and narratives during wartime, the postwar period, and the 1980s and 1990s. For each of these three periods I have written one chapter using oral testimony in a conventional manner, a second chapter about the making and remaking of the legend, and a third chapter comprising 'memory biographies'.

The initial, oral history chapters (1, 4 and 7) draw upon the testimony of all my interviewees and outline the main features of their experiences as Anzacs during and after the war, and into old age. They note significant aspects of the life experience of working-class Australians who fought as soldiers in the Great War, some of which correspond with the legend of Anzac. Yet these chapters also reveal the complex and multifaceted nature of the Anzac experience, and suggest ways in which it sometimes contradicted the legend.

In the chapters about the making and remaking of the public legend (2, 5 and 8), I explore how influential individuals and organisations drew upon the Anzac experience and, through processes of selection, simplification and generalisation, moulded a national legend. How was an official legend created and, more importantly, to what extent did it gain the support of the diggers and of Australians at home? My argument is that an official or dominant legend works not by excluding contradictory versions of experience, but by representing them in ways that fit the legend and flatten out the contradictions, but which are still resonant for a wide variety of people.

I explore the relationships between experience, memory and the legend in more detail in the memory biographies which comprise chapters 3, 6 and 9. These 'memory biographies' explore the particular ways in which veterans composed their memories of the war in relation to the legend, and in relation to their own shifting experiences and identities (the writing of memory biographies is discussed in Appendix 1). They explain how men of war related to the range of masculine and national identities available to them during key periods of their lives — enlistment, battle, behind the lines, repatriation and old age — and how they sought to compose comfortable ways of remembering the war and identifying themselves as soldiers and veterans.

This book includes the memory biographies of three diggers whom I interviewed in 1983 and again in 1987 — Percy Bird, Bill Langham and Fred Farrall — chosen because they had very different experiences of the war and postwar periods, and very different ways of remembering the war and relating to the Anzac legend. The following thumbnail sketches introduce the main features of the life stories of these three men, and evoke the variety of their experiences.

Percy Bird grew up in the Melbourne port of Williamstown, and was engaged to be married and employed as a clerk with the railways when he enlisted in 1915. He fought with the 5th Battalion on the Western Front until he was taken out of his unit to do clerical work behind the lines. A gas attack damaged his throat and he was sent home in 1917. After a brief period of rehabilitation he married his fiancée and returned to his job with the railways, where he rose through the ranks of the Auditing Department. Running parallel to a stable family and working life, Percy was active in ex-servicemen's organisations and was a keen participant in Anzac Day and other forms of war commemoration.

While still in his teens, Bill Langham left home and school in rural Victoria to work in the stables at Melbourne's Caulfield Racecourse. He was under age when he enlisted in the AIF, and despite parental protests he joined an artillery unit of the 8th Brigade and served on the Western Front, where he was a driver with a horse team that transported guns and ammunition. He was wounded in the head just before the Armistice. Back in Australia, Bill battled to get a pension and adequate employment training, and was in and out of work throughout the inter-war years. Yet he shared Percy Bird's enthusiasm for the soldiers' club and veterans' reunions, and for occasions of Anzac remembrance.

Fred Farrall grew up on a smallholding on the New South Wales Riverina, but like Bill Langham was keen to leave the country. In 1915 he left home on a 'Kangaroo' enlistment march to Sydney, and in the summer of 1916 he joined the 55th Battalion on the Somme. Fred was not a confident soldier. Most of his mates from the enlistment march were killed or wounded, and he was himself wounded and ill on a number of occasions, and eventually suffered a form of neurosis or shell-shock. Upon his return to Australia Fred decided to stay in Sydney, but he was in a poor physical and emotional state; he struggled to find work and had a nervous breakdown in 1926. At about this time he became active in Labor politics and gradually established himself as an active pacifist and socialist, and a leading figure in the trade union movement. Unlike Percy Bird and Bill Langham, Fred Farrall was alienated from all things Anzac and was virtually unable — at least until his later years — to speak about his wartime past.

I am not arguing that these three men were typical Anzacs, or even that between them they cover the range of possible digger experiences. They did, however, have extremely varied experiences as soldiers and veterans, and thus serve the aim of this book, which is to explore the relationship between the Anzac legend and the lives and remembering of the Anzacs. These three men also had very different ways of remembering their lives, and of composing their life stories, as the following sections explain.

Percy Bird

Percy Bird rang me as soon as he received my initial letter asking if I could interview him, and was eager to talk. We conducted our first interview in September of 1983 at the house in Williamstown in which he had lived for much of his adult life. He was a widower and lived alone, an old man of ninety-four who was physically frail but had an energetic and lively mind. When I arrived he handed me twelve pages of wartime anecdotes, which he had prepared specially for me, titled 'The 5th Battalion, 1916 and 1917, France'. He then related these same stories with great glee, and sung some soldiers' songs in a quavering but beautiful tenor voice. He told me that he had talked and sung about his youth and wartime experiences to local school children and at recent veterans' reunions. When I conducted the second interview with Percy in 1987 he had moved into a home for elderly ex-servicemen and war widows. It was on the other side of town but near to his daughter. In that interview he repeated most of the same anecdotes and told me how much his fellow residents enjoyed his stories and his singing.

There were two outstanding features of Percy Bird's remembering. Firstly, a primary aim of his remembering was entertainment, and he derived enormous satisfaction from his performances and the positive response of audiences. Secondly, these performances drew upon a fixed repertoire of anecdotes. The written stories that Percy gave me, and the stories that he wanted to tell regardless of my questions, were short, discrete anecdotes, loosely arranged in approximate chronological order but also prompted by cue words in a previous story or a question, or by the established sequence of the performance. One section of Percy's written story, which begins when his battalion was returning to the Somme for the winter of 1916-17 after a spell in Belgium, conveys this anecdotal form particularly well:

> On the way back to the Somme we marched about fourteen miles the first day. Marched about twelve miles or so the next day. Came to a village and passed outside the village for about four miles when we came to a big field. The Colonel told us the village we passed had an epidemic or something there so we were sleeping out in the open, but we would all receive canteen parcels. They all thought they would be receiving cigarettes but all we got were tins of acid drops.

> It rained all night and we were wet through and had to march twelve miles to the next place where we stopped for three nights.

> There was a big Brewery in the village.

> In our peregrinations around the country we were always on the march and the English Tommies went round in buses. So a yarn went round that while we were on the march a Tommy Regiment called the West Ridings were in the vicinity and an English officer hopped out from somewhere and yelled, 'Are you the West Ridings?' Back came the reply, 'No, we're the bloody Australians walking'.

> A couple of days before Christmas we came out of the line and C and D companies were billeted in tents between the villages of Dernancourt and Méaulte where there was a big Railhead. A number of Tommy permanent base men and German prisoners were also there. On Christmas Day the Germans sang a number of Christmas carols. It was wonderful.

On one occasion when the men were working one morning at the Railhead the Tommy captain in charge said to them, if they unloaded the two trains that were there then they could knock off at twelve noon and have the afternoon off. They had the job finished at twelve noon but another train came in and the captain said they would have to unload that. They told him he had promised them the afternoon off and they were going to have it. He went to our Captain and told him the men wouldn't work in the afternoon. Our captain said to him, 'I believe you told them that if they emptied the two trains that were there at the time by twelve noon they could have the afternoon off, so take my advice and give them the afternoon off and you will get a lot more out of them whilst they're with you'. So he did what our Captain said.[11]

The form and content of Percy's remembering are significant in a number of ways. Most of the other men I interviewed told their stories as an unfolding life story with a smooth, sequential flow. Percy's remembering was more akin to the anecdotal style of the stand-up comedian. Percy's anecdotes also combined his own story with that of the men of his battalion — it was usually the story of 'us' rather than 'I' — and were presented as a history of the battalion. At first reading the anecdotes appeared to be simply descriptive, but a closer reading showed that each story had a punchline or 'tag' which helped Percy to fix it in his memory, gave it a purposeful theme and made it a 'good' story. Some of the main themes through which Percy articulated his war memory were the humour of trench life, lucky escapes from enemy shells, his successful participation in army concerts, and the nature and effectiveness of the Australian soldiers. In Percy's remembering the war was never horrifying or disillusioning and there were obvious silences; for example, about enlistment and his feelings in the line. Extensive questioning sometimes drew him to discuss these aspects of his war, but he was always eager to return to his own, standard stories. Through his memory biography, I will explore how Percy Bird came to compose his remembering of the war in this form and with these meanings.

Bill Langham

When I first met Bill Langham in 1983 he was in his late eighties but was alert and remarkably healthy. He lived in his own house in the western suburb of Yarraville. The west had been Bill's home since his return from the war, and he had been an active participant in its clubs and societies. He shared his home with his wife of fifty years and a retired son. His wife showed tolerant amusement at Bill's 'yapping on' with me and got on with other activities in and out of the house during the interviews. They were both still very active in various Yarraville musical associations, and Bill was relatively content with his life as an old man. He also seemed comfortable with memories of his youth and a successful working life. The war retained a tangible presence, with a photo of Bill in uniform on the mantelpiece and occasional meetings with other diggers, but it did not have the powerful emotional charge that some of the other men I interviewed were still struggling to cope with. Bill's war experience had been incorporated into a more general sense of his life, identity and value system, which included contradictions and occasional doubts, but which on the whole worked well for Bill's peace of mind and personal fulfilment.

One key thread of this identity, which was influenced by his war experiences but also shaped the way he remembered the war and his life, was Bill's image of himself as a successful battler, the epitome of the 'little Aussie battler', a resonant character in Australian working-class and national folklore:

> That, see my life I reckon it was changed altogether when I came back. I was, I was a real roustabout, anything that went I did. Didn't matter what it was I put my hand to, and I'm not skiting when I say that whatever I tried to do I made a success of.[12]

Bill related his own success story to a more general story of Aussie battlers and pioneers; like the 'cockies' in the district where he grew up 'who started with nothing' and 'they struggled but they worked and they worked long hours and they made a success of it'. Bill's identity and world view was not simply an enterprise ideology for self-made men. It was a class-specific story with a clear sense of the hurdles faced by working-class men and was imbued with the belief that his own ability to earn a decent living was due to the struggles of the union and Labor movement. In Bill's view of the world working people were distinct from, and ill-treated by, the powers that be, which included the Royals — who 'get a lot of money out of the country and they get it from the people' — and the 'big nobs' who started the war and 'looked to the poor ordinary citizen to fight it for them'. Yet in Bill's world view this populist ideology did not include an analysis of the structural causes of inequality and oppression, and served primarily as the background to his identity as a battler who had overcome life's obstacles, including the war.[13]

Bill's identity as an old man also influenced his remembering. In the interview he made frequent comparisons between past and present. Sometimes these were explanatory contrasts between schooling or prices, and even about his own changed attitudes, but often the contrasts were value-laden and critical of social change. Thus the streets were no longer safe, and modern youth — the target of most of Bill's comments — was not as tough or well disciplined as his generation and lacked its bush and military training. These contrasts were made by many of the men I interviewed, although they were more common in Bill's remembering than most. They reveal his ignorance of a youth culture with which he had relatively little contact, and the anxiety of an old man no longer able to defend himself on the street and conscious of the loss of his own physical strength. These contrasts acted for Bill as an important reminder of his own young manhood, just as in turn memories of youthful masculinity were a source of emotional pride and strength in old age.[14]

Another significant context for Bill's remembering was the interview itself and our relationship. Bill enjoyed having his memory 'jogged' because his wartime experiences were important to him: 'I don't want to forget the mates I made, I don't want to forget the things I did'. With most of his digger mates long gone, the interview was a rare opportunity for Bill to reminisce about the war. Once he began to talk he derived a great deal of pleasure from relating and performing his story to me, often accompanying words with mimicry and play-acting; his strutting depiction of a self-important officer was sharper than any verbal description. Like Percy Bird, Bill remembered in vivid detail and showed little sign that his memory had been affected by mental or physical deterioration.[15]

Yet unlike Percy Bird, Bill had not sought out audiences for the performance of his war memories, and did not feel any strong need for public affirmation of those memories. Nor did he perform one particular, fixed version of his war. In fact, the most striking feature of Bill Langham's remembering was its candour and openness. He did not relate a stock of anecdotes or deliver a well-rehearsed autobiographical monologue. Rather, his remembering was reflective and discursive, and sometimes self-questioning. He decided that he should tell me stories that he preferred not to relate to other audiences or to dwell upon when he was alone, and he did not appear to be consciously withholding any memories. Although the pain of the memory of one particularly gruesome battle story was clearly evident on Bill's face, he tried to tell the story in a matter-of-fact manner because he felt it was an important part of the history of the war.

He also developed an active relationship with my questioning in which he would ask me to clarify what I meant by a question (rather than focusing on a keyword which sparked off particular memories), or would paraphrase a clumsy question more succinctly ('I think what you're trying to get me to say is whether it's [Anzac Day] trying to glorify war or not'). He was quite willing and able to disagree with me when I played devil's advocate or expressed an opinion with which he did not concur. Bill's self-awareness was also apparent in the relationship between his remembering in the two interviews. He played his copy of the first tape several times between the two interviews, including on the morning of the second interview. In the new interview he sometimes referred to comments he had made on the earlier occasion, but, unlike Percy Bird and Fred Farrall, he knew which stories were on the first tape and was careful to avoid repetition. He rarely contradicted his earlier version, but did provide many new stories as well as fresh angles and complexities about the first set of stories.[16]

Bill's remembering was not entirely flexible and independent. In his stories he often matched his past with his identity as a successful battler and with the identity of the Anzac legend. Although Bill's wartime and postwar experiences might easily have facilitated an oppositional stance to the war and the legend, his participation in various public practices of remembrance led him to generalise about his experiences in terms of the legend. Yet unconventional experiences were not scrubbed from Bill's memory. Because of the relatively unproblematic nature of Bill's identity as a soldier and a man, he did not need to compose a fixed, unitary version of his war. His remembering comprised a multifaceted layering or patchwork of stories and understandings, which derived from the complex and often contradictory range of his experiences and from the different public ways of making sense of those experiences. Although certain experiences were highlighted in his remembering, the many facets of Bill's war memory remained accessible enabling him to relate them in appropriate circumstances. Even during the interview different contexts prompted different ways of remembering the war. Bill's memory was not neatly composed into a single, seamless account and, as a result of this, of the three memory biographies his was hardest to write. In it, I try to evoke the complexity of Bill's remembering while also showing how he came to emphasise certain memories and identities.

Fred Farrall

I first heard about Fred Farrall through a friend who knew him from the Melbourne Labor History Group. I attended a meeting of the group so that I could meet him and his comrade Sid Norris, who was also a digger and a socialist. I wrote at the time:

Fred, eighty-five, sprightly and alert, had on his lapel that night his 1919 returned from active service badge. He had also been a member of the Communist Party; of course I wanted to interview him![17]

We met for our first interview in July of 1983. Fred lived by himself in a small house in the inner Melbourne municipality of Prahran. His de facto wife Dot had died four years previously to the day, and Fred had gradually learnt to fend for himself around the house. On the walls of the lounge room where we conducted all of our interviews hung a pictorial history of Fred's life. Filling one wall were a dramatic portrait of Lenin and pictures of Marx and of Fred with Tom Paine, the only living Australian to have met Lenin. On the wall above Fred's armchair were photos of Fred and Dot as Mayor and Mayoress of Prahran and, superbly framed, a photo of Fred as a soldier next to his beautifully inscribed discharge certificate.

From our introduction at the meeting Fred had gathered that I was interested in the experience of Australian Great War veterans who were also political radicals, and he now welcomed the opportunity to tell his story of the war and its aftermath to a historian who promised a wider audience. It was not the first time that he had told his life story as history. Apart from the discussions of the Labor History Group, which had resulted in articles by Fred for their *Recorder*, he had talked about the war to Melbourne college students and to the producers of a film about the AIF hero Albert Jacka. He liked being interviewed because he believed that he had a story with an important political message to tell, and because in his old age talking was almost his only way of being politically active ('all I can do now is talk').[18]

Over the next seven years I made many visits to Fawkner Street for taped interviews and untaped discussions, and to bring Fred copies of tapes and transcripts (he was delighted by the transcripts, but immediately set out to 'correct' his grammar and prose), and articles I had written based on the interviews. When the tape recorder was off, as we made a cup of tea or a meal together, Fred shed his interview persona and opened up about other aspects of his life, and asked me about my life and thoughts. When I went to England we maintained a correspondence which was partly sustained by Fred's interest in my attempts to publish a book about radical diggers. Fred read my manuscript drafts with approval, and was especially interested in my prefatory stories about Hector and Boyd Thomson and my personal connection with what he regarded as the tragedy of the war. Over the years and despite our differences — I did not share Fred's uncompromising pro-

Soviet socialism — we developed a friendly and trusting relationship. We said goodbye in June 1990, with Fred close to death.

Fred Farrall was an obvious choice for a case study, for a number of reasons. Firstly, I know more about Fred's life and memory than about any of my other interviewees. Many hours of tapes, combined with notes from untaped discussions, provided a wealth of material about Fred's life and thoughts. Secondly, Fred's life is relatively unusual because he was a digger who became a political radical. Through his biography I can trace why and how he took that divergent path, and analyse his relationships with the dominant Anzac legend and the more radical tradition of the Labor movement. Thirdly, Fred's identity as a digger and his remembering of the war went through several different phases. His memory biography shows how his identities and remembering changed over time through interactions between personal identity and public affirmation that were very different to those experienced by Percy Bird and Bill Langham.

Finally, the particular forms in which Fred remembered and related his life are instructive about the ways in which memories and identities are composed. Fred Farrall was a storyteller who used a particular narrative style to make specific meanings about his life. Like Percy Bird's anecdotes, though unlike Bill Langham's remembering, Fred's story did not change much between tellings; in 1983 he related stories to me in much the same form as he had told or written them in other recent historical contexts, and in 1987 he repeated many of the same stories and found it difficult to respond in new ways to my popular memory interview approach. At times I became frustrated by the difficulty of breaking into the monologue to ask a question, and by the lack of dialogue or discussion. More than any of the men I interviewed, Fred had his story well and truly composed; he wanted to 'improve' the transcripts of the interview to produce a more polished text and to iron out inconsistencies. In comparison with Percy Bird's anecdotes or Bill Langham's multi-layered remembering, Fred Farrall's story was deliberate, detailed and sequential, a well-rehearsed unfolding of the transformations of his life, often superbly told through tensions and twists towards climactic punchlines.[19]

Most importantly, Fred's stories composed a particular meaning in which his war and his life was a process of conversion. Stripped to its essentials, Fred's narrative is as follows. The naive and patriotic farm boy goes to war as a willing recruit but unwitting sacrificial lamb. He becomes a frightened, inadequate and disillusioned soldier on the Western Front, and a confused and traumatised veteran upon his return to Australia. After several years in a personal wilderness he discovers in the Labor movement supportive comrades and a new, socialist way of understanding his life and the world, and regains his self-esteem as a man. He articulates his disillusionment about the war in political terms and thus redefines the war as one stage on the way to his enlightenment.

Many of Fred's stories were framed in terms of this development towards conversion. The narrative is often interrupted and explained by an ironic reflection that situates a particular incident within the larger pattern that he made of his life. One example shows how Fred's remembering worked in this way and conveys his skill and style as a storyteller. The example is the opening section of a much longer story in which Fred takes his audience from his parents' farm to the Western Front:

> The war broke out in 1914. Well of course in August 1914 I was then actually fifteen years of age, and as I'd wanted to be a jockey I wasn't very robust, in fact I was very small. So my father then began to make some changes, politically speaking, because the Labor Party were not very enthusiastic about the war. Although there was a Labor government, Andrew Fisher, who pledged Australia's support to the last man and the last shilling, as politicians can easily do. But there was a fair bit of opposition from other sections of the Labor Party and my father became a staunch supporter for the war. When the landing was made in Gallipoli, of course, we all had to have it read to us from the papers after tea at night. It was sort of ... almost something like a religious service and we listened to it and we believed it.
>
> The war went on and 1915, in September 1915, I'd reached the age of eighteen. What did I say? Did I say earlier that I was fifteen when the war ... I was sixteen when the war broke out. In another month's time I was seventeen then, so I'd reached the age of eighteen and having given away the idea of being a jockey I began to build up my physique somewhat. Previous to that I'd, you know, endeavoured to keep my weight down, but after that denial on the part of my father to do what I wanted to do I just sort of grew up more. The physical standard for joining the army in 1914 and early 1915 was very, very high and I had no hope anyway at being able to measure up to that sort of thing, that standard. But after Gallipoli, or while Gallipoli was on, and they'd suffered a lot of casualties, they lowered the standard considerably here.

When we were in the harvest field in November 1915, we were hay-making and Dad had one or two or three men working for him in the harvest as we used to have. They would be swagmen that would be picked up in Ganmain and brought out to do some work, you see. So I was working with one of them, Bill Fraser, and I said to Bill one day, 'I'm thinking about going to the war, Bill'. Well he was an Irishman and he gave me some pretty good advice and a bit of a lecture while we were picking up sheaves of hay and putting them in stooks where they belonged. He told me that I should stay on the farm. He said, 'You should remember', he said, 'that next to your life the most valuable thing you've got is your health. You stay here on the farm and look after it because it's worthwhile keeping'. And finished his advice by saying 'Let the rich men fight their own wars'. Well, I didn't take Bill's advice. I had listened to the Prime Minister and his 'last man and last shilling to defend the Empire', the Premier of New South Wales, the Archbishop. I did what they said to do. I enlisted.[20]

This excerpt is a typical example of Fred's storytelling technique. The listener's interest is sustained by the tensions between different characters and courses of action. The narrative works as a story precisely because it is framed by Fred's retrospective vision of his life. The ironies are resonant because we know, as Fred came to know, the consequences of his enlistment, and that Fred should have listened to Bill Fraser's advice.

That doesn't mean that the details about how Fred felt and acted at the time are invented, merely that the way in which he composed his story of enlistment caused him to highlight certain experiences and make sense of them in particular ways. Like all stories it is meaningful, and the meanings came from the present as much as the past. At the same time it excluded alternative senses, and either reworked contradictory experiences to fit the narrative or ignored them.

Nor was the larger story of Fred's life as a process of conversion an invention. Fred was a simple farm boy, he was a traumatised soldier and veteran, and he was transformed by political activism and understanding. The way in which Fred composed his memories in these terms made sense of those actual experiences. But Fred highlighted conversion as the primary theme of his autobiography because it enabled him to fit his war service and political radicalism into one coherent story. The identity of the soldier as unwitting victim fitted into his overall sense of his life, and the public sense of the Labor movement that sustained it. Because his wartime inadequacy thus gained political purpose and meaning it was, to a certain extent, easier to live with and remember. Yet, as the memory biography that commences in chapter 3 will show, for many years Fred did not have a comfortable or coherent sense of his experience as a soldier, and even the political analysis of conversion did not resolve the personal traumas of Fred's war.

Part I

Making a legend

CHAPTER 1

THE DIGGERS' WAR

The diggers' war was shaped by the distinctive origins and nature of the Australian Imperial Force. When the German army marched into Belgium in 1914, the Commonwealth government offered to send an Australian contingent of 20 000 men to support the British cause. The existing military forces could not make up a fraction of that number and a call was made for volunteers, who rapidly filled the ranks. Unlike other Allied forces the AIF remained a volunteer force throughout the war, despite two attempts by the government to introduce conscription by referenda. The diggers regarded themselves as citizen soldiers and baulked at the traditions and regulations of the traditional British army; and the Australian government felt obliged to resist British pressures to introduce the death penalty for desertion, which operated in the rest of the British army, into the AIF.

Nevertheless the AIF was not an independent national force. Australian units were part of the British army and subject to the orders of British High Command. Even within the AIF many staff and regimental officers were, at least in the first years of the war, British regular army soldiers. In November of 1917, after a protracted political campaign, the five Australian infantry divisions were finally amalgamated into one army corps, staffed and commanded by Australians, although the Australian corps was still answerable to British headquarters.

Australian soldiers fought at Gallipoli and on the European Western Front, while members of the Light Horse also saw action in the Middle East. By 1918 the AIF had achieved a reputation as an effective fighting force, but at the cost of terrible casualties. Just under forty per cent of Australian males between eighteen and forty-four enlisted, and of the 331 814 who had served overseas or were undergoing training by November 1918, about sixty-five per cent were casualties (the highest rate in the British army) and 56 639 had died.[1]

The legend that was made of the Anzacs is familiar to most Australians, and can be summarised as follows. At Gallipoli, and then on the Western

Front, the Anzacs proved the character of Australian manhood for all the world to see and, through their victories and sacrifices, established a nation in spirit as well as in name. The Australian soldier of the legend was enterprising and independent, loyal to his mates and to his country, bold in battle, but cheerfully undisciplined out of the line and contemptuous of military etiquette and the British officer class. The Australian army suited his egalitarian nature: relations between officers and other ranks were friendly and respectful, and any man with ability could gain promotion. According to the legend these qualities, fostered in the Australian bush, discovered and immortalised in war, typified Australians and Australian society, a frontier land of equal opportunity in which enterprising people could make good. This was the nation that 'came of age' at Gallipoli.

In many respects this familiar story intersects with the remembered experiences and attitudes of the diggers themselves. Yet their testimony records a war experience that was much more complex and multifaceted than the homogeneous identity of the legend, and which sometimes even contradicts the legend. The following account of the diggers' war contrasts key features of the legend with the experiences of the men I interviewed. It is not intended to prove that the legend was 'false' — some of its features are corroborated by veterans' memories — but rather shows how Anzac legend-makers, including the diggers themselves, articulated a legend through processes of selection, simplification and generalisation.

Gone for soldiers

Traditional accounts of Australian motivations for enlistment focus on dutiful patriots and innocent adventurers. In Charles Bean's official history of the events of August 1914, the men of the Australian outback 'became alert as a wild bull who raises his head, nostrils wide, at the first scent of [...] an old friend in danger'. In more recent times, an advertisement for the Anzac Seventy-fifth Anniversary Commemorative Coin eulogised the motivations of the Anzacs in the following terms:

> They fought for what they believed in. They fought for freedom. They fought for their country. They fought for us. They fought for our children.[2]

Historians now question the extent and significance of patriotic motivations for enlistment. Richard White argues that there were other more private or self-interested motives, and that working-class recruits were less likely to

be inspired by patriotic duty than middle-class recruits; studies of Kitch-
ener's army make similar claims about the motivations of British working-
class volunteers. The stories of my own interviewees reinforce these more
complex explanations of enlistment by the so-called 'generation of 1914'.[3]

Although there is debate about the degree and depth of patriotic feeling
amongst working-class Australians in the years leading up to war, it was of
undoubted significance in the early lives of many of the men I interviewed.
Stan D'Altera grew up in a poor, factory-working, Yarraville family which
enjoyed few material luxuries, but when war broke out his mother scraped
up enough money to buy the newspaper so they could read about Australian
exploits: 'Oh I was patriotic like, we'd had all this patriotic stuff at school
[...] Britons never shall be slaves'. Stan's older brother who 'was mad on
going to the war' enlisted immediately and served at the Gallipoli landing.
In May of 1915 Stan followed his example:

> I wasn't eighteen. I'm working on the lathe, like, next to another chap,
> and ... everyone said go to the war, and I said to him, 'Why don't you
> enlist?' He said, 'I'll enlist if you do'. I was apprenticed, not of age. 1
> went straight up to Victoria Barracks and enlisted. We left the factory
> and I had to get my father's signature. Well I forged that.

Stan's father found out and had him discharged, but within two days the lad
had run away to Ballarat and enlisted under a false name, and before the year
was out he was on Gallipoli.[4]

Stan D'Altera was one of several interviewees who twisted their parents'
arms or lied so that they could enlist under age. Alf Stabb grew up among
the seven children of a bricklayer and a home-working mother in the inner
eastern suburb of Prahran. He was working for the railways when war broke
out:

> I wanted to go, straight away [...] Just had the urge [...] Well see, I'd
> been soldiering all my life and the war broke out and of course your
> mates, they were all enlisting. They were going away and you were still
> stuck. My father wouldn't let me go ... see, if you were under twenty-
> one you had to have your parents' permission to go. My dad wouldn't
> let me go. Well he wouldn't give a reason but I think he was afraid that
> if I got killed or something like that, it'd be on his mind.

One Sunday in February 1916 a boyfriend of Alf's sister arrived at the house
with the intention of enlisting the following day, and Alf persuaded his
father that Vic would look after him if they enlisted together. After a short

time in camp Vic left the army because the route marches were affecting an old injury: 'He said, "You can cook something up to get out". I said, "It's taken me too long to get in to want to get out"'.[5]

Alf's story reveals the genuine enthusiasm of many young recruits, and shows that for some the years of cadets and compulsory military training (required of boys aged twelve and over, since 1911) had whetted their appetite for the real thing. It also highlights the encouragement or pressure of mates, and of a general public that was unsympathetic to young men who stayed at home; Alf was one of many eligible men who received white feathers in his letter box. For James McNair these public attitudes were an indirect cause of enlistment. A clerk with the Melbourne Post Office, he was convinced that he had 'no hope' of being accepted into the army because he was too scrawny. In January of 1916 he went to the Town Hall to get a volunteers' badge which would prove that he had at least tried to enlist:

> I thought that's all I'd get. So, [laughs] 'You're in'. I signed the form, came home, I told the mater, oh she cried. I said, 'I've joined the army, mother'. Oh well [...] 'That's it. Got to put up with it'. I was neither disappointed or pleased.[6]

Underlying the enthusiasm of youths like Alf Stabb and Stan D'Altera was the anticipation of adventure; indeed, for a generation brought up on deeds that won the Empire there was a fine line between adventure and patriotism. For others the war seemed to be an opportunity for excitement and the chance of a lifetime. Jack Flannery had grown up in a family of potato pickers who travelled around the north west of Tasmania looking for work. After the Gallipoli landing he decided to 'go to the war' and persuaded his father that unless he got permission he would catch the boat to Melbourne to enlist: 'To be truthful, I thought we were going to have a good trip, see the world and have a good trip'.[7]

E. L. Cuddeford 'thought it would be a pastime, like all the lads who enlisted'. He had left a farm which his father managed in outback New South Wales to serve an engineering apprenticeship in Sydney, and decided to enlist at a carnival held in Parramatta Park early in the war:

> I had a girlfriend with me. I remember that quite well. We were up in the hurdy-gurdy, went round and around and around. We got way up the top, the brakes weren't much good, and raining cats and dogs. We were stuck up the top all the time, they'd go down the bottom, they couldn't hold her, she'd go up the top again. We were out in the pouring

rain. That's how I remember that one. And I decided from there on that I would enlist. I think it was that night I made the move, there was a chap around that night called people who wanted to enlist.[8]

The forgotten side of the great adventure is that many young men were only too keen to leave tedious, exhausting or unfulfilling working lives. Ern Morton had grown up in Dookie, where his father was an instructor at the Agricultural College, and went straight from school into farm labouring work. Within a week of the outbreak of war he and a mate who was working on the same property had enlisted in the Light Horse: 'I think it was patriotism … I felt it a duty that everyone go to the war'. Yet he was also delighted to leave a working life which 'had no future in it' and to travel for the first time to the big city where he met 'chaps from all walks of life and … you soon got to have a different outlook on life'.[9]

For others, enlistment was an alternative to poverty and unemployment. Sid Norris grew up with eight brothers and sisters in a slab hut with a mud floor, near Gulgong in New South Wales. He killed rabbits for food, and foxes and possums for their hides, 'because father used to drink a lot and he never left any money'. Sid left home and went on the track looking for labouring work as soon as he could. He was not keen to go to war, but when he was laid off a mate lent him the fare to get to town to enlist: 'Well there was no work. I had no money. I never enlisted for any reason for King and Country. That wasn't in it'.[10] Sid Norris's story — and others from the city and the bush — recalls the poverty of many working-class Australians in the first decades of the twentieth century. Low and intermittent wages, large and sometimes single parent families, bad housing conditions, and prospects of an uncertain future in factory or farm labouring work meant that enlistment provided an attractive alternative for many young men.

The war could also serve as a respite from domestic problems. Although most of my interviewees were too young to be escaping from family responsibilities, Jack Glew went to war 'because I couldn't hit it with me dad, because the way he treated me'. His father ran a wood and coal yard out of Geelong. Every evening Jack went to the forest to load up cut wood, and then in the morning he would bring it to the yard before going to school, 'where I was the biggest dunce because I was that darned tired'. Eventually he ran away from home and worked as an onion weeder, and when his family discovered his whereabouts and brought him home he enlisted to get away.[11]

For older men like James McNair and Percy Bird, who were settled into work or family responsibilities, the decision to enlist was more difficult.

Charles Bowden's story highlights the pull between family and military responsibilities, and the tragic consequences of the pressures of duty and mateship. Born in 1888 he had worked in mining and other mechanical jobs on the Victorian High Plains before finding employment with the railways in 1911. He married the following year, and by the end of 1916 he had a two-year-old son and had secured his position in the railways through study at night school:

> I was working down at Port Melbourne, and this old Doull [the foreman], he was a warmonger, you know. As soon as I mentioned the war to him, going to the war, he bounced right and gave me a kit bag and some other things to take to the war with me. There was another chap in the office, a fellow called Maddigan, Chris Maddigan, he was a clerk and I was a store-man. He came in to me in the store and he said, 'Look, I'm going to enlist today, how about coming with me?' I said, 'I'll think about it'. So he got leave from the old foreman, who was only too happy to see us go to the war.

The two men went to the recruiting office in South Melbourne, but Maddigan would only enlist if he was offered a commission.

> So of course, phsst, out. They didn't want him. So I go in after him. They ask me a few questions, give me a few tests and then sign on the bottom line, and I'm a bloomin' soldier [laughs]. Anyway, after I'd become a soldier I didn't fancy it too much, see, and I wanted to get out of it, but I couldn't. I tried all sorts of manners and means to get out.

Charles's wife was appalled at what he'd done, and when he went away she rarely wrote because 'she was bloody well disgusted with me'. He regretted leaving her and the child because 'she was a very fine looking woman and in that she was up against all sorts of obstacles, you know'.[12]

Wild colonial boys

Going to war was an extraordinary adventure, and a time of excitement and discovery. Among the most vivid and pleasurable of the veterans' stories are those of the voyage through exotic new worlds to the northern hemisphere. Lesley David was on one of the first Australian troop-ships to travel via the Panama Canal in 1917. Before the war he had worked as a clerk with the railways in Melbourne, but from an early age he had dreamt of going round the world as a wireless operator on a ship. Lesley delayed enlistment until he turned twenty-one, as his parents had died and he was looking after his younger brother and sister, and then joined the Wireless Corps so that he could 'study and see the world at the same time'. Lesley recalls with relish the delights of tropical Tahiti, the hospitality of Americans in Virginia and the shore-leave pleasures which he shared with the officers after being given a typing job in the ship's Orderly Room.[13]

For many of the younger recruits this excitement was laced with confusion and trepidation. Albie Linton had grown up with a family of fourteen in the Tasmanian bush, and then worked in a North Melbourne factory before he enlisted with mates from his football team. For Albie and other young recruits based in training camps around Cairo, Egypt was a source of exotic bewilderment:

> Oh, it was a really different kind of life to what we had been used to leading. Actually, we were lost, out on your own like that [...] in the way of living and getting around. Their livelihood was altogether different to ours. Their streets were different, their marketing, and their markets were different to ours. Course, you've got a different view when you're eighteen and nineteen, in those things, to what you've got now.[14]

Ted McKenzie, who had been an apprentice coach-maker in Richmond before the war, recalls that 'we were thrilled with it':

> It was quite a turnout for us young blokes anyhow. It was something different. We had the pyramids and all that to go around. Cairo was a place where you could be ... quite interested and see quite a lot of things.[15]

Stan D'Altera was also curious about ancient monuments, but his strongest impressions were of the soldiers' brutality to the Egyptian people, and of the brothels that attracted 'a lot of the blokes, you know, older than me'. Contempt for the local population ranged from casual violence — the brutal elements among the Australians would 'wack 'em over the jaw for anything' — to a more calculated imperial policing which made young Stan question the school-book histories of justice in the 'Great British Empire':

> [...] about once a week, all spruced up, you know, our uniforms neat, boots looked highly shined and shining bayonets, we'd march through the native quarters of Cairo and give them the impression that's what they'd be up against if they revolted.[16]

Most of my interviewees glossed over such behaviour and portrayed their Egyptian adventures as 'a good time' of youthful innocence and enthusiasm. Perhaps the younger recruits were particularly careful in this strange new world, or perhaps as ageing veterans they were careful to sanitise their stories for a modern audience with different attitudes to race and behaviour. The historical record shows that the violence and racism of some of the Australians in Egypt, and their local unpopularity, continued throughout the war.[17]

The record also shows that many Australians were enthusiastic visitors to the Cairo brothels. In the year ending February 1916, almost 6000 men from the AIF were treated for venereal disease and over 1000 were returned to Australia. Ern Morton recalls that 'having no experience of city life or of anything like that it was really degrading. I could never, never force myself to associate with any of it'. While several of my interviewees shared Ern's disgust, others described the brothels of Egypt, France and England in vivid detail, though very few were explicit about their own sexual experiences. By contrast, chance meetings with young French peasant women and blossoming relationships with English 'girls' are common stories recalled with evident pleasure. These types of relationships with women provided a memorable contrast with the stark experiences of the trenches and the everyday male world of the army. The stories reveal how men's wartime relationships with women were easily polarised into the stereotypes of foreign whore and feminine rose, and suggest that this distinction may have been used by men to sustain their own moral self-respect.[18]

The difficult relations between the Australians and the local population in Cairo also signalled a tension between the Australian soldiers' attitudes to discipline and those of the military authorities. This tension would be a sore point within the AIF, and between the AIF and British headquarters, throughout the war. Both in Egypt and in Europe, where the bulk of the Australian force was based from 1916, absence without leave, drunkenness, and disrespect to officers were common AIF offences, despite repeated attempts by the Australian staff to improve the behaviour and image of their troops. At one point the proportion of Australians in punitive detention was about nine times that of the British army as a whole. British headquarters was appalled by the 'the extreme indiscipline and inordinate vanity of the Australian forces'. The diggers' disdain for military etiquette — a memorandum of the First Brigade recorded that saluting at officers 'gradually became extinct' towards the end of the war — particularly infuriated British officers.[19]

Many diggers relished the life and identity of — to use Albie Linton's tag — the 'happy-go-lucky army'. For example, Stan D'Altera describes his good fortune when he was given permission to transfer to his brother's battalion on the French coast at Étaples:

> I was cashed up, I'd had a good win at a two-up school. I had me papers. It took me three weeks to find them. I went round having a good time, [laughs], going to what they call the pub, the estaminet, and having some good feeds and that. I was really AWL [absent without leave], but I wasn't AWL really, officially.[20]

Alf Stabb recalls that 'you wouldn't go AWL in France [...] but if you got the chance to slip away in England well you'd just nick off for a few days and if you were lucky you'd get back and if you weren't lucky you'd get caught [laughs]'. Even in France the men of his unit enjoyed themselves when they could, as on one occasion in 1918 when they came upon the evacuated town of Corbie:

> That's one place where we had a ball. We had feather beds, [laughs], sheets and pillow slips. It didn't go on for long. About three or four days and the MPs [Military Police], they came along. There was a brewery there and we raided that for grog. It was the best show we had up there all the time we were there, I think.[21]

Not all the Australian soldiers were comfortable with this way of life. Bill Williams was unusual among the men I interviewed in that he came from a relatively well-off family and had worked in his father's property agency before the war. Now he found that he was 'different' to the other soldiers:

> I don't know that I was ever one of them in some ways. I'm not a snob, but I felt that I was more literate than most of them, probably I was. I think I had different tastes. Beer, booze, never appealed to me. It appealed to a lot of our fellows. I don't know that I could ever have the idea of chasing women who had been ... well, as the fellow said, preloved [laughs]. I couldn't face going to brothels like a lot of them did, and that sort of thing, and I suppose they thought I was what we'd call in those days a wowser ... I didn't feel that I was very happy about being different.[22]

Percy Bird and Fred Farrall also felt uncomfortable and even excluded by the diggers' larrikin behaviour, but for most of my interviewees it was a feature of the AIF in which they took great pleasure and pride.

Albie Linton explains that the 'happy-go-lucky army' was different because 'we didn't have the professional soldier instinct' of the British army, and because 'the outlook of our officers was totally different to them, because we were all volunteers [...] They had to do what they were told but we could argue the point a little bit with our officers'. The interviews are full of similar contrasts between the distinctive forms of discipline that Australian officers used with independent diggers, and the rule-bound British army with its autocratic officers and servile Tommies. Ern Morton explains this difference in terms of an Australian egalitarian ethos: 'in Australia you never took any undue interest in a man because he was in a different position to you, he was an Australian and he was a cobber and that's the end of it. That's the same way it was in the army'. Others agree with Alf Stabb that the Australian officer was 'one of the boys' because he came from the ranks.[23]

There is evidence that these attitudes were prevalent in the Australian army.[24] Yet for all their pride in the distinctive forms of AIF discipline, the diggers were still bound by army authority. Bill Harney was a cattle drover before he enlisted, and he contrasted life in the army with 'the old days of the cattle station where the man rounded up cattle and used his own initiative — that's all gone. You're just a big cog in the machine'. Jack Glew had joined the army to escape the discipline of his father, but found that it was another 'hard old life':

> Nothing that was any good that you would like, because you had to obey orders. Do this and do that, and if you didn't do it, by Jesus, they'd fix you up and put you in the clink. If there was another war I wouldn't go to it anyhow [laughs]. I'm buggered if I would.[25]

Sometimes, despite their loyalty to the egalitarian AIF, old diggers included Australian officers in their criticism. Ern Morton explains that as the Australian officers associated with English officers 'and saw how things were run in a regular army, some of them changed'. He recalls that one 'very great cobber' who was promoted out of the ranks didn't like 'to be seen associating with me' in front of other officers. Jack Flannery states that in extreme cases Australian officers were shot by their own men:

> They don't last long if they're bad. I know one bloke, he got bumped off, red-headed bloke. But, oh gosh, he was a cow. He was a cow out of the line. Oh, a private was only a dog to him. Like a dog. So, got rid of him.[26]

Because the Australian divisions served within the British army and were subject to British military discipline, such extreme criticisms were more often directed at British authority figures — especially staff officers and military policemen — and thus deflected away from Australian leaders. Yet veterans' memories remind us that even within the AIF there were significant tensions between Anzac officers and digger mates.

Sailors and non-combatants

For an Australian brought up on soldiers' stories, one of the surprising findings of the oral history project was that many veterans had not experienced war in the frontline. Men who served in the navy had very different war stories, and for these veterans their memory of the war seems to have been affected by the Anzac legend's emphasis upon the infantry experience. Bill Bridgeman had been encouraged by a mariner grandfather to join the navy before 1914. As an able seaman on *HMAS Sydney* (the escort for the first Australian troop-ship convoy which later became famous for knocking out the German cruiser *Emden)* he spent much of the war in the miserable conditions of the North Sea. Though he was very proud of his war service, Bill Bridgeman always felt that the Australian soldiers had got too much of the credit.[27]

Even within the army there were significant differences in the war experience. An army in the 1914–18 war depended upon a large administrative and support staff. Although the AIF had a higher proportion of fighting soldiers than most armies, there were still many Australian soldiers — including about a quarter of the men I interviewed — who did not see service in the line. They tended to be older men who had gained appropriate work experience before the war, and who now joined support units or were plucked out of their platoons to take on administrative duties. The war experiences of these men, which rarely figure in the public accounts of the Anzacs, were markedly different from those of other soldiers. They often felt uneasy or guilty about the relative safety and comfort of their service, yet they also knew that survival and good health depended upon their special status, and were grateful for training and education opportunities.

James McNair joined the 14th Battalion at Albert on 14 August 1916. After six weeks with the battalion on the Somme and at Ypres he was spotted by an old workmate from the Melbourne Post Office. This friend was now on the staff of First Anzac Corps, and he offered James a job keeping the records of the Assistant Adjutant-General. James was hesitant at first — 'I can have a rest when I'm with the battalion, here you're working nine in the morning till ten or twelve at night' — but he accepted when he was also promoted to Lance Corporal:

I was there pushing a pen till the end of the war — I never saw any more action [...] But he saved my life ... nearly everyone I went away with — you know, you were a little bit of a clique in the tents, in the camps [...] Oh, a lot of the chappies I went away with were killed in action or taken prisoner of war.[28]

Charles Bowden joined a unit which drove trains carrying soldiers and supplies between the French ports and the railheads of the Western Front. Though he had not really wanted to join the army he was quite happy in the job, which took him around the country and was relatively safe. Charles made the most of the opportunity to teach himself French, and after the war the army put him through night school at Pitman's College in London. Yet he always felt that 'we weren't what you might say, real soldiers, we never had any arms or any ammunition or anything like that ...'[29]

Like James McNair, Leslie David had also been a clerical worker before the war. Upon arrival in England he was ordered to work in the Camp Orderly Room, where he remained until the Armistice. He was 'very disappointed' that he was not allowed to do similar work closer to the action in France, yet he benefited from the work experience and from army-sponsored training at Pitman's College. For Leslie David the war 'was mainly a pleasant experience':

> You see, I was fortunate. I never got into any strife in France and so on. And six months in London, that was a pleasant experience; I made some good friends there. But generally speaking my war experience couldn't be regarded as anything but pleasant. I was a lucky one.[30]

End of innocence

For men in front-line units the war was a very different story. The outlines of the story of the AIF are familiar to most Australians. The soldiers of the Australian First Division landed at Gallipoli on 25 April 1915, and, after a bitter struggle with the Turkish defenders, established a toehold on the peninsula and gained lasting fame. The Gallipoli campaign became a trench war of attrition punctuated by bloody offensives, and after a bitter winter the emaciated Anzac survivors were evacuated from the peninsula in December. In March of 1916 the AIF, more than doubled in size by reinforcements but leaving the Light Horse units to fight in the Middle East, moved to France. After a short spell in a quiet sector of the line, the

Australian divisions joined General Haig's infamous Somme offensive. Although they made a name for themselves at Pozières by succeeding where earlier British attacks had failed, the Australians suffered terrible casualties from the artillery bombardments that were a feature of war on the Western Front. In 1917 Australian losses numbered 55 000 as the AIF spearheaded several of the step-by-step offensives at Ypres, which were eventually and disastrously bogged down in the mud of Passchendaele. The following year, Australians played a significant role in halting the German spring offensive, and then contributed to the vital breakthroughs in August which shattered the German armies of the Western Front and turned the tide of war.

At the time, Anzac publicists extolled the military triumphs of the Australians on the Western Front, which they regarded as final proof of the fighting qualities of the AIF, and a fitting climax to the national trial by fire. Though it is important to recognise that even for the Australians most of the battles of the war were strategic and human disasters, there is considerable evidence to support claims about the comparative military effectiveness and success of the AIF. The relatively new armies of the Dominions were less confined by military tradition than British units and better able to adapt to the demands of twentieth-century warfare. The high proportion of citizen-soldier officers in the AIF was significant in this regard; the Australian Corps commander, John Monash, exemplified the new breed of modern generals for whom war was fought like a modern business, and who emphasised appropriate training, planning and logistical support. Furthermore, within the huge British Army the relatively small and homogeneous AIF was a cohesive military unit, and, inevitably, the glowing reputation of the AIF became partly self-fulfilling as new recruits leant from old hands and were encouraged to live up to the Anzac tradition.[31]

Some of the men I interviewed highlighted the military qualities and effectiveness of the AIF. E. L. Cuddeford believes that it was initiative and independence which made the Australians such effective soldiers. He recalls that the diggers did not trust British units to defend their flanks, and explains why he thought the Tommies were inferior:

> I hated the British Tommy as a soldier. I always said he was a very good soldier, and a very game soldier, if led, which he wasn't. He was led a lot of times at the point of a revolver. They didn't treat him properly at all, didn't know how to handle the man.[32]

Jack Flannery concludes that 'there was only one soldier in the world equal to the Australian, and that's a Kiwi. And that's a big order isn't it?'

He had no fear. You could be a lieutenant or a captain or a major. I'm a lieutenant. We're leading a battalion into action. We get skittled. There's always a private, there's always someone in the mob that'll take over. They're what made it so good.[33]

Yet the fighting ability of the Australian soldiers is not the dominant feature of the war memories of diggers from the ranks. Veterans also emphasise the appalling conditions of trench warfare, the losses of friends, and their own feelings of vulnerability, confusion and fear. They recall the ordeal of constant shell and rifle fire, the stench of the unburied dead on Gallipoli, or the mud of France and Belgium, churned up by unceasing rain and bombardment, rotting the feet of the soldiers and deep enough for wounded men to drown in without trace. They recount images of mates caught on the wire, of battalions lining up reduced to the size of platoons, of treading on corpses in narrow trenches.

Ern Morton had joined the Light Horse, but he left his horse in Egypt when he was sent to Gallipoli after the landing:

We scratched little holes in the side of the hill, for protection. Then for three and a half months it never went out of my mind for one second, there was constantly shell-fire and rifle-fire, machine-gun fire. Continuously for three and a half months. It was a terrible ordeal for a lad of nineteen to go into [...] We weren't on Gallipoli long before we were up in the front lines, of course, and used to throw what we called 'egg bombs' at one another [...] One of these egg bombs landed on my cobber that I enlisted with. I'd been living with nearly all my life. Landed on his head and completely blew his head off. I was standing alongside of him. That was a shock ... [34]

Albie Linton arrived in Egypt just as the Australians were evacuated from Gallipoli. Among them was his older brother, who was in 'very, very poor condition, under-fed [and suffering from] dysentery'. As a member of the 31st Battalion of the new AIF Fifth Division, Albie's initiation into battle coincided with the first major Australian action on the Western Front, the ill-conceived attack at Fromelles:

We went over the bags, about 15 000, between half past six and seven in the evening, and we fought our way across to Fromelles. About twelve or one o'clock in the morning we got the order to retire to our own lines. See, because we had no hope of taking it. It was too well defended. They'd flooded us out, the barbed wire, we had to cross through barbed wire, and then they flooded open dams on us, and they flooded no man's land, and we had to get back through water and barbed wire that you couldn't see [...] I felt rotten. Wouldn't you? The man that said he wasn't frightened, well he's a liar [...] I think at the Roll Call next morning there were about five or six thousand answered the Roll Call out of about fifteen. That's how bad it was.[35]

Offensive actions like these were an infrequent part of a soldier's life in the war of attrition on the Western Front. Units alternated between the front line, reserve trenches and billets behind the lines, with occasional periods of leave in England. Life in the trenches was often a monotonous routine, yet the strongest remembered impressions are of appalling physical conditions and the terrible emotional and physical impact of artillery bombardment. James McNair only spent six weeks in the trenches, but that period is stamped in his memory, and he conveys the main themes of his experience in a few lines. In the heat of summer, the decomposing bodies dug up by shell-fire produced a terrible stench. Then rain turned the trenches into a quagmire and the diggers were 'up to our bloody thighs in water and weighing like blazes'. Worst of all was the shelling:

You can't get that blasting ... oh it's a terrific feeling [...] at times in the trench there, the shells can come over at the front, at the back; oh your ears would ring. See, you can be just a few yards away and not get a scratch. But oh, the detonation![36]

Embattled manhood

The diggers suffered from fear of death or mutilation, from the trauma of appalling sights, sounds and smells, from loss and guilt, and from extreme discomfort, exhaustion and illness. These stresses often produced disorders such as combat fatigue, nervous exhaustion and shell-shock. Some historians of the Great War argue that the experience of trench warfare precipitated a psychological crisis for many soldiers. Feminist historians suggest that this was a crisis of masculinity. Unable to live up to prescriptions for military masculinity that required them to be bold and enduring, but equally unable to escape from the trenches, soldiers took the psychological escape of nervous collapse.[37]

Statistical evidence suggests that Australians were no better off in this regard than other Great War soldiers. Indeed, one Australian study argues that diggers who bottled up fears so that they could live up to the Anzac reputation of bravery and self-control were especially prone to nervous collapse. Digger memories confirm the statistics. Fred Farrall recalls that after being wounded and dazed during the 1917 battle for Polygon Wood he began to show signs of a nervous condition which manifested itself in a stutter and extreme anxiety. Though few interviewees are so frank about the emotional effects of war, most remember mates who could not cope.[38]

There were ways of getting out of the line. Veterans recall that they were glad to receive a minor wound which took them away from the front (a minor wound requiring hospital treatment in England was known as a 'blighty'). As the pressures of war mounted, some Australians took more extreme measures. In the last year of the war incidents of self-inflicted wounds and desertion increased dramatically throughout the AIF. The Australians had the highest rate of desertion in the British army, and although the absence of a death penalty in the AIF may have been partly responsible for these statistics, there is little doubt that some diggers were desperate to stay out of the trenches.[39]

Alf Stabb joined the 29th Battalion and served in many of the battles on the Western Front. He recalls that after living on inadequate rations through the bitterly cold winters of 1916 and 1917, 'we were really sick of it'. He also remembers his own extraordinary good fortune, and describes several occasions when he left groups of mates to undertake a particular task and returned to find them all blown away. 'Well there must have been somebody looking after me up top, that's what I reckon, because it was uncanny.' Alf's luck continued, and in April 1918 he survived when a shell landed at his feet. The men around him were badly wounded but Alf suffered only minor injuries to his right hand. The regimental doctor bandaged him up and commented, 'You were lucky Stabby, this is a good one'. Back at the Casualty Clearing Station Alf was put with two other men who had hand wounds similar to his own. He could not understand why the three of them were being ignored by all the orderlies. Eventually a doctor arrived and questioned the three men in great detail. After some time he spoke to one of the nurses about Alf, 'Shift this man down below, you mixed him up. He's not a self-inflicted wound'. It was only then that Alf realised that his hand wound had made the doctor suspect that it was self-inflicted. He was sent to a hospital near Paris, but assumes that the other two men were court-martialled: 'It was getting near the end of the war, I suppose, and the fellows had had it and just couldn't take any more'.[40]

On Gallipoli, Ern Morton remembers men who would 'go to any lengths to get out of the army' and off the peninsula. He recalls one man who pestered him for days asking for his arm to be broken, and others who rammed picks into their legs while they were tunnelling underneath no man's land. One common ruse was to put condensed milk on your penis and claim that you had venereal disease. Ern was evacuated from the peninsula with a wound on the night before the August offensive. When he recovered he was sent to France with a unit of reinforcements for the Second Division Machine Gunners. In France, Ern became an outspoken campaigner against conscription amongst the diggers, because he felt 'very, very strongly against forcing people to go and fight in a war for the benefit of others'. In 1918 Ern's brother deserted from the army and went to Ireland, and Ern recalls in vivid detail the occasion which provoked his own rebellion:

> One of the most momentous experiences I had in the war, I think, in the whole of the war, was when we were going to hop over one day, and there was a German officer. Only a young chap, be in his late teens I suppose, and he had been mortally wounded. And of course with the machine-gun I didn't have a rifle, I had a pistol. He prayed to me to kill him. Put him out of his misery. He spoke English better than ever I'll speak it in my life, probably educated in an English university, and it flashed through my mind then, we've been taught all these years about these heathens we're fighting and they've got to be exterminated at all costs, and here's a man that could speak English and asking me to put him out of his misery. I turned against war. I was probably gradually drifting that way but that was the end. From that time on, and that was late in '18, I think, I never did anything in the war that ... all I did was keep out of it. I wouldn't fire a machine-gun, wouldn't shoot at anybody. I thought to myself, 'I'm not going to have anything at all to do with this'.[41]

More commonly, digger disillusionment was articulated as resentment towards their army commanders. The stubborn commitment of the British High Command to a war of attrition, and British generals' mismanagement of most major battles, were sources of bitterness and derision. It was also widely believed within the AIF that the Australian forces were being used to excess. As E. L. Cuddeford recalls, 'any tough corner, the Australian troops were pushed into it. The British troops were kept out of it. We objected to it, while we were pushed into everything dirty and rough and the British troops weren't there to help us'.[42]

But criticism was not only directed at the British. Fred Farrall relates two telling incidents that occurred when his unit was stationed behind the lines in the last weeks of the war. Fred and two others were sent to the Australian Army Corps compound at Flexicourt to bring back some men from their unit who had gone absent without leave and had served a sentence in 'the clink'. When they arrived at the compound they noticed a body of 1st Battalion men who were being held within a barbed wire fence. After talking to the guards, Fred and his pals discovered 'that the 1st Battalion had walked out of the Hindenburg Line and declared the war "off", or in other words they went on strike so they were all put in the clink'. In fact 119 members of the 1st Battalion had refused to return to action after their relief had been postponed. These men believed that they were not getting a fair deal but they were arrested, court-martialled and found guilty of desertion.

Fred Farrall remembers that the men of his own battalion were also getting restless in these last weeks of the war. It was a Sunday and the battalion, by now reduced from a thousand to three hundred men, was on church parade when their general arrived on his rounds. He spoke to the men in 'the usual glowing terms of how good we were, and then dropped a bombshell by saying that they had hopes that the war was coming to an end, but that if it was not successful then we would be back in the front line again':

> Well that's as far as he got. It was easy to get clods of dirt which were aimed at his horse, if not at him, and this was something that wouldn't be dreamt of in days gone by. But the situation developed to an extent where the general thought that it would be better to remove himself. So he did. That was the frame of mind that the soldiers had got into. Generally speaking they didn't want any more frontline or talk about it, any more than the 1st Battalion didn't want any more of the Hindenburg Line.[43]

Yet in common with most other units of the British army, the majority of the Australian soldiers did stick to their task, not with any great enthusiasm but because there were few realistic alternatives. Some veterans recall that their endurance was sustained by a sense of duty or responsibility. Alf Stabb recalls that 'it was a job that you knew you had to do and you couldn't squib it':

> You had to go on with it. Of course, we were volunteers, we weren't conscripts and we just went on with it. But nobody was … well everybody was more or less frightened I suppose. You could put it that way, nobody was a hero. You kept your head down when you possibly could.

Sid Norris remarks that 'you didn't care much, you developed that way that you didn't care, everybody was looking for a blighty'. But he concludes that 'the majority thought they had to win the war, they had to finish, you see, they had that at the back'. Some veterans recall that they were motivated by loyalty to the AIF battalion that had become their family and home. In particular they proudly relate the story of digger resistance to the merging of many of the depleted battalions in 1918, and use the incident to explain the bonding and spirit of their battalions.[44]

Soldiers also used a variety of more personal coping mechanisms to help them endure life in the line. While many longed for the temporary relief of blighty leave (in England) or even a blighty wound, others sought solace in alcohol, gambling or prostitutes, as well as less frowned-upon pastimes. Some individuals turned to God or other forms of spiritual succour; after several miraculous escapes Ern Morton and Alf Stabb began to feel that they were fated to survive, and that someone was looking after them. Most men adopted a fatalistic attitude as a psychological defence against anxiety and vulnerability, and expressed that fatalism in the distinctive language and humour of the trenches. Jack Glew had joined the Australian artillery in France as a horse driver, and recalls that 'I didn't seem to care [...] if I'd got knocked I wouldn't have cared. I'd seen so much of it'. A. J. McGillivray was an infantryman, and comments that he 'had great faith right to the end. Don't know why it was, we seemed to be so jovial and that under such conditions'. None of these coping mechanisms were unique to the Australians. Studies of British and European veterans show that they relied upon much the same strategies for survival.[45]

The company of mates was the most important physical and emotional support, both in and out of the line. War, like other isolated stressful situations, encouraged men to form close and supportive relationships. Comradeship helped men to cope in appalling conditions and to enjoy their life out of the line. Comrades could share equipment and skills, or sleep together for extra warmth and security. In the line a good friend could save your life, and most soldiers believed that they had a better chance of survival if they helped each other. Because soldiers' friendships were lived in extreme conditions and were likely to be cut short, they were often very intense, and comradeship became revered as an almost sacred bond. Though Australian mateship is highlighted in the Anzac legend, comradeship was not unique to the AIF. Many personal accounts from the European armies in the Great War single out comradeship as the most positive aspect of the war experience.[46]

Bill Langham recalls the material and emotional support of mates as the most positive feature of his war experience. In the cold of the European winter the six men of his artillery team slept close for warmth, so that 'when you turned, you all had to turn together'. When one of the men received a food parcel it was shared without question. One man, fifteen years his senior, became like a brother to Bill until he died of bronchial pneumonia.

> You had so many good mates … made you forget homesickness. When you go to war you find real mates. They, they'll die for you. They will too. You don't think, you don't think of yourself. You think of the other fellow. As long as he's all right, don't worry … about me. He thinks the same about you.[47]

Mateship provided a vital support network, but it also exerted a powerful pressure on soldiers to maintain a particular code of manhood and not let their mates down. Albie Linton recalls that 'you couldn't turn your comrades down, you had to be with them, in it'. Fred Farrall consistently refused to leave his front-line battalion for the relative safety of his brother's support unit, because that would have meant leaving his mates. The creed of mateship was a double-edged sword.[48]

Digger culture and identity

A man's comrades also served as the main forum for the articulation of feelings, attitudes and identities. They provided a place to grumble about the food, about officers or about army life in general. Among his mates a man developed particular ways to talk about experiences such as battle. And within this small community, soldiers identified ways in which they were distinctive in comparison with civilians and with soldiers in other units or armies.

Although in many ways the life of the diggers was similar to that of soldiers in other armies, their articulation of the experiences outlined in this chapter was shaped in particular ways by Australian attitudes and customs. Perhaps most importantly, my interviews bear out the claim that for many Australian soldiers the war was a potent experience of national self-recognition. Australia was too far away for the soldiers to return home for regular leave. The battalion became home and digger mates became a soldier's family, so that despite the separation from Australia the men's most intimate contacts were fellow Australians. In their training, and during life in and out of the line, they began to see themselves as part of a distinctive unit from a faraway country, identified by their Australian kit and the famous slouch hat, with their own common experiences of place, culture and even humour. The Australian vernacular was an important part of this identity; for example, military comradeship was defined using the language of 'mateship', with all its associated understandings of relationships between men and with authority.

The Australian soldiers savoured the military reputation that they had gained at Gallipoli and in subsequent battles, and liked to think of themselves as better than British soldiers. Stan D'Altera remembers that by the time he was invalided home in 1917 he had become 'a mad Aussie, and I'd proved that Australians were the best at anything, best in the world'. The diggers were proud of their status as citizen soldiers in an army that did not have conscription and did not use the death penalty, and resolutely maintained an informal attitude to military authority. They also became homesick, a feeling that was reinforced by the miseries of the Western Front, and developed in contrast an idealised and often pastoral image of their country.[49]

For all these reasons they enthusiastically identified themselves as Australians and adopted the affirming labels of 'Anzac' and 'digger'. Yet whereas the term 'Anzacs' was invented by head-quarters staff and used by publicists to denote the Australian and New Zealand soldiers, the term 'digger' was coined by the Australian other ranks on the Western Front in 1916 and 1917. They preferred the latter term because it was of their own making and because it signified their own distinctive culture. Most of the men I interviewed referred to themselves as 'diggers' and not 'Anzacs'. Though digger culture asserted a common national identity, it was primarily the identity of the Australian other ranks, and it did not necessarily carry the patriotic meanings that informed the language of 'Anzac'.[50]

The digger culture of the other ranks was defined and reproduced in codes of behaviour about and attitudes to mateship (no true digger would leave his mates), authority (diggers did not salute officers) and life out of the line (diggers were drinkers, gamblers and womanisers). Australian soldiers had to match their own behaviour against the public identity of the diggers. Though some relished the diggers' manly comradeship, for others the prescriptions were uncomfortable or exclusive. My interviews include several examples of men like Bill Williams who would not or could not live up to the standards of the digger culture, and who 'didn't feel […] very happy about being different'.[51]

Digger culture was articulated through storytelling, rumours and songs, which used distinctive language and slang; many soldiers' publications included glossaries of 'diggerisms'. In turn, this oral culture was worked up and crystallised for a wider digger audience, and occasionally for a civilian audience, in writing. The most important forms in which the soldiers wrote for and about themselves were the trench newspapers and annuals. There were a great number of digger papers throughout the war, with numerous contributors writing and drawing for a large readership. The papers varied from handwritten and stencilled sheets of gossip and verse to sophisticated, printed newspapers with news, photographs and articles, edited by journalist soldiers. Their content also varied, but on the whole they served as outlets for gossip and rumour. Though they were typically humorous in tone, they included bitter-sweet reflections on the war and mild criticisms of officers and the army, and thus provided a safety valve for the soldiers' discontent. They were also sentimentally Australian. Writers often adopted the popular Australian literary styles of the bush ballad and urban larrikin verse, although they usually excluded the more rough and critical elements of digger oral culture.[52]

Thus the Australian soldiers did not just have a legend created by others about their experiences; they were actively involved in fashioning and promoting their own collective identity. Nevertheless the making of digger culture and identity was not an independent process. Digger codes of behaviour were often in conflict with army regulations, and even in the AIF behaviour that overstepped the mark was sternly punished. Although the diggers' oral culture expressed the attitudes of the men of the ranks, it also drew upon the ways in which Australian soldiers were represented by others, in letters from home, in press reports and official citations, and in the steady stream of Anzac books that were published after the Gallipoli landing; indeed, official and journalistic writing was often intended to counter troubling aspects of digger culture. The writings of the diggers were also regulated by the military authorities. The Anzac annuals and several of the main trench newspapers were controlled by AIF staff officers and were primarily intended to bolster morale. Even the most simple productions were monitored by a supervising officer and were censored to exclude material that might stir discontent or undermine discipline.

The making of a digger culture and identity was thus influenced by, and in constant interaction with, the making of a more official legend by the army and by war correspondents and commentators. The next chapter explores how one influential commentator, the official war correspondent Charles Bean, responded to the diversity of Anzac experience and moulded a national legend.

CHAPTER 2

CHARLES BEAN AND THE ANZACS

Charles Bean is widely regarded as the most influential of those who contributed to the creation of Australia's Anzac legend. There were other important figures, such as C. J. Dennis, who wrote the best-selling account of Ginger Mick's transformation from urban larrikin to Anzac hero; Bean stands out, however, because of his unique role as official war correspondent and historian, his wartime proximity to the Anzac experience, and the enormous output of his writings about the war.

The scope of Bean's Anzac endeavours was prodigious. As Australia's official correspondent he was based with the AIF and wrote regular reports for the national press. He also edited a series of Anzac annuals, was responsible for some of the AIF newspapers, and published his own 1916 press reports in *Letters from France*. Apart from his writing he was active in various AIF political campaigns, including efforts to persuade the soldiers to vote for conscription, and attempts to boost the status of the AIF and establish a more independent Australian force. Anticipating his future task as war historian, he gathered oral and documentary evidence about Australian participation in the war, and created an Australian War Records Section which was responsible for collecting documents and relics for his proposed war museum and archive. To the same end he sought the appointment of Australian war artists and photographers, and was the dominant figure on a committee that met in London to vet their work. After the war Bean became general editor of the twelve volume Australian official history of the Great War, wrote the six volumes about the AIF, and had a leading role in the creation of the Australian War Memorial.

Bean's Anzac writings have had immense influence upon the ways in which Australians understand their participation in the Great War. Patsy Adam-Smith records in her own 1978 best-seller, *The Anzacs*, that 'for those researching this war one Australian stands out beyond all others [...] This writer is indebted to the man, Bean, as is anyone who searches for reality in the study of that time'. As this acknowledgement implies, for the most part

Bean's war writings have been regarded as a realistic and truthful account. Yet examination of the evolution of Bean's representation of the Anzacs, in both his journalism and his history, shows how he constructed a particular version of the experience of Australians at war.[1]

My argument is that Bean's Anzac legend-making provides a superb example of the 'hegemonic' process whereby a legend was created, not by excluding the varieties and contradictions of digger experience, but by using selection, simplification and generalisation to represent that complexity. Bean's representation of Australians at war was a result of the interaction between his pre-conceptions and his experiences with the AIF; his Anzac account was bounded by the limitations of his official roles and fashioned into an evocative narrative. He produced an idealised version of the Anzac experience which, nevertheless, captured and expressed key elements of digger culture and identity, and was resonant and appealing to many Australian soldiers.

Chapter 5 assesses Bean's history-writing in the context of postwar Australian society and politics. This chapter focuses on Bean's wartime writing, exploring the preconceptions he brought to the war and the ways in which they were remoulded by his Anzac experience; the constraints of writing at war and the influences of civilian and digger readers; and, crucially, how Bean's writing was used by Australian soldiers as they sought to comprehend and articulate their war.

An English Australian

Charles Bean's Anzac legend was influenced by a world view shaped during his pre-war years in England and Australia. Bean was born in the New South Wales country town of Bathurst in 1879. His father Edwin, who came from a British imperial family with East India Company connections, emigrated to Australia in 1873 to be a teacher in a private school. In 1877 Edwin Bean married Sarah Butler, the daughter of a Tasmanian solicitor, and together they moved to Bathurst where Edwin was to be headmaster of All Saints' College, and where Charles Bean apparently had an idyllic rural childhood. Twelve years later the family returned to England, where Edwin Bean was appointed headmaster of Brentwood School in Essex and Charles became one of his first pupils. In 1892 Charles enrolled at Clifton College in Bristol, and then, in 1898, he won a scholarship to read Classics at Oxford.[2]

In his education and family life Charles Bean was imbued with the values of service, honour, patriotism and valour which comprised the public school ethic of imperial England. He also had the martial upbringing typical of a boy of that imperial epoch and class. Bean later recalled that, like many Australian and English children, he was 'brought up on tales of Crécy and Agincourt, Trafalgar, Waterloo, the Indian Mutiny and the Crimean, Afghan, Zulu and other British wars'. In his teens he was thrilled by trips to Waterloo and to his father's Volunteer Force training camps, and after one visit to Portsmouth Bean developed a detailed interest in the British navy which inspired him to read every book he could find on the subject. He joined the School Engineer Corps at Clifton and the Oxford University Battalion of the Oxfordshire Light Infantry.[3]

Bean's martial and imperial enthusiasm was fuelled by his pride in the heritage and values of the Anglo-Saxon race. A few years before the war he defended the 'pure old Cross of St George' against critics of 'flag wagging', and explained 'what that flag means to me [...] whether it has five stars upon it or a maple leaf':

> It stands for each and every one of these ideas — for generosity in sport and out of it, for a pure regard for women, a chivalrous marriage tie, a fair trial, a free speech, liberty of the subject and equality before the law, for every British principle of cleanliness — in body and mind, in trade or politics, of kindness to animals, of fun and fair play, for a politeness [...] that will be made good in real life by real sacrifices if need be, for the British Sunday, for clean streets and decent drainage, for every other canon of work and sport and holiday, and a thousand and one ideas, wrung out by British men and women from the toil and sweat and labour of nine hundred years [...] that made the Anglo-Saxon life worth living for the Anglo-Saxon.[4]

The passage marks Bean as a man of liberal ideals, but it also signals an ignorance of social divisions and living conditions in Britain, and a very English chauvinism, which were born of Bean's own limited experience. It reveals a concern with moral and physical cleanliness which would recur in Bean's Anzac writings, and it evokes an interest in national character which Bean sustained when he returned to Australia. In 1902 Bean failed the examination for the Indian Civil Service. He took a degree in law instead, and in 1904, twenty-five years old and eager for change, he sailed for Australia.

Bean settled in Sydney, where he dabbled in teaching and the law, including a stint as the associate of a circuit judge whose duties included tours of outback New South Wales. He began to write newspaper articles about his impressions of Australia, and in 1908 joined the *Sydney Morning Herald* as a junior reporter. The editors were impressed with his background knowledge and skilful reporting, and offered him special assignments to report on naval and military affairs and on life and work out in the country. By 1914 he was the successful author of two books on naval subjects and two books dealing with impressions formed on his outback travels.[5]

Like other English immigrants of his generation, Bean's pre-war life and writings display a creative tension between Australian and English identities. He idealised Australia as an Anglo-Saxon oasis in the Pacific, and was convinced that the transplanted race needed to be protected from the oriental threat. He shared the ideas of contemporary social Darwinism, which assumed that there was an innate relationship between race and moral and cultural traits, and he was convinced that the English were pre-eminent because of their superior characteristics. Bean's racial attitudes would change, but the notion that the character of each individual exemplified distinctive national traits would remain the central explanatory tool of his life's writing. This idea dovetailed with a typically Victorian personal philosophy which assumed that an individual of sound character could determine his (rarely her) own fate, regardless of personal privilege or economic power. For Bean, individuals and nations were the main actors of history; race and physical environment were its driving forces.[6]

In fact, Bean gradually began to judge the Australian character as a racial improvement. In his travels around rural New South Wales he concluded that the Australian environment, and bush life in particular, brought out the very best in the Anglo-Saxon race. In the struggle to tame their harsh environment Australians had developed qualities of resourcefulness and independence which were stifled in the cramped industrial cities of Europe. Bean's thinking coincided with a popular notion about the progress of the British race in the frontier lands of the New World. Although there was some anxiety about the work discipline and social behaviour of colonials — especially in Australia with its convict origins — the 'Coming Man' of the White, imperial frontiers was idealised in adventure fiction and boys' culture, and became the most resonant symbol of Australian character.[7]

Bean also believed that Australian social and political relationships were a distinct improvement on the British model. For example, the rigour and isolation of bush life promoted 'mateship', the quality of sticking to your mates through thick and thin. In the decades preceding the First World War the meaning of 'mateship' was contested in Australian political culture, with the mateship of workers espoused by radicals during the strikes of the 1890s opposed to a mateship which included men of different classes. Writing almost twenty years after these strikes, and with an idealised notion of class and society, Bean portrayed shearers and labourers as mates with their sheep station bosses, and thus contributed to a conservative appropriation of the language of mateship. Freed of the feudal shackles of the Old World, Australia was, according to Bean, a more egalitarian society. As he later wrote in his history: 'Probably nowhere were the less wealthy folk more truly free, or on such terms of genuine social equality with the rich, in dress, habits and intercourse'.[8]

Bean's ideal Australian may have been an improved breed of Anglo-Saxon manhood, but his Coming Man was not politically independent, let alone republican. In the pre-war period Bean easily accommodated his rediscovered Australian identity with imperial loyalty. As a journalist he approved the militarisation of Australian society, which included the introduction of compulsory military training for boys and the creation of an Australian navy for national and imperial defence. In 1911 his book, *The Dreadnought of the Darling*, prefigured Australian participation in the impending war by stretching the concept of mateship to define the imperial relationship. If ever England needed help, he wrote, it would be found:

> [...] in the younger land, existing in quite unexpected quarters, a thousand times deeper and more effective than the more showy protestations which sometimes appropriate the title of 'imperialism', the quality of sticking — whatever may come and whatever may be the end of it — to an old mate.[9]

When war was declared in 1914 Bean was convinced of the rightness of Australian participation, and that the extraordinary enthusiasm for enlistment would dispel anxieties about the imperial loyalty of the young race.

In September 1914 the British government invited each Dominion to include an official correspondent with its military contingent. In Australia the Federal government asked the Australian Journalist's Association to nominate a man for the job, and Bean was delighted when he won a ballot of Association members by a few votes. Over the next four years, Bean's experiences with the Australian soldiers would alter his attitudes to war and national identity. However his perception of those experiences, and the legend he created, was influenced by his perspective as an upper middle-class English Australian.

Baptisms of fire

Bean's preoccupations with an ideal and truly Australian identity are evident in the way he handled his first major task, which brought him into immediate confrontation with the larrikinism of Australian soldiers. Having sailed to Egypt with the first Australian contingent in November 1914, he became embroiled in the controversy over the bad behaviour of the Australians in Cairo. Bean's own view, expressed in a booklet of advice for the soldiers about the culture and history of their temporary home, was that 'even the humblest Britisher here in the East' should maintain the reputation of the race for 'high principle and manliness'. But, as he later recalled, 'leave-breaking, desertion, attacks upon natives, robbery, and disease began to reach such a pitch as to destroy the great name' which the Australians should have been earning in their training.[10]

The Australian commander, General Bridges, decided to discharge several hundred offenders and ship them back home, and asked Bean to send an explanatory despatch to Australia. The despatch, dated 29 December 1915, began by arguing that it would be 'a deceit upon the people of Australia if it were reported to them that Christmas and the approaching New Year have found the Australian Imperial Force without a cloud in the sky'. He praised the physique, bearing and potential of the Australians, but confessed that there were problems in the AIF caused by 'a leaven of wasters':

> There is only a small percentage — possibly 1 or 2 per cent — in the force, which is really responsible for the occurrences about which Cairo is beginning to talk; the great majority of men are keen, intelligent young Australians who you will meet enjoying their hours of leisure in front of the cafes, or in the museum, or the zoological gardens, or the postcard shops, dressed as neat as any of the other soldiers in town, and behaving themselves in a way which any rational Australian on a holiday would behave [...] But there is in the Australian ranks a proportion of men who are uncontrolled, slovenly, and in some cases, what few Australians can be accused of being — dirty.

He suggested that the older veterans of the South African war were leading young soldiers astray, and concluded that it was necessary to send the miscreants home to preserve the country's reputation.[11]

Bean's despatch was printed in most Australian newspapers and provoked an outburst of indignant correspondence. Some letter-writers were appalled that such wasters had infiltrated the noble AIF, while others condemned

Bean's criticisms of Australia's bold volunteers. Bean's portrayal also caused great resentment among the soldiers at Mena Camp. In a letter to his mother, AIF Private K. S. Mackay wrote that 'the stories are true to a certain extent but not enough to warrant such attention', and concluded that Bean had 'done himself and Australia more harm than all the men Australia sent to this or any other war. The men will not forget him either'. Bombardier Frederick Rowe wrote dismissively to his family about Bean's report that 'it is lies from start to finish and he has got himself in very hot water', and described 'indignant meetings all over the camp'. He enclosed a poem by a mate from his unit which concluded:

> Let me tip yer mister Critic
> Don't take walks along the Nile
> Else perhaps yer taste its waters
> While the boys look on and smile.[12]

The tension between Bean and the soldiers at Mena Camp is significant in several ways. Firstly, it signals the differences between Bean's understanding of the soldiers' behaviour and how some of them thought about themselves. Bean could not comprehend that any 'genuine' Australian soldier would want to visit a brothel; his language reverberates with moral and physical disgust for the 'unheard of vileness' of Cairo low-life and for the dirty soldiers who could not be true Australians. Bean's attitude was influenced by the 'rational recreation' movement of nineteenth-century England, with its ideal of moral manhood and support for healthy and controlled recreation as an alternative to the popular working-class pleasures of gambling, gaming and drinking. Lack of experience or empathy for working-class life and pleasures caused Bean to project his own values upon the diggers. In turn, the Mena affair reinforced Bean's personal shyness and distance from the men he so admired, as he recorded in his diary in 1918:

> I have been shy of these men — have done my work from outside as a staff officer as it were — I don't know if I should have mixed with them more if the unpopularity I gained at Mena had not made me shrink from living among the men — anyway I am too self conscious to mix well with a great mass of men.[13]

The language of Bean's Cairo report is also revealing. For an official correspondent he was remarkably frank about Australian misbehaviour, but he also used a narrative strategy of distinguishing between the majority of 'genuine' Australians and an alien minority, which subsequently became a defining feature of his historical writing about Australians in battle. Yet the public outcry over the Mena affair was a lesson to Bean and the Australian command, and thereafter criticisms of Australian misbehaviour were either censored or played down by Australian publicists, including Bean, to avoid tarnishing the image of the AIF or undermining recruitment.

More generally, the Mena furore reflected and articulated tensions between the official ideal of a soldier in the British army, and the behaviour and identities of the Australians. Within the British army Australian misbehaviour out of the line, and the diggers' disrespect for military authorities, remained notorious throughout the war, just as the diggers remained suspicious of official prescriptions for their behaviour. Caught between his official role and his admiration for the men of the AIF, Charles Bean would continue to struggle, personally and in his writing, to make safe sense of the disturbing behaviour of some of the Anzacs.

Bean's other way of coping with Anzac larrikinism was to emphasise its positive features. Like many of the Australians, Bean disliked the British military sticklers who thought the salute was more important than the soldier. In his diary, and occasionally in his public reports, he began to characterise the issue of Australian behaviour out of the line not as a problem of Australian indiscipline, but rather as representing a positive development away from the authoritarianism of the British army. Even amidst the troubles of Cairo he believed that the 'strong positive virtues' of the Australians would outweigh their 'strong positive vices', and that the diggers would be loyal and well-disciplined in their own way when it really mattered. Like many other Australians, Bean keenly anticipated that the Australian qualities of the men, and hence an Australian national identity, would be triumphantly realised in battle.[14]

These expectations shaped Bean's reporting of the Australian initiation at Anzac Cove, though in fact the first press report about the Anzac landing at Gallipoli, rapturously received in Australia on May 8, was filed not by Bean, but by the flamboyant British war correspondent, Ellis Ashmead-Bartlett. His report made the most of the mythic ingredients of the landing — the untried soldiers of the new nation, the impregnable cliffs, and the classical setting of Asia Minor — and praised the Australian soldiers:

> There has been no finer feat in this war [...] These raw colonial troops in these desperate hours proved worthy to fight side by side with the heroes of Mons, the Aisne, Ypres, and Neuve Chapelle.

Undaunted by the cliffs of Gallipoli, the 'race of athletes' had demonstrated that 'colonials were practical above all else'. Australians had been 'tried for the first time, and had not been found wanting'.[15]

The lavish approval of the fighting qualities of the Australians set the tone for subsequent stories of the landing. Ashmead-Bartlett's report had an immediate and influential impact in Australia, in part because he could not be accused of boasting, and because it finally disproved fears about the military discipline and effectiveness of the AIF. Here was an Englishman announcing to the world that Australian soldiers had proved themselves among the best in the Empire, with their own distinctive, practical qualities. Following Ashmead-Bartlett's lead, Australian politicians and newspapers declared April 25 the national baptism of fire, and claimed that the young country had proved itself a worthy partner of the Empire.[16]

Bean landed at Gallipoli on the first morning and remained on the peninsula for most of the campaign. Though to his chagrin his own report of the landing was held up until May 13, it also received wide coverage; by May 18 it had been printed together with Ashmead-Bartlett's report in a pamphlet for New South Wales schools titled *Australians in Action: The Story of Gallipoli*. Bean's account was less florid than that of Bartlett. He concluded that the first attack up the cliffs — 'like a whirlwind, with wild cheers and bayonets flashing' — would go down in history, but he also stressed that it was no reckless charge, but was directed from the front by junior officers trained to act on their own initiative. Though in his journalism Bean had not yet fully crystallised his assessment of the Australian fighting man, he was already articulating the main themes of the Anzac legend in terms of his pre-war preconceptions. After only a few days on Gallipoli he wrote in his diary that 'the wild pastoral independent life of Australia, if it makes rather wild men, makes superb soldiers'. For Bean, Gallipoli set the standards of the AIF, such as indifference to fire by soldiers and non-combatants (he described Australian soldiers working on the beach 'careless of any fire, in the good old Anzac way') and close relations between officers and men, resembling, from Bean's perspective, the relationship between manager and workers on a sheep station, or prefects and boys in a public school. For Bean, Gallipoli dispelled the doubts of Egypt and proved that the Australians responded to good leadership in battle, and could hold their heads high as soldiers.[17]

Conversely, events on Gallipoli, and subsequent battles on the Western Front, reinforced Australian bitterness about British command. Though Bean's war correspondence was characteristically discreet about this bitterness, his diary was seething with criticism. He identified the main cause of the problem as the British class system. For example, he wrote in his diary that the defeats of the British army in April 1918 were not only the fault of Generals Haig and Gough. They had a deeper origin:

> [T]he real cause has been as plain as an open book since Suvla Bay [a British military fiasco at Gallipoli] — it is far deeper than the failure of this or that division or general. The real cause is the social system of England, or the distorted relic of the early middle ages which passes for a system, the exploitation of the whole country for the benefit of a class — a system quietly assumed by the 'upper class' and accepted by the lower class.[18]

According to Bean, in this system the British upper class did not need brains or ability to gain command, and British soldiers were not solid yeomen but were rather the feeble products of exploitation, industrialisation and urban decay. The war experience shattered Bean's pre-war idealisation of British society and confirmed his belief that Australian social and political conditions were a great improvement and had produced a better army. Bean argued that because Australian officers and men came from the same social background and treated each other as equals, they were able to work together in an effective military partnership. He also concluded that the informal discipline system of the AIF encouraged initiative and loyalty, and produced skills of battle rather than of the parade ground.[19]

Bean remained a staunch imperialist — he hoped that Australia would be a model for 'a great empire [...] young, beginning, active and thinking' — but his English identity faded and he increasingly identified himself as an Australian. His reports fondly described 'the familiar old pea-soup overcoats and high-necked jackets and slouch hats of the Australians', and he came to share the diggers' apparent preference for dirty old Australian tunics, and their disdain for British uniform issues: 'The feeling is extraordinarily strong — and I have it too [...] I *hate* seeing them go into a British tunic. It seems to me the hallmark of a different being — a more subservient less intelligent man'.[20]

Bean and other Anzac publicists like Keith Murdoch and Will Dyson became forthright, nationalist advocates for the diggers and the AIF. Angered by the lack of recognition in the British press for the AIF military successes of 1917 and 1918 (the War Office often assumed that the Australians were British and, therefore, that there was no need for distinction), Murdoch and Bean organised a series of visits to the AIF front-line by influential politicians and writers. They also campaigned for the creation of an Australian national army, and were delighted by the establishment, in November 1917, of an Australian army corps staffed and commanded by Australians. As a result of this amalgamation, the victories of the Australian Corps in 1918 could be unequivocally credited to the Australians, and Bean celebrated these victories as an epic national achievement to rank with the landing at Gallipoli.[21]

Writing at the cutting edge

When Bean was appointed official war correspondent the Australian government anticipated that he would also produce a history of Australian participation in the war. This dual role of correspondent and historian influenced the way Bean gathered information and wrote about the Australian soldiers. He did not think of himself as an ordinary reporter, and believed that he needed to collect detailed, first-hand evidence for his reports and for the national history; he was usually based with the AIF divisions in order to have ready access to military resources and information.

When, for example, the war correspondents were instructed in June 1915 to transfer from Gallipoli to a base on the nearby island of Imbros, Bean wrote a memorandum to the ANZAC commander General Birdwood arguing that he needed to remain on Gallipoli because 'the category of news which my duties require me to obtain [...] has no relation to that required by journalists responsible to newspapers'.[22]

This national responsibility complemented Bean's personal commitment to thorough investigation of the facts of an event so that he could 'record the plain and absolute truth so far as it was within his limited powers to compass it'. It also suited his interest in individual experience and the ways in which individual character, moulded by race and nationality, influenced the outcome of events. Bean liked to focus on the exploits of individuals and small groups because he believed that wars were often won or lost at the 'cutting edge' of battle. Furthermore, Bean believed that because he was responsible to the Australian nation and people and not to any newspaper editor, his job was not to report the activities of other forces or the general trends of campaigns, but was to describe and assess Australian participation in the war. As he wrote in the same memorandum, what he could not get at Imbros were:

> [...] the details as to the life, scenes, bearing of men, scenes that will stir Australian pride (there are plenty of such details told to the British people of their soldiers) — which is what the nation I represent wants to hear.

Thus Bean's perception of his job led him to explore the character and fighting qualities of his Anzac heroes, which he perceived in national terms and which became the main themes of his journalism and history.[23]

Driven by these perceptions of his role and by a keen sense of duty, Bean was a diligent and brave war reporter. He spent much of his time in or near to the front-line, determined to observe every major action of the Australian forces. At Gallipoli, in particular, he shared the privations and dangers of the soldiers. He was recommended for a decoration for bravery after assisting wounded soldiers — thus regaining the respect of the men he had criticised in Egypt — and was wounded while covering an Australian offensive. Despite the pain and discomfort he remained on Gallipoli so that he could continue his work. In addition to his own observations, Bean amassed the information he required by interviewing Australian soldiers and staff about every battle. Later he described his work at Gallipoli, which marks him as one of Australia's first oral historians:

> Day after day I would walk or climb to some part of Anzac, sometimes of Suvla, and seek from eyewitnesses accounts of events in which they had often been principal participants. They would send me on from one to another, submitting to the closest examination carefully jotted down by candlelight, sometimes until two or three in the morning.[24]

Figure 2   *Will Dyson's wartime illustration, 'Captain Bean typing despatches after a hard day at Polygon Wood.' (c. 1917, crayon, 25.7 x 17.8 cm, Australian War Memorial [9931]).*

Then, late into the night — though he was often cold, exhausted and unwell, and the light was so poor that he frequently wrote on top of other words — Bean transcribed his daytime interview notes and observations into the diary that subsequently became his chief record of the war.

Bean's diary is an invaluable source for exploring his wartime writing and the ways in which, over time, he gradually articulated and fashioned his account of the Anzacs. The diary has been the subject of historical debate, and it does need to be used with caution. Bean himself noted that because the diary was often written in a stressful situation, his observations and statements were sometimes confused and inaccurate. Bean was also careful about what he wrote in the diary because he was concerned that it might fall into the wrong hands. More importantly, close examination of the copies of Bean's diaries held at the Australian War Memorial reveals that, far from being a consistent chronological series of day by day accounts, each 'diary' contains material written at different times and in very different contexts.

Take, for example, Bean's diaries of the Anzac landing at Gallipoli on 25 April 1915. Diary number 4 contains brief handwritten and shorthand notes which Bean made on the spot during the day. He stopped writing in the early afternoon — it was too difficult and dangerous — and that night recommenced with a slightly more detailed account of the rest of the day. In this diary many of the faded pencil marks have been subsequently retraced in ink; pages have been torn out, and additions and corrections have been made with different writing implements at later dates.

Diaries 5 and 6 also record the events of the landing, and are the diaries that Kevin Fewster edited and published in 1983 as *Gallipoli Correspondent*. They draw upon the notes in diary 4, and upon interviews with participants which Bean collected in subsequent months and recorded in another set of notebooks (catalogued by the Australian War Memorial as diaries 25 to 28). Textual evidence suggests that the bulk of diaries 5 and 6 were written by Bean towards the end of 1915 and early in 1916. He continued to amend and delete material — certainly in 1920 and 1952 — as he came upon new information or developed new understandings of the events of April 1915. By comparison with diary 4, diaries 5 and 6 are well-worked narratives of the Anzac landing — often switching between past and present tense — which were probably written with a view to publication, or at least as the basis for subsequent historical writing. Contrast, for example, the following passages as Bean observes the landing from his ship.

7.22 Our men seen on top of ridge. (Helio working from further hill face of it 7 [o'clock].) Men quite plainly visible in large numbers — entrenching slightly behind Hill top — walking in quite unconcerned. All wounded in boats. [Diary 4]

Ten minutes later someone sees men upon the skyline. The rumour gradually spreads round. At 7.17 I heard of it [in 1921 Bean recorded in diary 4 that his own watch was thirteen minutes fast by comparison with the Corps Diary]. Through the telescope you can see them, numbers of them — some standing full length. Others moving over it. Certain ones are standing up, moving along amongst them. Others are sitting down, apparently talking. Are they Turks or Australians? The Turks wear khaki, but the attitudes are extraordinarily like those of Australians. Just below them on our side of them a long line of men is digging in quietly on a nearer hill. They have round caps, I think clearly you can distinguish that round disc-like top. They are Australians! And they have taken that second line of hills! [Diary 5][25]

In short, the 'diary' which is usually quoted in studies of Bean's work was a carefully constructed narrative written after the event with a view to publication. The line between 'private' and 'public' accounts is blurred, and diary passages clearly need to be checked for their provenance and for their intention at the time of writing.[26] Despite these qualifications, Bean's diary, which is often an impressively frank and self-questioning document, is still the best source for studying the development and articulation of his wartime thinking, and his handling of the contradictions of the Anzac experience. It provides a useful contrast to Bean's contemporaneous war correspondence, and reveals how his press reports were affected by the dictates of propaganda and censorship, by loyalty to the Allied commanders and cause, and by Bean's own idealisation of the Anzacs.

British propaganda and censorship were superbly well controlled during the Great War. The guidelines for literary propaganda were drawn up in London just after the outbreak of war: correspondents were expected to write 'interesting' articles and books with small doses of propaganda that were not too apparent and would not detract from their appeal; they were also required to depict the soldiers as 'cheerful and happy'. The basic rules of censorship — that reports must not provide information for the enemy, needlessly distress the bereaved, or criticise the military conduct of the war — ensured that these guidelines could not be breached.[27]

The nature of propaganda and the degree of censorship largely depended on definitions of what was distressing or critical. Bean vehemently opposed one official view that lies were justified if they helped the war effort, and argued that the destruction of public confidence in the government could 'conceivably do far more harm to a nation than defeat'. Bean's criticism of this form of propaganda was directed specifically at British military censors and was fuelled by a personal and national grudge. The censors often used trivial excuses to block his own reports, and Bean believed with some justification that they were unsympathetic to the national role of an official correspondent from the 'colonies'. Yet Bean's anger was largely confined to his diary. He was much too obedient to flout the rules of censorship, and criticised correspondents like Ashmead-Bartlett who tried to make uncensored reports.[28]

Bean's own loyalties were more directly influential than censorship in the construction of his wartime account of the diggers. As a confidant of many of the Australian commanders, he refused to betray their trust and the inside information that they provided. When he was critical, it tended to be about the way the Australians were being used rather than about the war itself. He maintained his belief in the Allied cause and would not write reports that might provoke doubt about that cause. As Kevin Fewster remarks, when Bean had to choose between truth, devotion to Empire and loyalty to its military leaders, 'he invariably chose to keep his criticisms to himself'.[29]

By his own admission, Bean threw a cloak over 'the horror and beastliness and cowardice and treachery' of war, and the Australian public received a highly selective version of the soldiers' experience. In his Gallipoli diary Bean admitted that soldiers — 'even Australian soldiers' — frequently ran away; that they needed to be threatened by their officers; that almost all dreaded the front and some shot off their fingers to escape it. Very little of this was revealed in Bean's war correspondence, which celebrated the Australians' successes or, at worst, argued that they were 'putting a good face upon it under conditions which [...] were sheer undiluted misery'. For example, although Bean's report of the disaster at Fromelles admitted 'very severe losses' (Bean was castigated by the British censor for that admission) he included none of his diary criticisms of the mismanagement of the battle, and instead focused on the few soldiers who had reached the enemy lines, characterising the Australian efforts as 'worthy of all the traditions of Anzac'.[30]

Within the limitations of censorship and personal loyalty, Bean was a far more rigorous investigator than many other war correspondents, and had a much higher regard for the truth. While others laced their stories with information from official communiqués or with the accounts of hospital heroes, Bean was wary of the official line and despised the exaggerations of 'Cairo correspondents'. On 26 September 1915, in one of the most revealing passages of his diary, Bean exploded against this type of journalism:

> But what wretched cant it all is that they talk in the newspapers [...] I can't write about bayonet charges like some of the correspondents do. Ashmead-Bartlett makes it a little difficult for one by his exaggerations, and yet he's a lover of the truth. He gives the spirit of the thing: but if he were asked [...] 'Did the first battle of Anzac really end with the flash of bayonets all along the line, a charge, and the rolling back of the Turkish attack', he'd have to say 'Well — no, as a matter of fact that didn't occur'. Well, I can't write that it occurred if I know that it did not, even if by painting it that way I could rouse the blood and make the pulse beat faster [...][31]

Bean knew that any initial Australian eagerness for battle was by this time long gone, but that journalists were still being misled by 'the false literature of other wars' or were 'bent on sustaining the national determination in this one'. Experiences at Gallipoli and later on the Somme persuaded Bean to curb 'any tendency to glorify war', and to depict 'as far as possible the suffering and misery of the war'. As a writer at the cutting edge, Bean had realised the inadequacy of what Paul Fussell has termed 'high diction' — the romantic rhetoric of war that men of Victorian and Edwardian generations, including Bean himself, had grown up with — for describing the soldier's experience of modern warfare. Bean's remedy, which suited his journalistic style, was to describe battle in plain, simple and 'Anglo-Saxon' prose, with a minimum of rhetorical flourish.[32]

Yet despite this avowed intent to record the plain details of battle, the facts of Bean's war correspondence were still selected and articulated according to his Anzac ideal, and were fashioned to make a good story. Although Bean had realised that some men could not cope with the 'suffering and misery of the war', his main impression, and the impression he recorded in his diary and conveyed in his journalism, was of the heroism of the men who stuck to the task. The worse the conditions of war, the more heroic was the ordinary Australian soldier. He wrote in his diary that for soldiers to go into Pozières 'in spite of their natural state of mind and do all they would is a hundred times finer than the heroics that have been written in the past', and he elaborated on that theme in his report to Australia about the men of Pozières:

Steadfast until death, just the men that Australians at home know them to be; into the place with a joke, a dry cynical Australian joke as often as not; holding fast through anything that man can imagine; stretcher bearers, fatigue parties, messengers, chaplains, doing their job all the time, both new-joined youngsters and old hands, without fuss, but steadily, because it *is* their work. They are not heroes; they do not intend to be thought or spoken of as heroes. They are just ordinary Australians doing their particular work as their country would wish them to do it. And pray God Australians in days to come will be worthy of them![33]

Despite his criticisms of the exaggerated heroics of much war writing, Bean did not reject the idea of heroism. He simply redefined it by arguing that the soldiers who endured the horrors of trench warfare demonstrated their own brand of heroism, just as impressive as that of military fiction. Significantly, he defined this heroism of endurance in terms of Australian national character. Thus Australians were more likely to stick to the task than soldiers of other armies, as he wrote in his diary and reported to Australians back home:

> [...] the actual truth is that though not all Australians, by any means, do their job, there is a bigger proportion of men in the Australian army that try to do it cheerfully and without the least show of fear, than in any force or army that I have seen in Gallipoli.[34]

Convinced that the qualities of the Australian soldiers were the reason for their military successes, and determined to convey their experiences in those terms in order to sustain Australian pride and resolve, Bean was clearly not just writing matter-of-fact description. His prose was carefully framed to make particular meanings about the war and about the achievements of the Australian soldiers.

One of the best examples of the literary and didactic aims that were implicit in much of Bean's war journalism is his account of the Anzac evacuation from Gallipoli. The evacuation took place over several days in December 1915, and appears to have been unnoticed by the Turks, resulting in negligible Anzac casualties. This was a remarkable military achievement, and Bean knew that it would make a good story. On 23 December he recorded in his diary a conversation with a fellow correspondent, who complained that a rearguard action would have made for better news. Bean dissented: 'I say that battle stories are almost commonplace nowadays; and the spectacle of our whole position gradually left bare to the Turks [...] is as good as any battle story'.

Bean described the despatch he wrote as one 'on which I had poured out more care than anything of which I have written here — the only chance one has of even attempting to rival Bartlett's work'. Though it was delayed and curtailed by the censors, Bean's report gave the Australian Gallipoli campaign a dramatic conclusion and confirmed its 'historic' significance:

> Three miles away from me, across a grey, silky sea, lies the dark shape of the land. Eight months ago, just as the first lemon-grey of dawn was breaking over that long lizard-shaped mountain, I watched such signs as were visible of the landing of the Australian troops at Gallipoli. Now, as night falls gradually down upon the same historic scene, I am watching for the signs of their departure [...][35]

As historian Richard Ely comments about this passage, 'Bean knew very well how to *frame* a narrative to enhance its power to move'.

Although the report about the Anzac evacuation had a self-conscious literary tone, Bean was wary of romanticising war and usually concealed his literary art by framing his representations of the Anzac ideal in evocative but simple prose. But his new style of war reporting, which appeared to depict the 'ordinary heroism' of Australian soldiers in plain language and without glorifying war, was not always appreciated by Australian newspaper editors and their civilian readership. In September 1915, Bean discovered that the Melbourne *Age* and *Argus* were not publishing some of his reports because they were of insufficient interest. He was incensed by the papers' preference for the 'wild, sensational inventions' of the 'Cairo correspondents', and retorted that his reports had 'merely the interest that I risk my life hundreds of times over on the spot itself in order that they may know that every word is as true as it can be'. The charge of 'insufficient interest' plagued Bean throughout the war (in 1916 the *Argus* compared Bean's writing to that of a bank-clerk's ledger and claimed that readers wanted less accuracy and more spirit) and his reports were not always printed by Australian newspapers.[36]

As the war dragged on, civilian Australians gradually began to comprehend something of the misery of trench warfare, and Bean's war correspondence achieved more thoughtful praise. In October 1916 the *Bendigo Advertiser* quoted Bean's account of the Australians at Pozières — 'ready to weep like little children' but still 'doing their job' — and praised his evocation of the 'great souled heroes'. By portraying some of the difficulties of life in the line while still praising the Australian soldiers' forbearance, Bean's reports began to touch a powerful nerve in Australia, where people were loathe to regard the losses of loved ones as a futile waste. Although the codes of propaganda and censorship made it difficult to bridge the gap between home and front, and ensured that Australians read a sanitised version of the soldiers' lives, Bean's correspondence attempted to reduce this gap; an aim that would become his main preoccupation as a historian.[37]

By the middle of 1917 Bean had recognised that his rigorous historian's approach was increasingly detracting from what was expected of him as a war correspondent, and he appointed an assistant official correspondent so that he could spend more time on historical research. Bean also realised that he was trying to cater for two very different audiences, and was writing as much for the men of the AIF as for people in Australia. Living among the diggers, Bean identified with their experience and tried to write about it in terms that they would accept. Some Australian soldiers had mixed feelings about Bean's writing. Late in 1918 the Australian war artist Will Dyson, one of Bean's closest friends, overheard a group of diggers discussing Bean's despatches:

> [...] some said 'I reckon he does the right thing in sending them the dinkum story'; others said, 'That might be all very well for the historian, but they reckoned the war correspondent ought to put a little more glory into it'.

Although some soldiers wanted 'a little more glory' in Bean's journalism, many of them were critical of the usual journalistic cant which so grossly misrepresented their experiences to Australians back home; one poem penned at Gallipoli scorned the language of '"deathless heroes — lasting glory", and the other foolish fuss'. After Gallipoli many of the soldiers appreciated Bean's writing precisely because he wrote from first-hand knowledge of their experience. Lieutenant Noel Loutit of the 10th Battalion wrote to his father from Gallipoli that 'Captain Bean gave a better account [of the landing] than Bartlett did. *Bean was up with us*'.[38]

Although some of the diggers wanted more public criticism of the mismanagement at Gallipoli and in subsequent campaigns, Bean's simple but affirming image of ordinary men who were just doing their job was very appealing to those who wanted to find a positive meaning for their own experiences, and to believe that their mates had not died in vain. The secret of Bean's success as a war correspondent was that he was able to construct a version of the soldiers' war that, while constrained by censorship and informed by his reverent Anzac ideal, overlapped with the Anzacs' own articulation of their experiences and fulfilled deep emotional needs. Bean's writing was thus more influential among the Australian soldiers than that of most other correspondents, in that it provided interpretative categories with which they could make sense of the war and of their Anzac experience.

Bean's *Anzac Book*

The Anzac annuals that Bean edited were even more successful than his correspondence in their articulation of the Australian soldiers' experiences. The first and most popular of these was *The Anzac Book*. In November of 1915 Bean was invited to join a committee to create an Anzac Annual. He soon became the energetic editor, changed the title, and sent a circular to all Anzac units asking for contributions 'to make it worthy of Anzac and a souvenir which time will make increasingly valued'. Although the proposed annual was publicised as a New Year's entertainment for the troops on Gallipoli, it seems likely that its initiators on ANZAC staff, perhaps including Bean, knew of the plans for evacuation. It would thus appear that the book was intended to serve as a commemorative souvenir for the soldiers, and for a wider Australian audience; its aim being to construct a positive account of military defeat.[39]

According to Bean, the response to his request for contributions was 'enormous'. The front cover of the final product stated that it was 'Written and Illustrated In Gallipoli by the Men of Anzac', and enthusiastic Australian reviewers believed that it was the authentic voice of the troops; a Sydney *Bulletin* writer commented that 'there must have been almost as many poets as fighters at Gallipoli'. But Bean had exaggerated the extent and representativeness of the contributors. From over 36 000 men in the Anzac zone, only about 150 contributed to the book, and that included Bean and several other correspondents.

These contributors may have been a cross-section of Anzac troops — though the requirement of artistic or literary skill probably narrowed the sample — but the material that was published did not represent a cross-section of Anzac experiences and opinions. In the archives of the Australian War Memorial, historian David Kent recently discovered a file of soldiers' manuscripts that Bean excluded from *The Anzac Book*, together with a set of original manuscripts which differed markedly from the edited versions that were published. They show that Bean was a very selective editor, partly to maintain a consistent literary standard, but also to project a particular image of the Anzacs. There were both overlaps and tensions between Bean's Anzac ideal and the more varied and complex ways in which the soldiers represented themselves in writing and drawings. The

Figure 3 *Front cover of* The Anzac Book.

following, illustrated study of Bean's editing of *The Anzac Book* shows how he worked to define the Australian soldiers in a particular way, and to exclude characteristics that did not match his ideal or the requirements of the censors.[40]

Some contributions to the book, such as David Barker's front cover illustration (Figure 3), affirmed the conventional, heroic stereotype of the battered but unyielding warrior, and thus implied that Gallipoli was a triumph rather than a failure. But Bean also allowed contributors to express the disdain for romantic journalism and official lies — the 'tosh' of 'press pudding' — which he shared (see, for example, the letters 'R' and 'V' of 'An Anzac Alphabet' in Figure 9). Poems and cartoons, like those in Figure 4, explicitly contrasted 'The Ideal and the Real' of the soldiers' experience, and debunked newspaper images of 'heroic colonials' as a 'race of athletes'. Bean himself contributed to this debunking in an ironic drawing of an Australian soldier 'returning from the field of glory at Helles', although another of his own illustrations, portraying semi-naked Anzacs, shows that he admired the sun-bronzed, athletic Australians and was equally willing to represent that image in the book (see Figures 4 and 5).

Many contributions showed more vividly than any newspaper report the physical discomforts endured by the soldiers on Gallipoli — the dirt, flies and lice, intense heat and bitter cold — and the ways in which the men made the most of their difficult situation. In this regard, *The Anzac Book* was typical of the trench publications of Australian soldiers, in that it constructed an image of the quintessential digger: a tough man who shrugs off discomfort or pain with ironic grumbles and grins; a reluctant soldier but casual under fire and scornful of its consequences, fun-loving and cheeky, a man who disdains military swagger (see Figures 6–10).

Although the irreverence and sarcasm that is apparent in these illustrations sometimes rubbed against more respectable official images, the less reputable aspects of Anzac behaviour were kept within acceptable limits by Bean's editing. For example, Bean limited the representation of Anzac larrikinism by rejecting several manuscripts that portrayed the boozing, brawling, swearing and racist Anzacs in Cairo.

He also kept criticism of officers and authority within acceptable boundaries. For example, the first cartoon in Figure 10 expresses the Australian soldiers' contempt for upper class officers, and shows that it was possible to criticise authority in *The Anzac Book*. The second cartoon shows how such criticisms were bounded by the further implication that in the army — or at least in the egalitarian Australian army — even upper class twits

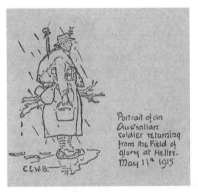

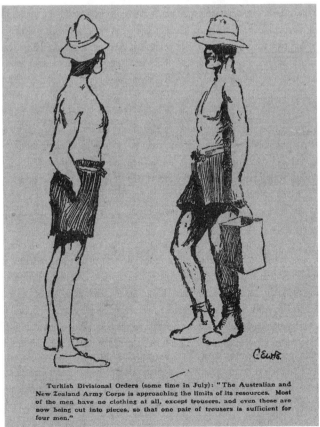

Figure 4   *(Top) Cartoons from* The Anzac Book.
Figure 5   *(Bottom) Illustrations by Bean from* The Anzac Book.

could discover their true qualities as leaders, and win the respect of their men. Along the same lines, Bean excluded manuscripts that contrasted the unequal comforts or sacrifices of officers and men, and included material which implied that the AIF had done away with class differences. Despite the odd cartoon which ribbed staff officers, frequently British staff (see the letters 'I' and 'P' in 'An Anzac Alphabet', Figure 9), extensive contributions from Generals Hamilton and Birdwood promoted the respectable military viewpoint of unity and loyalty; Birdwood was even photographed swimming in the sea just like one of the men.

Though he was more than willing to include images and text which showed the difficulties faced by the men of Gallipoli, Bean rejected poems that depicted Australian losses in bitter or negative terms. In one soldier's poem he deleted the line 'We ain't got no Daddy now our Daddy's killed and dead', and inserted 'Simple words that bring her memories o'er the boundaries of the dead'. By editing out the plain language of death and converting it into elegy, Bean ensured that death and anguish were justified as dutiful sacrifices for the greater good. Similarly, Bean excluded manuscripts that showed the savagery or monotony of battle, and the anger, despair or terror of Australian soldiers. In the published *Anzac Book* such feelings are transcended by noble action: in a story with one of the few references to fear, the coward redeems himself in heroic death. As David Kent concludes: 'When fear was mythologised in this way there was no room for contributions which accepted it as a fact of life at Gallipoli'.[41]

Bean's selective editing of *The Anzac Book* must be partly attributed to his keen awareness of what was acceptable in wartime, and to his desire to show the Australian soldiers in the best possible light. It may also have been influenced by an unconscious projection of his own Anzac ideal. Although there were still minor tensions in the text between Bean's ideal and the self-image of some contributors — for example, writers and illustrators were less respectful of authority than Bean might have liked — that self-image was carefully edited to sustain and promote the legend of ordinary Australians who had displayed characteristic qualities of bravery, humour and endurance in the most trying circumstances.

In Australia *The Anzac Book* was a monumental best-seller, with sales of 100 000 copies by September 1916, and it became one of the most influential early versions of the Anzac legend. Civilians seemed to like its depiction of the stalwart Anzacs, and reviewers were convinced that *The Anzac Book* was 'one of the most complete studies in the psychology of our Australian brothers'. The evidence of the rejected manuscripts, and of oral testimony,

"AT THE LANDING, AND HERE EVER SINCE"

Drawn in Blue and Red Pencil by DAVID BARKER

"Gawd help the first bloomin' Turk I see to-night"

The Aus: "Are you wounded, mate?"
The Victim: "D'yer think I'm doing this fer fun?"

Figures 6, 7 and 8   *Cartoons from* The Anzac Book.

shows that the book was not such a complete study of digger psychology, and that some of 'our Australia brothers' had a very different understanding of their experience at war. Yet by all accounts *The Anzac Book* was very favourably received by the Australian troops; by November of 1916 AIF members had ordered 53 000 copies.

These massive sales were partly due to Bean's energetic and efficient distribution work. Australian soldiers were able to buy a copy by direct debit of two shillings and sixpence from their pay, and for an additional sixpence they could consign a copy to an address in Australia. Bean set up book-selling committees in each unit, and sold thousands of copies by his own efforts. He also organised for each Australian division to have 3000 copies to sell to new recruits so that they could learn what sort of soldiers they should be.[42]

The success of the book can also be attributed to the fact that the men liked the image it portrayed. Australian soldiers believed *The Anzac Book* to be their own collective attitudes and identity expressed in their own words and pictures. The image it conveyed was certainly preferable to the lies of official communiqués or the exaggerations of the press. *The Anzac Book* admitted the discomforts of their life, and highlighted the ironic humour that was one of their main ways of coping with danger and discomfort. It also showed the Australian to be a distinctive kind of soldier, and this was a notion that the diggers were beginning to enjoy, even though they sometimes gave their nationality more unofficial and even oppositional meanings.

The Anzac Book was Bean's most effective wartime rendering of the Anzac legend. While Bean's portrayal of Anzac manhood was more wholesome and virtuous than the actual behaviour of many Anzacs, his positive account of Australian military manhood was affirming for soldiers who wanted to feel that their manhood had been sustained, that their actions had been valued, and that their efforts were worthwhile. Although Bean's articulation of Anzac identity was more homogeneous and nationalistic than the varieties of Anzac experience, the Anzacs did not reject his definition of Australian military manhood because they shared a substantial common understanding of what it meant to be an Australian soldier.

Some men may have recognised the selectivity of the supposed self-portrait of *The Anzac Book*, but the influence of the book among the troops and back in Australia would help to ensure that contradictory experiences were defined and felt as abnormal and marginal. Indeed, Bean's wartime writings, as both journalist and editor, helped Australian soldiers to artic-ulate and generalise their experience of war in terms of an Anzac legend, and

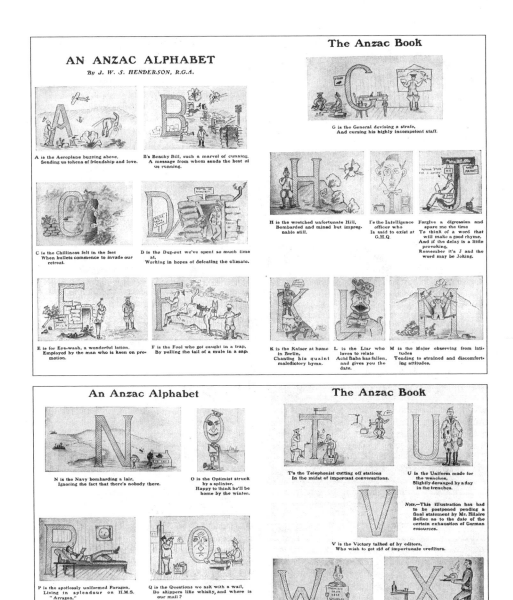

Figure 9 *'An Anzac Alphabet' from* The Anzac Book.

at the same time worked to silence or filter out contradictory understandings. The following chapter explores this relationship between experience, personal understanding and collective identity through the wartime memory biographies of Percy Bird, Bill Langham and Fred Farrall.

Young Officer: "Haw, haw, no shave?"
Australian; "He, he, no ―― razaw!"

Figure 10   *Cartoons from* The Anzac Book.

CHAPTER 3

MEMORIES OF WAR

Percy Bird

Percy Bird was born in the western Melbourne port suburb of Williams-town in 1889, the youngest of seven brothers and sisters. His mother's family had run a boarding house and his father was a boilermaker with the railways and a Mason, with no political or union interest that Percy could recall. Percy proudly explained that according to family tradition his paternal grandfather's family had owned a shipping line in the old country, and that his grandfather had eloped to Sydney and sold his boat and worked on the harbour. The story highlights the ambiguity of Percy's class position and identity. Though the Birds were 'not a bit wealthy', Percy was careful to define his family as 'near enough we'll say to the middle class'. This contrasts with most of my other western suburbs interviewees, who defined their backgrounds and social position as 'working class'. The Birds aspired to the middle-class lifestyle and values of church, association and education, and Percy adopted these ideals and subsequently sought professional training and positions with professional status, albeit limited and subordinate, in his army and working life.[1]

In our first interview Percy described an idyllic childhood in Williams-town. He went to the local State school, was prominent in several sporting teams, and sang as a soloist in the Presbyterian Sunday School choir. He was also an enthusiastic member of the Williamstown Boys' Naval Brigade and participated in its band, in public military displays and camping holidays. Percy enjoyed remembering these aspects of his childhood; he had already helped researchers into the histories of baseball and the Boys' Naval Brigade. In the second interview he admitted that when he read the first transcript he realised that 'he didn't have such a wonderful time' because he was the youngest and had to run messages for the family and neighbours. Percy's explanation hints at a general unease about his childhood which is usually hidden by the enthusiastic remembering of organised youth activities. This childhood unease about private life and relationships was repeated in adult life, where public activities remained the most affirming for Percy and became the mainstay of his identity and remembering.[2]

In 1904 Percy left school and, on his mother's advice, attended a business college. The following year he started work as a clerk with the Victorian railways. He still lived with his family in Williamstown, where he was active in local sports teams, the church choir and the Australian Natives Association branch. Unlike most of my other interviewees, Percy was already in his mid-twenties in 1914, and had been in regular employment for almost ten years. He was also engaged to be married to Eva Linklater. In the interview he didn't recall any strong personal response to the outbreak of war, and explained that he delayed enlisting until July of 1915 because his father was dying. I asked him why he enlisted:

> Oh [laughs]. Be like all the others [laughs]. I wanted to enlist like all the others, you know. Well, like lots of the others, I should say, because I thought I was ... well, I was ... should enlist. Being a member of ... being an Australian.[3]

When I listened to the tape I was struck by the awkwardness and discomfort of Percy's reply. Enlistment was not one of Percy's standard anecdotes (unlike some other veterans for whom it is an important signpost in memory), and the awkward form of the account of enlistment contrasted sharply with his usual, confident storytelling. Enlistment did not become one of Percy's stories because at the time it required a difficult choice between competing demands. In the memories of most of my interviewees the story of enlistment is highly significant and fraught with contradictions. It reveals a struggle to make sense of a decision that may have been difficult at the time, which sometimes had disastrous personal consequences, and for which public regard has shifted dramatically, from wartime enthusiasm through to doubt, ambivalence and even opposition.

On the one hand Percy was influenced by his perceived duty as an Australian which he had learnt in the Boys' Naval Brigade and the Australian Natives Association, and which became a predominant motif of recruiting rhetoric. He was also affected by the subtle pressure of mates who had joined up; later in the interview he remarked that he 'had to go' because his pals were going. On the other hand, the family trauma of his father's illness and the commitment to a new fiancée were countervailing pressures that made enlistment difficult. Enlistment represented a choice between two different prescriptions of masculinity, between the family man and the independent soldier adventurer. The pressures to join up and be a soldiering man were stronger, and to justify the decision — especially necessary in relation to his family and fiancée — the explanations of duty and mateship became the main features of Percy's account. In turn, these explanations were consolidated by wartime public approval; though in later life Percy would be further troubled by public questioning of the worth of Australian participation in 'a European war'. Enlistment never became a favoured public story for Percy because it recalled the old tension and hidden pain. It is perhaps significant that he never discussed Eva's feelings about his decision, or how she and his family were affected by his absence.[4]

Following his difficult decision to enlist, Percy Bird was sent to Egypt with a unit of reinforcements for the 5th Battalion of the AIF. On the ship his clerical skills were discovered and he was made Orderly Room Sergeant for the duration of the voyage. After a period of training in Egypt and at Étaples, Percy joined D Company of the 5th Battalion at Bonneville in July 1916, just after the battalion suffered heavy casualties in the first battle for Pozières. After a short spell in the line at Pozières he went with the battalion to Belgium for six weeks, and then returned to the Somme for the winter of 1916–17. In February 1917 he was reappointed as an Orderly Room Sergeant for the battalion. The new job at battalion headquarters took him out of his company and, for the most part, out of front-line fighting, although he still went up the line on occasional tasks. The line also came to him on one occasion at Lagnicourt in May 1917, when the battalion headquarters was almost overrun by the enemy. While at Lagnicourt he was gassed, and in August he had a life-saving operation on a gland in his neck. He ended up in a hospital in Weymouth, and from there was shipped to Australia on the day before the Melbourne Cup horse race in 1917.

There were a number of key issues in Percy's experience and identity during the war. Like every infantry soldier he had to cope with life in the front line. Though his experience of the front was not as severe as that of many other diggers — he was only in a fighting unit for about six months, and in all of that time D Company never went 'over the top' — he did endure the bombardment of Second Pozières and the miserable conditions of the Somme winter of 1916–17. Most of Percy's stories about the front refer to the difficulties of getting through the mud and finding the line, and the lack of food because ration parties were bogged or lost:

The 1916–17 winter was the worst for thirty years, and the mud was shocking. Well, we were up the line one time for three days and three nights and you know what we had to eat? Bread and milk. Sodden bread, Tommy cookers, condensed milk […] they ducked out with a couple of dixies to get some shell-hole water and we mixed it with the condensed milk and heat it, you see, and the captain said to me — there were about twenty of us in this dugout — he said, 'How much?' I said, 'Two mouthfuls each'. Went round, I said, 'Fill a mouthful'. And that's what we did [laughs].[5]

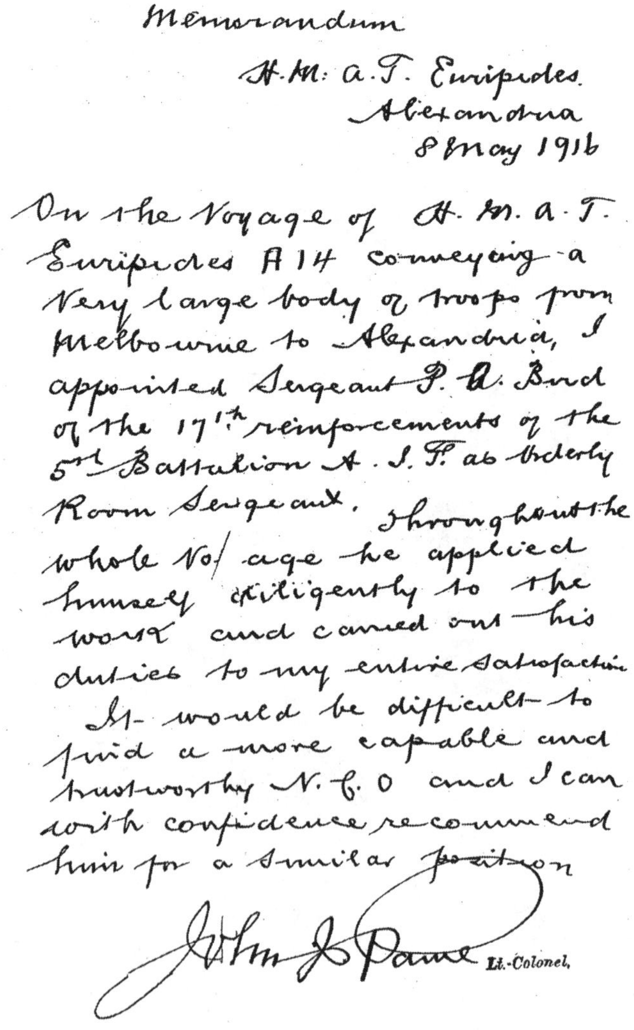

Memorandum

H.M.A.T. Euripides.
Alexandria
8 May 1916

On the Voyage of H.M.A.T. Euripides A14 conveying a Very large body of troops from Melbourne to Alexandria, I appointed Sergeant P. A. Bird of the 17th reinforcements of the 5th Battalion A.I.F. as Orderly Room Sergeant. Throughout the whole Voyage he applied himself diligently to the work and carried out his duties to my entire satisfaction. It would be difficult to find a more capable and trustworthy N.C.O and I can with confidence recommend him for a similar position

John J Paine Lt.-Colonel.

Figure 11 *Letter of commendation for Percy Bird, 1916. (Kath Hunter)*

The 5th Battalion history confirms that these were important features of the battalion's experience of that winter on the Somme, in which even getting to the line was a miserable experience of 'wading, crawling, wallowing in mud for hours through the darkness'.[6]

Humorous anecdotes helped Percy and the men of D Company to cope with these physical hardships (the stories are still his main way of remembering the winter of 1916–17), but the conditions did not cause major emotional problems for Percy. In contrast, silences and repressions in his remembering suggest that he did not cope so well with the artillery bombardments that were the other main feature of life on the Somme. The battalion history describes the scene when Percy first went into the front line at Pozières' 'Death Valley' on 14 August 1916:

> The whole of the shell-rent surface was torn into the ghastliest commixture of decaying dead, tattered clothing and broken equipment [...] Stinking corpses, or portions of them, everywhere exuded their foul gases [...] forming a dreadful paving on which perforce the men walked.

The job of D Company was to dig hopping-off trenches for other battalions, all the time under intense shell-fire and suffering severe casualties, unable to respond in any way to the bombardment. According to the history it was 'a time of nerve-wracking passivity for the Fifth'. Yet every time I asked Percy what it felt like to be under shell-fire he changed the subject back to one of his standard stories about concert parties behind the lines or getting bogged in the mud. He volunteered no stories about the smells and sounds of the trenches, about his own feelings under fire, or about the mutilation and deaths of his mates. The most that he could say in the first interview, before he changed the subject, was, 'I think we were all frightened but we all stuck together'.[7]

In the second interview, perhaps because there was greater trust between us, and because I was expressly trying to guide him out of his set sequence of stories, Percy expressed a few more clues about those feelings. He said that he did not like watching the 'Anzacs' television series because it brought back painful memories of dead pals. After I had repeated the question about his feelings several times, he rushed through a set of disturbing memories that had not been in his previous written or spoken remembering — of watching helplessly while another battalion was 'knocked to ribbons', and of two NCOs being blown up just after he walked away from them — before changing the subject again.[8]

The way in which Percy told these stories, and avoided telling them, suggests that like many others he was extremely distressed by these experiences in the firing line. Though he denied that his nerves had been affected, he did remark that 'we were glad to get away. I will admit that'. None of Percy's stories were positive about his worth as a fighting man, and that role was virtually excluded from his remembering because he could make no safe or comfortable sense of it either at the time or subsequently. The evidence for the traumatic effects of bombardment lies in the silences of his remembering. Disturbing experiences and feelings were either repressed from conscious memory or pushed aside into a 'private' drawer of Percy's memory, from where they came out only under pressure, in response to probing questions, or by association and in dreams, but never in his public stories.[9]

Repression of the most disturbing aspects of bombardment was only one way in which Percy coped with the experience. The refrain 'we were frightened but we all stuck together', which Percy repeated several times, testifies to the physical and emotional support provided by the men of D Company. In sticking together the soldiers developed their own language, stories and songs which made the experience easier to live with. Percy's oral skills ensured that he was actively involved in that process of collective narration. Many of his standard stories were told and enjoyed during the war. Thus stories about lucky escapes (an Orderly Sergeant who has a shell go between his legs and bury itself in the ground while he is asleep) and the language they employed ('Fritz' lobbing over 'a couple of shells'), made sense of the experience of bombardment with humour, bravado and a touch of fatalism. Percy's stories did not make sense of the war in terms of bitterness or disillusionment, partly because most of his war experience was in relative safety behind the lines, and partly because he enjoyed good relations with his superior officers and had an uncritical attitude towards the military authorities and their decisions. His own experiences were almost always generalised in terms of the positive, collective experience of the unit, and formed a repertoire of affirming stories. Over time the stories and soldiers' songs came to stand for the experience of the trenches and provided Percy with comfortable ways of remembering.[10]

Percy's masculine and military identity was also troubled by his redeployment to clerical duties behind the lines. Both on the ship and at the beginning of 1917, Percy protested that he did not want to leave the company, and he recalled the move as 'unfortunate'. It was 'unfortunate' partly because it took Percy away from his friends in D Company, but also because Percy felt guilty about leaving his mates, and inferior in his non-combat role (a wound was a valid excuse, and Percy had no qualms about admitting that he was glad to get away to Australia after he was wounded). To compensate for feelings of guilt and masculine inadequacy, Percy seems to have revelled in the dangers of his life behind the lines, and to have highlighted them in his memory; indeed, there is far more military action and excitement in these stories than in his account of life in the line. He also developed a positive alternative identity for himself by taking pride in his competence as Orderly Room Sergeant and in the compliments of senior staff officers.[11]

Even more affirming for Percy's military identity were the experiences of his life with the battalion out of the line, when front-line and support soldiers shared the common identity of the battalion. At these times Percy came into his own as a performer and enjoyed the acclamation of the soldiers as 'Birdy', a star of battalion concerts. For example, Percy represented D Company in a battalion concert competition in which prize money of one hundred francs was offered to the company with the best singers:

> So I sang that night and I got an encore, and then when it was all over they had a committee passing the votes and everything. So the next morning the captain we had, oh thrilled to bits, we got the three first prizes and the best effort. Seventy out of a hundred [laughs].[12]

This was the story that Percy told when I first asked him about life on the Somme, and he subsequently repeated it with great gusto. Performance out of the line was the main way in which he gained affirmation from the men in his battalion, and the most positive aspect of Percy's identity as a soldier. It became a central feature of his remembering of the war because in comparison with other aspects of his war experience it was entirely positive, and because the recreation of wartime performance — in stories punctuated by song — was popular with postwar audiences.

Percy also enjoyed his reputation as a scrounger of food; both officers and other ranks often came to him to share the food that he had acquired from local villagers. But he was less comfortable with other aspects of digger culture. Because of his job Percy mixed closely with staff and battalion officers and, in contrast with almost all my other interviewees, in his remembering he highlighted occasions when he sought and received their praise. Though Percy edited out of his remembering any tensions between officers and men in the 5th Battalion, even the battalion history recorded such tensions, and other sources note the diggers' renowned antipathy to men who were overfamiliar with officers. It may be that some of the battalion's other ranks frowned upon Percy's comfortable relationship with his superiors.[13]

More obviously troubling for Percy were the differences between his social standards and what he perceived to be the prevailing standards of behaviour among the diggers. Because of his Presbyterian upbringing Percy didn't drink or smoke, and he felt that he was unusual in this regard. In the interview he emphasised his differences in a recollection that his one 'vice' was lollies, and that he supplied the men with sweets when they ran out of cigarettes ('Got any lollies, Birdy?'). Percy also felt that he was different in his attitude to women. He claimed that he 'never worried about the women folk' because he was engaged, and because he disapproved of the other fellows who 'used to go to see the women there for certain purposes [...] certain ones, unfortunately, they were caught, certain diseases'. He was also rather disgusted by the men's jokes about sex and masturbation:

> [...] where we were sleeping and that, somebody would yell out, 'The old squire's been foully murdered'. And of course, they'd all 'What? Again?' This seemed to be one of the little jokes. Poor joke I thought it was.

On one occasion at a new battalion billet when Percy was looking for a quiet place to read, he came upon a 'young lady' in a 'lovely big place'. They swapped coins and talked, and for the next two nights Percy returned to sit with her. Yet he refused to tell his mates where he had been because he thought the relationship would be misunderstood and his character smeared.[14]

These stories suggest that despite his skills as a singer, story-teller and scrounger, Percy sometimes felt like an outsider among the diggers, excluded by his own ideal and practice of moral manhood from the more larrikin masculinity that he perceived to be predominant. When prompted by my questions, he expressed ambivalence about the diggers' behaviour and concern that the digger reputation was not true for men like him, but on the whole he preferred not to discuss these aspects of life out of the line. This contrasts with some of the other interviewees who relished the drinking and gambling and highlighted it in their memories.[15] Throughout the interview Percy referred to the Australian soldiers as 'Anzacs' or 'Australians' but never 'diggers', perhaps because he felt uncomfortable about the digger identity of the other ranks.

In contrast, the common unit identities of D Company, the 5th Battalion and the AIF were comfortable and affirming for Percy. The company and battalion identities were reinforced by the bonding of life together, in and out of the line, and by the inter-unit rivalry of battalion and brigade sports meetings and concerts. Thus Percy proudly identified himself as a 'Don

Company' man. The memoir that he wrote for me about his war was an account of the 5th Battalion in France, because at the time and in memory the identity of the battalion was affirming for Percy, and because he had shaped his personal war story in terms of the battalion's story.

National identity was also significant for Percy, and this, too, he used to help comprehend and articulate his experiences. Percy's sense of his Australian identity was drawn, to some extent, from personal experience. For example, while loading supplies for English and French officers he noted that they treated the diggers with less respect than their own Australian officers, who were willing to negotiate about rights and responsibilities and got a better job as a result. He observed that, in contrast with his own battalion, English units often left their billets in a mess, and he overheard remarks by English officers who were surprised at the quality of an Australian drill, and by some French women who felt that they were 'safe now, Australia's here'.[16]

The specific, national meanings of these and other similar anecdotes were articulated within Percy's unit, and in newspapers, books and even official commendations. The story about the French women was a common, apocryphal tale within the AIF, and 'everybody reckoned the Australian soldiers … they made a big name for themselves, and they were the best soldiers to get anywhere'. Public opinion about the Australians thus helped Percy to articulate his experiences in a particular, nationalistic way, and in turn led him to highlight in his memory experiences that made sense in terms of this national pride. Although Percy could not relate his trench experiences in terms of the Anzac hero — the gap between his experience and that aspect of the legend was too great — he could enjoy the more general and official collective identity of the AIF. The Australian national identity worked for Percy because, like the identities of the company and the battalion, it was an affirming and inclusive identity that did not necessarily distinguish between Australians in and out of the line, between officers and others ranks, or between different standards of behaviour. In those terms, being an Australian was something Percy had in common with his fellows, which also proudly distinguished them from other soldiers.[17]

Yet during the war the affirmation of Percy's experience through national identification was undermined by tensions between his own experiences and attitudes and the Anzac and digger prescriptions for masculinity. Only when the war finished and those tensions were no longer part of lived experience, would Percy's war memory and identity become less troubled.

Bill Langham

Bill Langham was born in 1897 in Axedale, a small town about forty miles from the central Victorian provincial city of Bendigo. The Langham family of thirteen had to survive on the meagre income Bill's father's made from quarrying and woodcutting work — about a pound a week — and Bill's memory of his childhood is summarised in the familiar language of his generation and class: 'We were a good happy family, but we were very poor, and I don't mind admitting how many a time I went to school with a bit of dry bread, in those days'. Apart from enjoying the outdoors' freedom of a country boy, Bill loved to read, especially history, and he was awarded a framed Merit Certificate when he completed the last two grades of primary school within six months. He accepted a place at Bendigo High School but, as he recalled, 'the war settled me as a student. That came along and I went to the war and there was no more education'. This claim that the war caused Bill to sacrifice opportunities for further school and university study was repeated in both interviews. Yet further questioning revealed that Bill actually ran away from high school and home before the outbreak of war because he'd 'had school'. He got a casual job picking up fleeces in a shearing shed near Bendigo, and then headed to Melbourne to work in the Caulfield racing stables. He had been a keen rider as a boy, and now hoped to follow in the footsteps of a brother who was a country jockey: 'course the war intervened and altered all that'.[18]

The war did disrupt Bill's plans and aspirations, but in his remembering he highlighted that disruption and played down other factors for a number of reasons. The war provided an obvious and socially acceptable explanation for change and loss. Bill's explanation allowed him the sympathy and understanding due to a victim of circumstance, and mitigated his own agency in the 'sacrifice' of his education, improved job prospects and enlistment. Finally, the disruption of war became the necessary backdrop for Bill's conclusion that he was able to overcome his wartime sacrifices and make a success of the rest of his life. In his remembering Bill satisfied the need for an affirming memory by emphasising his active role in the successes of his life, and by implying that he was a victim of fate in less positive circumstances.

Bill's various accounts of his enlistment in 1915 show that while he tried to compose an affirming memory of the event, his remembering was influenced by other, public accounts. In our first interview, Bill's initial explanation for joining up was that it was a very personal, spur of the moment decision:

Oh, I don't know. I think it just came suddenly. I used to pick up the papers and see fellows that I knew and mates that had been knocked over [...] One of my very good mates, he was a bit older than I was, but he was a lovely fellow and I read in the paper one morning where he'd got knocked, so I thought, 'Well, I'll go and have a go at it'.[19]

In response to a follow-up question he denied that he felt under any pressure to enlist, yet at the time and subsequently there were a number of public influences that affected his decision and how he remembered it.

Firstly, even his initial explanation suggests that the decision was not entirely 'personal', and that mateship — feelings of guilt and inadequacy because he was not already with his mates — was a motivation for enlistment. Secondly, Bill recalled that there were a 'lot of people' who said after the war that he only went away because he thought he was going to have a good trip. Bill was very clear that that was 'fartherest from my mind, because [...] I'd read a lot of history in my day, when I went to school, and I realised that war wasn't a picnic'.[20]

Thirdly, Bill described how he was 'a real Royalist in those days' and used to love reading British and Empire history. This reading, and the celebration of Empire Day at school, confirmed some degree of affection for the 'old Dart' which had been the home of his forefathers. These feelings, and their exuberant public declaration during the early years of the war, may have been a motivating factor behind the decision for which his mate's death was a catalyst. He may also have eagerly anticipated a proud martial identity, which he had relished as a cadet, and which he certainly enjoyed after he enlisted: 'We used to march, we'd have a band and we used to march in from Royal Park to the city. Oh, it used to be lovely, striding out behind that band, you know. You thought you were king'.[21] For Bill this personal pleasure and pride seems to have been more important than any deep-rooted patriotic sentiment. Bill never said that national or imperial sentiment affected his decision, and of Empire Day he recalled that his main interest was in the spread of food that was provided for school children on the day.[22]

Fourthly, he recalled that 'we used to hear bad stories about them [the Germans], about carrying babies on their bayonets and all that'. He claimed that the stories were 'a lot of hooey', but it is difficult to tell if he was so critical when he enlisted, and whether or not they might have been another background influence on his decision. In response to a question in the second interview asking how he felt about joining up, Bill remarked:

Oh that's pretty hard, that's a pretty hard question. I just thought I
was doing the right thing anyway. By the things that we'd been told I
suppose when the war first started, they, the press reports and things,
they gave us stories of German soldiers marching along with babies on
their bayonets. Well, that sort of got under your skin. You know. You
thought to yourself well you want to try and put a stop to this, if that's
true.[23]

The quote suggests that when he enlisted Bill may have shared some of the
mystification of the barbaric enemy, and that his rejection of the stories
as 'hooey' was at least partly due to subsequent events during the war, but
that he had shifted his criticism back in time so as not to associate his own
enlistment with an attitude that had become publicly unacceptable.[24]

Finally, in the context of a discussion on recent feminist Anzac Day
protests, which Bill denounced as ungrateful to the men who had made
sacrifices without which 'we wouldn't be living under the conditions we're
living under now', he remarked that if he lived his life over he would 'do the
same again' and enlist (my emphases, in italics, show how Bill slipped from
singular to plural when he ascribed general motivations to himself):

Because *I* thought it was right, *I* had to do it. *We* only went, *we* went
away, *we* tried to save the world from a fate worse than we've got now,
I think. But I don't think *we* succeeded somehow. That's the trouble.[25]

The context of these comments influenced the explanation of his motivation
to enlist. In discussion of a movement that he perceived to be critical of
soldiers like himself and other comrades who were 'crippled, maimed and
ruined for life', Bill needed to assert the worth of the men's sacrifices and the
validity of enlistment. He adopted the standard rhetoric used by newspapers
and Anzac Day speakers to denounce the feminist protests because it
matched his own indignation.

In another context, however, with different associations and emotional
needs, Bill articulated the worth of the war and the validity of his enlistment
in different terms. A meeting with some German prisoners, which he
described in an account of his experiences on the Western Front, 'sort of
altered my attitude altogether', and made him realise that 'we shouldn't be
fighting fellows that didn't want to fight'. At the time and in his remembering
he blamed the 'big nobs', sitting in their comfortable office chairs back home,
for sending ordinary men to fight their war for them. Criticism of the 'big
nobs' was reinforced by a feeling that they had not rewarded the soldiers

sufficiently when they came home. This same anger about the neglect of ex-servicemen by civilians after the war also informed Bill's anger at the feminist opponents of the 'diggers' day', but resulted in a very different assessment of the value of war service.[26]

Bill's account of his enlistment thus comprises several layers of memories, in which the contradictions derive from the complexity of his decision and his own shifting attitudes and identity. Those shifts are in turn related to the ways in which public versions of enlistment and the war have changed over time, and to the various contexts and associations that were operative when Bill was remembering.

Once Bill Langham had decided to join up he worked hard to be accepted. His father refused to sign the consent forms, which were necessary because Bill was under age, so the lad lied about his age at the Bendigo Recruiting Depot. He then had himself posted to a training camp in Melbourne to avoid being discovered and so that he could visit the big city for the first time. He was transferred to Maribyrnong Camp where he joined recruits from every State in the artillery unit of the Eighth Brigade of the Third Division of the AIF. The unit sailed to England via South Africa — Bill had a vivid memory of one exotic day in Durban — and then spent several months training with the new division. From there he was sent to France and saw his first action at Messines in the British campaign of 1916.

Bill's job was to help drive a team of horses loaded with fresh ammunition for the divisional artillery, and to move the guns when necessary. The artillery driver's work engendered a special relationship with his horse:

> He came first, your horse. He came before you or anybody else. Because you couldn't get horses but you could always get men. Your horse was always number one. Anyway, he was the fellow that had to carry you through.

Grooming the horses and cleaning the harnesses took up a lot of time when the unit came out of action, and for that reason drivers were paid a shilling more per week than gunners and infantry privates: 'we reckoned we were millionaires too'.[27]

On 1 October 1918, Bill's luck ran out when an explosion tore a great hole in the side of his horse's head. A small piece of shrapnel lodged in Bill's skull, though the sight and stench of the horse's blood which drenched his clothes made his wound seem much worse than it really was. One eye was paralysed forever and, though this did not impair his eyesight, it required him to turn his head to look sideways. Hospitalised in London and Manchester, Bill was disappointed not to be with his mates, who wrote that they were having a lovely time in Belgium now that the war was over. Yet when he came out of hospital he said that he did not want to return to his unit — 'I've seen enough of that' — and was given a job guarding the AIF headquarters in London and escorting the AIF payroll. Another reason for declining the opportunity to return to his mates was that Bill had 'a little girl' up in Manchester. During Christmas Bill and a mate went absent without leave so that they could visit this girl. When he returned to London he was dismissed from his job and sent to camp on Salisbury Plain. There he reported his misdemeanour to the officer in charge, who happened to be Captain Jacka, the famous Australian Victoria Cross winner: 'Well', he said, 'The bloody war's over. Do you want to make anything out of it?' Two days later Bill was on a boat and on his way home.[28]

In general, Bill remembered that a soldier had 'a good life':

> I mean it's a good healthy job [...] you're fit and well. You get well looked after. You get well clothed. You get well fed [...] you learn some discipline, which is very badly needed in the world at the present time.

The last comment suggests that this satisfaction was at least partly retrospective, especially as he also recalled that at the time he didn't like taking orders.[29]

But there was a negative side: 'the only part that's not good is the war part'. Like Percy Bird, a main issue for Bill, at the time and in his memory, was how he fared in battle. In a number of ways the nature of Bill's experience in battle, and his ability to cope, was determined by his specific role in the artillery. Bill felt that he was lucky compared with the men in the infantry. Although the horse-wagon lines could be wet and miserable, they were relatively comfortable and safe when compared with the trenches. The artillery men rarely had direct contact with the enemy or knew if they had killed anyone, whereas 'the infantryman must have, specially a man that's been in bayonet charges, he must have some awful memories'. The artillery driving job was active and mobile, and sometimes even exhilarating, in comparison with the static, passive lot of the infantryman. As they could respond actively to danger and fear in battle, there was perhaps less likelihood of drivers developing battle-induced nervous conditions. Yet their work could also be terrifying and traumatic. Out in the open with no form of protection, the ammunition teams were at the mercy of enemy artillery who were quick to train their guns on an inviting target. Bill knew what an explosive shell or bullet could do to a body, and freely admitted that he was scared:

Percy Bird before he went to war in 1915, photographed in the backyard of the family home in Williamstown. (Kath Hunter)

Goulburn Training Camp 1916, after the 'Kangaroo' recruitment march from Wagga, with a very young-looking Fred Farrall holding his rifle (bottom right). (Fred Farrall)

Postcard from Fred Farrall, Goulburn Camp: 'Dear Mother & Father, This is a photo of last Friday's Guard. I am standing behind Lieut. Beattie. W.W. Hall is on the left [with cap on]. Les Hall is second from the end [sitting] on the right, from Fred, 1916.' (Fred Farrall)

Driver W. (Bill) Langham, 22942, 30th Battery, 8th Australian Field Artillery Brigade, pictured in Footscray in 1916 before embarkation. (Bill Langham)

A photograph taken by the author's grandfather, John Rogers, in the trenches at Gallipoli with Australian soldiers striking a typical Anzac pose. (John Rogers/Campbell Thomson)

Members of the Australian 2nd Brigade killed in the advance at Helles in May 1915, as photographed by John Rogers. (John Rogers/Campbell Thomson)

Captain C.E.W Bean with members of the Australian 6th Battalion in the trenches at Helles, Gallipoli. John Rogers is second from the right. (John Rogers/Campbell Thomson)

Above: Postcard from Fred Farrall, France: 'Dear Auntie, This is a dinkum War Service photo & pretty rough of course, but still a little like Fred.' (Fred Farrall)

Right above: Bill Langham (centre) with digger mates at Sutton Veney, 1916, celebrating Jack McCannon's (in front of Bill) twenty-first birthday. (Bill Langham)

Right below: Jack Beard, Bill Langham and Phil Macumber, Armentières, 1916. (Bill Langham)

Percy Bird with other wounded soldiers returning to Australia in 1917. Percy is slightly below left of the man in the back row with the chin bandage. (Kath Hunter)

Percy Bird in Melbourne after returning from the war, 1918. (Kath Hunter)

[There are] people who say you're not scared, but that's all bunkum. I don't give a damn who it is. Whether he's a VC winner or what he is, he's scared, at times, that's only natural. He gets used to it after a while, but you still, you still wonder whether your name's on one.[30]

There were also many gruelling moments in Bill's war. He told the following story with a particularly pained expression on his face:

One of the worst things I … things I don't like, I very seldom mention it, but when we made the big push on the 4th of August [1918] … we were taking up ammunition. We had to go through this cutting, there was only room for your wagon to go through, and your team. And you know, that was full of Germans, lying there. We couldn't go round. We had to go through. We had to go over them. That's always been a very … something that … oh, it's hard to explain. It hurts very much when you think of it, that you had to … you had to gallop over and pull your wagon over those dead Germans. It's an awful feeling you know.

Seeing and hearing 'the wheels of the wagons and things crushing them up' was traumatic because Bill was forced to break powerful taboos about the sanctity of the body. It also made him imagine that it was 'our fellows' (perhaps even himself) under the wheels, and showed how all soldiers were just bodies that could be mutilated.[31]

Terrifying vulnerability and participation in mutilation tested Bill's emotional well-being. Like Percy Bird he maintained his composure by relying on the support network of the diggers (he also shared Percy's love for music, and found comfort playing his harmonica). Bill highlighted the material and emotional support of mates as the most positive feature of his war experience: 'When you go to war you find real mates. They, they'll die for you'. Though he recalled that 'the odd one or two' didn't fit in with this mateship, there was little in his own character or attitude that might have excluded him from its support. Unlike Percy Bird, Bill felt entirely comfortable with the diggers' larrikin comradeship and disdain for military regulation. He enjoyed drinking and gambling with the men of his unit when they were out of the line. He also emphasised the pleasure of the company of women, although he was guarded in his references to brothels.[32]

Among his soldier mates Bill developed a language of fatalistic acceptance which he found to be the easiest way to articulate troubling experience — in comparison with Percy Bird's use of humour — and which helped him to maintain his emotional composure. He recalled that when he was expected to drive the wagon over the bodies he felt like disobeying the order, 'but where were you going to go? Still, it had to be. That's war'. Artillery barrages were regarded as a lottery of life — Bill recounted many lucky escapes — and you got 'used to it after a while'. Over time, and in memory, this language solidified into a form that maintained a safe and almost formulaic distance from traumatic events and feelings, and that generalised experience so that the hard and painful edges of particular moments were smoothed out. Thus the work of the ammunition teams was remembered and safely defined by Bill as 'some hectic times', and during the interview his feelings of the time were often expressed in the second or third person ('he's scared at times, that's only natural').[33]

This fatalistic language intersected with public depictions of the hero of endurance. Bill recalled that he liked Bean's war correspondence because it provided 'the true facts', and Bean's writing, as well as the culture of the diggers, shaped and affirmed certain general understandings about the war and caused others to be played down. Sometimes my questions, and Bill's genuine attempts at a full account, brought out the painful or repulsive particulars that were hidden in his memory — the damage caused by an explosive bullet through the penis, the sound of flesh and bone crushed beneath a wagon, the hole in his horse's head — but for the most part, and in most company, Bill preferred to use relatively safe and painless generalisations for these experiences.[34]

One other aspect of Bill's war experience threatened to undermine this emotional composure and his military resolve. He recalled that his attitude to the war was transformed by a meeting with German soldiers during the Passchendaele campaign. While Bill's team was waiting in the cold and rain by the wagon lines, a digger returned from the front line with some German prisoners. The artillery men took pity on the Germans, who were 'only bits of kids' and looked miserable, and offered them some stew and a hot drink. The guard, who resented having to bring the prisoners out through the slush, 'went crook about their generosity'.

> That altered my, sort of altered my attitude altogether, I realised we
> shouldn't be fighting. 'Cause one of them said to me [...] 'I didn't want
> to fight you, I was forced into fighting [...] we were happy to be taken
> prisoner. Soon as we could'. I realised, we were fighting fellows that
> didn't want to fight. I still got that attitude.[35]

This event had a dramatic, almost theatrical quality, which may have helped Bill to articulate incoherent feelings about the war, and which made it a good story and ensured that it stuck in his memory. Several of the men I interviewed described similar meetings with Germans as having a conversion effect, although Percy Bird's encounter with a group of singing POWs had no such effect, and was only significant to him because of the 'wonderful' singing. Bill's new attitude to the war was shared by at least some of the men in his unit; he also recalled that at one point 'the whole bloody AIF' almost went on strike because they kept getting pushed into the line and had no reinforcements. The experience of meeting German prisoners shattered the propaganda image of the demonic enemy and called into question the justifications for the war. It nurtured in Bill a critical view that perhaps the soldiers of both armies had been duped and used by the 'big nobs' sitting back home 'in their office chairs', and threatened to undermine his determination to keep on fighting.[36]

A number of factors sustained Bill's resolve at the time and are at odds with the conversion story. In the interview Bill recalled that a sense of duty was one motivating force. He explained that despite his change of heart about the war I kept going because we took an oath [...] when we joined the army ... I wasn't proud of the things we were forced to do, but I was proud that I did it and carried out to the best of my ability, what I had to do'. Bill did not mention that his choices were very limited, and it is difficult to judge the extent to which this attitude was a wartime motivation or a subsequent justification.[37]

At least as influential at the time were the perceived expectations of digger mates. The particular public of the other ranks sometimes helped the diggers to reject the behavioural prescriptions of newspapers and military officialdom. For example, although AIF leaders sought to persuade the diggers to vote in favour of conscription and then suppressed the voting figures which revealed considerable dissent, Bill and other diggers knew that many of them had voted 'no', and they were scornful of the official lies. Yet Bill's account suggests that the dominant digger attitude to behaviour at the front also overlapped with official expectations, especially with the hero of endurance promoted by Bean and other writers. Bill remembered that the diggers regarded deserters and men with self-inflicted wounds as laughable (one man took off his boot to shoot himself in the foot, and then put the intact boot back on again) or disloyal: 'well you're leaving your mates [...] to bear the brunt of it'. In an artillery unit the link between mateship and military resolve was especially strong; if a driver left the horse team his desertion affected his mates and was a cause for immediate official concern.[38]

The ways in which Bill used generalised, formulaic language to explain individual 'weakness' and his own resolve, suggest that he drew heavily upon the overlapping official and digger prescriptions for behaviour. At the time those prescriptions were affirming for men like Bill who did 'keep going', and they subsequently provided a positive self-image and a justification for participation in the war.

For Bill, the expectation of self-respect and resolve in battle was underpinned by his commitment to the code of appropriate masculine behaviour which was prevalent amongst the men of the AIF. Several of Bill's stories show how bravery was respected as a virtue in the AIF, and that cowardice (though not fear, which was 'natural') was not. He enjoyed the recollection of walking down the Strand with a VC winner and meeting two military policeman who expected to be saluted, but had to do the saluting when they saw the decoration ('you're saluting the honour, and they deserve it too'). By implication the military policemen, who were not combat soldiers, were lesser men.[39]

This code of appropriate military masculinity also had a specifically Australian quality. The digger was just an ordinary bloke 'until you got into action. And then [...] he proved what a good soldier he was', with 'lots of guts'. Bill argued that the Australians were 'star troops', and implied that in this respect the diggers agreed with their public acclaim. Yet when Bill made this national distinction it was entirely general. He did not relate it to his own experiences or ability, and the relevant passages are all in the third person. Nor did he provide specific evidence to confirm the generalisations; the only example he used was a story he had heard about some American soldiers who were decimated because they ignored Australian experience and advice.[40]

Indeed, there were aspects of Bill's own experience that might have caused him to question this identity. For example, the passage about the fighting ability of the Australians is followed by a story about Bill getting lost near the front line which implies that he was no great soldier. Yet in the interview Bill was not aware of any tension between his own remembered experience and the national generalisation. The story about getting lost was not given a more general significance because in the Australian army, and in Australian popular memory, there was no prevalent public narrative that he could use to articulate the experience in negative terms. Many of Bill's war stories — such as the story about driving over German bodies — appeared in his remembering as isolated descriptive incidents but were not generalised in more meaningful terms because they had not found shape and strength from alternatives to the predominant Anzac narrative, or because the paradigm of the war as a chaotic personal hell was less appealing for Bill than the more positive paradigm of masculine and national identity. The conversion story did attain a more general, critical significance in Bill's memory because both during and after the war it was affirmed by public narratives (my role as a historian of the post-Vietnam generation also encouraged Bill to emphasise this story in the interview). But it is discordant with the majority of Bill's war stories, which emphasised his military resolve and matched the Anzac legend's themes of endurance and national pride. At the time and in his remembering, the prevailing national definition of military manhood provided Bill with a powerfully validating sense of 'doing the job', and an affirming collective identity.[41]

Fred Farrall

Fred Farrall was born in Cobram, on the Victorian side of the Murray River, in 1897. His father ran a small carrier business until 1904 when he won a ballot for a block of farming land near Ganmain in the New South Wales Riverina: 'we moved up there in 1905 and established ourselves in the wilderness'. For several years the two adults and six children (Fred was the fourth child) lived in a bark hut and then in a pine log cabin, until relatives helped them to build a four room weatherboard house. It was an isolated and primitive life. The nearest neighbour was five miles away, and there was no local school until 1908. Fred only had about four years of schooling and recalled that 'we grew up just like brumbies'.[42]

The children worked hard to help get the farm established, but Fred did not enjoy farm life or work. In his spare time he loved to read newspapers about life beyond the Riverina, and he became a keen follower of far away sporting events. Fred was physically small and like many bush kids he could ride a horse. Although he had never seen a horse-race he decided that he wanted to be a jockey, but when the time came to think about employment Mr Farrall declared that no member of his family would be connected with horse-racing, and Fred stayed on the farm:

> Rather reluctantly because it was all hard work. Early morning to late at
> night and seven days a week. With not much time for any amusement
> or sport or anything else that might be about.[43]

Fred's political hindsight caused him to emphasise that the Farrall's were a political family. His great grandfather had been a follower of the English socialist Robert Owen, and some of the Australian Farralls supported the early Labor Party and the Victorian Socialist Party. Although Fred's father was not active in party politics he was politically minded, and Fred recalled that he gained notoriety among their conservative rural neighbours for supporting a Labor Party candidate in a State election and for paying higher wages to his farm workers. The Farrall children were also ostracised at school because their father backed the black boxer, Jack Johnson, in the 1908 world heavyweight title fight. In the interview Fred was ambiguous about the extent to which he was affected by this family political tradition. He described how an old school mate later remarked that 'We knew when you were going to school that you'd be involved in politics', and supposed that he was influenced by his father. But he also recalled that he was politically quite ignorant, and it is likely that the family tradition only became important for him, and emphasised in memory, later in his life when he was looking for political antecedents.[44]

Despite the family's socialist background, the Farralls were avid supporters of the Australian war effort in 1914. Stories about the Gallipoli campaign became sacred texts within the Farrall family and the children relished the 'religious' atmosphere of patriotism. Fred was inspired by Anzac deeds and recruiting rhetoric, although his decision to enlist may have been equally influenced by his distaste for farm life and a desire to escape. In our interview Fred did not refer to his more personal motivation for enlistment, preferring to emphasise his identity as a duped patriot and to show that a large proportion of AIF recruits were motivated by patriotism. This emphasis fitted more neatly into Fred's story of transformation from patriotism to disillusionment. The motivations of escape and adventure were less appealing to Fred the political activist who preferred to emphasise serious and principled motivations, even if they were mistaken.[45]

Fred joined a 'Kangaroo' recruiting march from Wagga to Sydney at Galong on 11 December 1915. His main recollection of the march was that it was interrupted by two strikes led by recruits from navvy gangs who were members of the Australian Workers' Union. The first strike occurred when the captain in charge of the march reneged on his promise that the men could go home for Christmas and Boxing Day. That strike was successful, but the second strike, an attempt to enforce another promise of three days leave in Sydney, was defeated by the arrival of an armed regular army unit, 'so we didn't see much of Sydney, until we came back in 1919, those that were left'. Fred emphasised this memory of the strikes because it highlighted the influence of trade unionists in the AIF and confirmed the ideal of the independent and anti-authoritarian diggers that he developed during and after the war. In the interview he did not portray himself as a radical or argue that he was greatly influenced at the time by these actions; he preferred to stress his own passivity and ignorance: 'I knew nothing about it, I just followed on like a sheep'.[46]

Fred had two photos of the men from the recruiting march, taken at Goulburn training camp, which suggest that the self-image in his memory was a fair representation of how he was and felt about himself at the time (see the first section of photographs in this book). In both photos Fred stands out as a very raw and awkward recruit amongst the older, tougher-looking diggers; in one it is striking that he is the only man in uniform. It may be that Fred had just come off duty, but it could also be that he did not easily share the men's relaxed disregard for military dress or the manly camaraderie that the scene evokes. In the 1980s the photos reinforced Fred's sense of inadequacy as a recruit and as a man. Fred reiterated this self-image in a story about Jack Carroll, a Ganmain Irish Catholic, who disregarded the qualms of the local Irish community about fighting an English war because, he said, 'If a fellow like Fred Farrall can go to the war, anybody ought to be able to go'.[47]

Fred Farrall's war experience was deeply troubling and, rather than providing solace, the nascent Anzac legend simply reinforced Fred's feelings of inadequacy and alienation. At the end of several months at the Goulburn training camp, on 14 April 1916, Fred boarded the *Seramic* in Woolloomooloo Bay with 3000 other recruits. After a crowded and uncomfortable sea voyage during which Fred was shockingly sea sick — Fred as a weakling is a recurrent theme — the boat landed in Egypt, where the men were separated into different units for some more training. Then, late in the summer of 1916, Fred joined the 55th Battalion of the Fifth Division of the AIF at Fleurbaix, where the division was recovering after the slaughter at Fromelles in July. In October the battalion moved to the Somme, where Fred suffered the winter until December when he was hospitalised for six months with trench feet. He rejoined the battalion a few weeks before the attack on Polygon Wood in Flanders on 26 September 1917. In the attack he was wounded in the leg by a machine gun bullet and struggled back to a Regimental Aid Post, from where he was sent to a Casualty Clearing Station to have the bullet removed, and thence to hospitals in France and London. He had another long stay in hospital, lengthened by a 'nervous condition that had developed that was probably more serious than the gunshot wound'. After more training on Salisbury Plain he eventually rejoined the battalion in France not long before the Armistice, and was then stationed in Belgium until he returned to Australia in the middle of 1919. In Sydney he spent another six months in Randwick Hospital being treated for his trench foot condition, and for rheumatism and a nasal complaint he had developed in the trenches. He was finally discharged from the AIF in January 1920.[48]

In Fred's narrative these bare facts are related in a chronological sequence and enlivened by a richly detailed portrayal of places, events, people, conversations and feelings. Fred's war story uses selection, emphasis and ironic reflection to make meanings about his experience. It is a confident, seamless narrative with few uncertain pauses or jerky shifts of direction; the opening story about 'our introduction to the Somme' filled the entire forty-five minutes of one side of a cassette, and it was not until I turned the cassette over that I felt able to interrupt Fred's flow and ask questions about how he coped and about relations in the AIF. In retrospect I can see that Fred was already beginning to discuss these and other issues in his own narrative, and that it was a mistake to interrupt before he had told his own story, in his own way, from beginning to end. Still, that uninterrupted initial story, and the ways in which Fred often used my questions to continue his own story and emphasise its themes in other long passages, do reveal the main issues that were significant for Fred during the war, and that were reinforced or reworked in his memory over time.

One absence in Fred's recollection of the war, and a corresponding emphasis, is particularly obvious. Fred's war story is pre-eminently located in the trenches of the Western Front. Although he spent a large proportion of his war in hospitals, training camps and billets behind the lines, these experiences are relatively insignificant in his remembering, especially in comparison with other veterans like Percy Bird. In Fred's story, hospitalisation and leave are interruptions to the real business of the war in the trenches. He emphasised his life in the line because it had an enormous physical and emotional effect upon him at the time and subsequently, and because in old age the main meanings that he wanted to make about his experiences between 1916 and 1918 were about the war in the trenches. As we shall see, Fred was not always so positive or talkative about that aspect of his war, and only became able to compose a coherent sense of it well after the war was over.

Mateship was one important theme in Fred's memory of the trenches. He began his story about his arrival at the Western Front by describing the deaths of several friends at Fromelles and, on the day before he left the quiet Armentières sector, of his mate Gus Stevens from Ganmain. He then explained how these friendships were forged on the recruiting march back in New South Wales:

> I was there on the night that Gus got wounded, of course. You see, when we were on the route march from Wagga to Sydney, people team up, according to their outlook on life, to a large extent, and on that route march was Harry Fleming, Gus Stevens who came from the same town, Les Hall and Bing Hall from Tumut, Bob Pettiford from Albury, Walter Allen from Coolamon that I'd known for many years, and Sam McKernan from Adelong. And we all seemed to have something in common, and others that had things in common teamed up too. We nearly always camped in the same tent and kept together, until we got to Egypt. Then there was some separation took place there. But out of all that lot, Harry Fleming, Gus Stevens, Walter Allen and Les Hall were killed. Bob Pettiford and Sam McKernan were badly wounded.[49]

The support of these friends had begun to help Fred to develop from farm boy to soldier, and to affirm a more positive self-image as a digger and man. Their loss shattered this new found security and identity. For the remainder of the war it was much harder for Fred to form close and supportive friendships, as he was often absent from the battalion and its personnel changed rapidly. When he returned to No. 9 Platoon in the summer of 1917 there were only three or four familiar faces: 'the thing changed and changed and kept on changing, until it'd be hard to know what ... how the fellows, you know, that you were with for a few weeks, what they were thinking of'. For Fred, mateship was something he had lost rather than gained during the war. Although he did not talk to me in detail about how he felt about the deaths of his friends, the manner and priority of the passages about their deaths evoke the grief, loneliness and even guilt that Fred felt at the time. As a way of coping with these feelings, he commemorated his friends at their funerals and recorded in his memory the exact dates, places and details of each of their deaths.[50]

In his wartime distress and confusion Fred did not compose a more general sense of the deaths of his friends. But in old age, with the hindsight of subsequent understandings of the war, their story was intended to convey the waste and futile sacrifice it entailed. Thus he told the story of Fromelles:

> It was the first time that the Australians had crossed no man's land on the Western Front and was possibly the most disastrous. And got nothing. They finished up the next day back where they started, with

half the division casualties. Those losses were never recovered, ever. I lost a mate that night who has never been seen or heard, nothing of him left at all. He was reported missing for several years, but Harry Fleming must have been blown to pieces in such a way that nothing of him was ever found that could be identified.

For Fred the benefit of this method of making sense of the loss of his mates was that it attained meaning as part of a political critique. Fred was able to direct his grief and anger at clear targets, and he showed little pain when he described and explained the deaths of his friends in these terms. Fred's political sense of mateship contrasted with the more nationalist meanings that it assumed in the Anzac legend; his story did not conclude that the deaths of his mates purchased a national tradition of loyalty and self-sacrifice.

The loss of many of his closest friends could not have come at a worse time for Fred, who needed all the support he could get when he went into the line. He did not cope at all well with life in the trenches, and his inadequacy as a soldier was a second major theme of his memory of the war. When the battalion first moved into the Somme it was:

> [...] a different front line to what we were accustomed to up near Armentières. The shell fire for a start, never, never ceased. Never stopped. It was one constant barrage from both sides. So that was pretty disturbing for a start.[51]

The bombardment caused terrible casualties and chaos, and it was almost impossible to evacuate the wounded or bury the dead. With the onset of winter it became bitterly cold and conditions got worse. The shelling continued to pulverise the men and the mud, and to uncover rotting corpses. For Fred, one of the most upsetting consequences of the shelling was that he sometimes had to walk on dead and decaying men. He was also 'scared stiff'. Fred's memory of his inadequacy as a soldier was evoked by his account of the attack on Polygon Wood in September 1917. Before the attack the Australians were subjected to 'a merciless barrage':

> The worse that I'd experienced. With little or no protection, only a shell hole, it went on and on. Even an hour seemed an eternity really. So in the darkness of all this I'd got separated from Panton [his lieutenant] and things were, generally speaking, in a bad way [...] When our artillery made their attack, I'd lost my rifle, as a matter of fact I'd nearly lost my senses, and I'd lost Panton, and so I just started across no man's land in a sort of haze.[52]

Hospitalised after being wounded in the attack, Fred also began to show the first signs of a nervous condition. At the time he received no recognition or treatment for this condition because he did not display the symptoms of a 'dithering idiot'. After the war his nerves deteriorated, and the condition would produce debilitating physical symptoms and culminate in a nervous breakdown.

Part of Fred's problem during the war was that he was neither able nor encouraged to express or resolve his feelings of anxiety, vulnerability and inadequacy. Digger friends might have been able to help but his closest mates were dead and, in comparison with Percy Bird and Bill Langham, he does not seem to have received a great deal of support from other soldiers (the fact that he refused to join his brother's non-combatant unit suggests that his relationship with the men in his unit was closer than he remembered). Although Fred's fears were shared by many diggers, his feelings of inadequacy were so acute that he could not show them.

Fred's negative self-image was also reinforced by what he perceived to be the expectation 'to put up with that sort of thing and carry on', and by comparisons with other soldiers who seemed to be unaffected by the barrage: 'I wouldn't be in the same street as a whole lot of them'. The Australians had a reputation for bravado — Fred recounted 'the casual everyday sort of way that the Australians went about doing what they had to do' — and this code of masculine behaviour encouraged Fred to bottle up his feelings and fears. The platoon and the battalion, which were so affirming for Percy Bird and figured prominently in his testimony, were barely mentioned by Fred in the interview because they did not provide the support he required. Whereas Percy Bird and Bill Langham talked with other diggers to transform collective war experiences into relatively safe stories, Fred Farrall remained silent because his experience was so deeply troubling and his masculine self-image so negative, and because he could find no empathetic public audience.[53]

In contrast, as an old man Fred was able to tell the story of his wartime 'inadequacy' and its effects. Recent general histories which showed that Fred's condition was relatively common, and which Fred read with enthusiasm, helped him to feel that the story of his un-manning was publicly acceptable. The more radical accounts of the soldier as a victim of the folly of war provided a political explanation of his experience. Thus, remembering the war no longer undermined Fred's manhood and identity. In contrast to his previous silence, and unlike Percy Bird, in our interview Fred focused on conditions on the Western Front and highlighted his own misery and inadequacy. He no longer perceived his trench experiences to be unusual and isolated; rather, he told his story to represent the general experience of soldiers on the Western Front and to make political points about what modern warfare is really like for its participants.

In our interview Fred also represented the war as the first stage in his political conversion. He recalled that this conversion began after the Polygon Wood attack, when he found shelter in a concrete pill box with a mixture of British and Australian soldiers and German prisoners:

> [...] lying there and sitting there and standing there quite peacefully. The Germans were as helpful as they could be and co-operative. A couple of miles further up they were carrying on with the job of killing one another as fast as they could. It seemed to me, that was the first time I began to think, 'Well, what's this all about? How silly can we be'. Course in the back of my mind was the old propaganda that had been implanted there years before that we were fighting for the British Empire, so it was the right thing to do. But side by side with that was of course, also planted, the thought that, well, we could get on very well with the Germans if we were left alone.[54]

Like other stories about Fred's conversion, it is difficult to judge the extent to which this seed of change took root at the time or lay dormant until its postwar political fertilisation. There is no suggestion that Fred expressed his disillusionment in any way during the war when, by his own admission, he was very confused:

> I just didn't know where I was. I had an idea that war and everything connected to it was wrong, but I had no clear opinion on what should be done about it, and I didn't find that out until, well that didn't mature or materialise until, probably, fairly late into the '20s.[55]

Although moments of personal doubt or collective disillusionment began this questioning at the time, it seems that the subsequent 'mature' understanding caused Fred to highlight memories of the diggers' opposition to the war and to generalise about it in terms of collective disillusionment (in comparison with Bill Langham who did not generalise his conversion experience in this way). Thus he recalled that of the front-line soldiers, many of whom he believed had enlisted for patriotic reasons, ninety per cent lost their enthusiasm and became anti-war, and he told two stories to support this view. He described a visit to an Australian Army Corps compound towards the end of the war, where he saw men from the 1st Battalion who were 'in the clink' because they had 'walked out of the Hindenburg Line and declared the war off ('or in other words, they went on strike'). He also recalled a church parade where the men of his own, depleted battalion hurled clods at a general who threatened to send them back into the line just before the Armistice: 'that was the frame of mind that the soldiers had got into, generally speaking. They didn't want any more front line, or talk about it'.[56]

Fred's stories about the disillusionment of the soldiers were not explained in the usual terms of Australian antipathy towards British commanders. Some of my interviewees would have highlighted the British identity of the general at the church parade, but for Fred national tensions were less important than the anti-war theme. Although Fred did share the common digger hostility to the British headquarters' staff, when that criticism was part of his remembering it was overlaid by the subsequent political critique of imperialism. Thus he noted that the British government charged the Australians for the use of wartime camps ('the Australians had gone 10 000 miles to help the home government, and that's how much they appreciated it'), and he cited recent historians' claims that British generals used the Australians in the most dangerous sections of the line. Although Fred's anti-British sentiments intersected with one theme of the Anzac legend, they were not primarily expressed in terms of Australian nationalism. Rather they were part of a broader political critique of imperialism and imperialist war. Fred's war story was both socialist and internationalist. He argued that the war was fought for the imperial interests of political and military leaders on both sides, and wherever possible he highlighted moments of resistance by the men in the ranks.[57]

Though national character was not a significant theme of Fred's war memory (he talked about the war for almost an hour before he was prompted by a question to consider national distinctions), when he did consider the distinctiveness of the Anzacs his analysis intersected with the Anzac legend but made very different meanings. He did not perceive the war in terms of the birth of the nation, and he believed that the diggers' 'national character' derived from the working-class, trade-union strengths of Australian society and the AIF. He explained that because of the influence of trade unionists in the AIF, and of Henry Lawson's writings about mateship, 'the Australians had a reputation for being great mates, one in all in, whether you were right or wrong', and the AIF was 'by far the most democratic' army in the war. This characterisation was developed by Fred in an interplay between wartime experience and postwar ideology, in which aspects of his experience became more prominent in his memory over time. He proved his characterisation of the AIF by using evidence from his own wartime experience (the Kangaroo march strikes, the diggers' refusal to salute officers, and their opposition to the war in 1918), together with citations from recent histories and generalisations from his trade-unionist ideology. The result was an Anzac tradition with a radical inflection, in which digger mateship was expressed against authorities and the war. The staff officers and military policemen, who were recalled by Fred as the main targets of digger hostility, were identified during the interview in terms of their roles rather than their nationality.[58]

Fred was more ambivalent in his remembering about the relationship between diggers and AIF battalion officers. He described a shift in his own attitude to officers during the war. He was an extremely obedient recruit who believed that officers were 'like Mohammed's coffin, halfway to heaven'. But, as the victim of a number of bullying officers, 'by the time I'd finished in the army, I'd got to the stage where I downgraded them terribly' (an opinion which he later considered to be wrong, because like everyone else the officers had a job they had to do). Yet he also recalled some officers who treated him well and who were regarded as mates, and that it often 'depended on the individual' rather than the rank. He contrasted his relationships with Lieutenant Pye, the son of a millionaire who mixed with the men and understood their problems, with Captain Wilson, who was disliked because he would not 'come down' to the ranks and use first name terms, and with Lieutenant Panton, an authoritarian bully whom Fred and two mates planned to kill.[59]

The ways in which Fred, as an old man, generalised about the relationship between front-line officers and men were affected by a number of intersecting but contradictory traditions. He used Bean's history to support a claim that many of the Australian officers were working men from the ranks, in contrast with British officers who needed 'a certain social background', but then stretched the notion of the egalitarian AIF officer much further than Bean. For example, he emphasised the story of the AIF hero Albert Jacka, a trade unionist who rose from the ranks, because Jacka treated the diggers as equals and shared their dislike for higher authority: 'he had all the characteristics that were frowned upon by the high command, by the generals'. Without recognising the tensions within his own analysis, Fred also highlighted the issue of class, and recalled that the relative equality between officers and men in the line was not perpetuated out of the line, where officers had better pay and conditions and more freedom, and 'lived a different sort of life':

But generally speaking officers were officers and they knew it, and the rank and file knew they were the rank and file. Just the same as employers look upon the workers.[60]

Fred's account of the 'democratic AIF' shows how even within one man's memory that label could be pulled in various political directions, using different ideological frameworks. The account shows how the variety and contradictions of that aspect of Fred's Anzac experience were smoothed out and generalised in his remembering through the use of both mainstream and radical traditions, and how easy it would be to plunder Fred's story to make one or other statement about the AIF.

Although Fred relished stories about the independent, anti-authoritarian diggers, he was critical of their larrikin reputation. Because of his isolated and rather puritanical rural upbringing, Fred was an innocent and straight-laced soldier:

You know I can say this, and I'm not the only one. That all the time I was in the army I never smoked, although they were dished out free. I never drank, other than a few spoonfuls of rum, at one time on the Somme. I was never in a brothel and I never gambled. I never played two-up, crown and anchor, or anything else. Or poker. None of these things.[61]

In our interview Fred was critical of diggers who did over-indulge in these pastimes, such as the Australians who got 'as full as googs, kicked up as much noise as they could, or laid in the gutter' on the Anzac Day march through London in 1919. In comparison with these men Albert Jacka was Fred's perfect hero because he was 'a man of very high morals who didn't drink, smoke or swear'. Yet Fred believed that among the diggers such morality was a handicap for men without Jacka's warrior status, because more larrikin standards of behaviour were the norm. Like Percy Bird, Fred felt excluded by that norm, and this may be another reason why he was unable to use the digger identity as a resource to help him overcome his feelings of military inadequacy.

The public reputation of the larrikin Anzacs also made Fred feel uncomfortable. He was aggrieved that the drunks at the London march were 'the ones that would be seen, possibly, as representing Australians'. In retrospect, Fred stressed — in terms very similar to those of C.E.W. Bean — that the 'doubtful characters' were a rather unsavoury minority in the AIF and that his own behaviour was more typical. He was thus especially critical of the 'Anzacs' television serial, which he thought played up the 'don't care type, larrikin type' that came to be best known in the AIF. He concluded: 'now a lot of people wouldn't ever believe that there was anybody in the AIF that didn't do that. But there would be a larger number that would be the same as me and some probably better'.

Whether or not larrikin behaviour was predominant among the diggers, Fred's insistence was partly an attempt to overcome his lingering feeling of exclusion and to define his own identity as closer to the norm. It may also have been fuelled by the moral responsibility that had become a major ingredient of his personal political philosophy. Refusing to salute an officer was admirable because it asserted a particular attitude to authority; getting drunk made no such statement and was politically irresponsible. The development of Fred's radical political philosophy, and the ways in which it helped him to compose his war experience in more positive terms, need to be situated in the context of postwar Australian society and politics. Fred Farrall's war story was thus influenced almost as much by his return to Australia as by his experiences in Belgium and France.[62]

The aim of these three wartime memory biographies has been to use the rich, personal detail of three Australian soldiers' stories in order to understand how the men comprehended and articulated their war experience, and to explore the interactions between personal war experiences and identities and the collective identities of the diggers and the Anzac legend. Even three lives can evoke some of the variety of the Anzac experience. For each of these men there were very different contexts, motivations and pressures for joining up, and they remembered enlistment in very different ways. War service in the artillery or the infantry, in the front line or as a noncombatant, made for contrasting experiences of war and different identities as soldiers. Within the AIF, Bill Langham relished the masculine comradeship of the diggers and participated in the larrikin lifestyle of digger culture, but both Percy Bird and Fred Farrall were ill at ease with that lifestyle.

The experience of battle and trench warfare had the greatest impact on all three men, and posed the most difficult problems for the composure of military masculinity. They dealt with those difficulties in different ways, depending on the nature of their combat experiences, the availability of material and emotional support, and the suitability of public narratives with which they could articulate combat. The particular public of the diggers was one important support network, and a resource for the recognition and articulation of combat experience. Yet while Bill Langham was sustained by his mates, and found in digger culture a language to make a comparatively positive and affirming sense of the war, Percy Bird suffered the loss of that resource when he was taken out of the line, and for Fred Farrall — who had lost his best mates in the early battles of 1916 — the apparent self-confidence and competence of the diggers simply reinforced his own feelings of masculine inadequacy.

The official Anzac legend that was fashioned during the war by Charles Bean and other publicists, and which drew upon digger culture and reshaped it in more respectable ways, provided another set of public meanings and identities which the Australians used to articulate their war experience. Percy Bird was uncomfortable with digger culture, but the proud Anzac identity of his unit and the AIF helped him to feel positive about his role and identity as a soldier. For Bill Langham, the wartime Anzac legend provided general interpretative categories to make sense of stories about the quality and effectiveness of the Australian soldiers, but it also made it difficult for him to articulate aspects of his war — such as his conversion experience — that did not fit the legend. For Fred Farrall, the gap between the legend's characterisation of the Anzacs and his own experiences was so great that he felt utterly inadequate as a soldier, and was unable to express his feelings or articulate a positive military identity.

The memory biographies also show that the process of composing military identities did not end with the war. The war identities and memories of Australian soldiers who survived to return to Australia would be shaped and redefined by their postwar experiences as returned servicemen, by the new publics of postwar civilian life and ex-servicemen's culture, and by changes in the definition of the digger and the legend of Anzac.

Part II

The politics of Anzac

CHAPTER 4

THE RETURN OF THE SOLDIERS

In his war histories Charles Bean concluded that 'the AIF on its return merged quickly and quietly into the general population', and that the repatriation of the diggers was on the whole a great success. General Monash directed an elaborate demobilisation scheme that successfully completed the massive task of shipping several hundred thousand soldiers back to Australia before the end of 1919. Federal government pension schemes, which had been developed as part of the recruiting campaign during the war, were comparatively well-planned and generous, and disability pensions were more liberal than those of most other combatant nations. There were schemes for vocational training, assistance to ex-servicemen wanting to set up in business, and soldier settlement on the land. Ex-servicemen were to be granted preference in public employment, and private employers were encouraged to follow the government's example. Returned men were offered special terms for the purchase of War Service Homes, and received, on top of any deferred pay they had owing to them, an interest-bearing bond or 'gratuity' of one shilling and sixpence for every day they had served between embarkation and peace. With the exceptions of the schemes to set up ex-servicemen in small business or on the land, these measures were, according to repatriation histories, relatively successful.[1]

Yet the memories of working-class ex-servicemen show that for many of them the battles of peacetime were as hard as any they had fought during the war. How a man fared in this postwar battle depended on how badly he had been affected by the war, how well placed he was for employment, and the degree of support he received from family and friends. The nature and success of a soldier's repatriation experience would, in turn, shape his relationship to ex-service organisations and Anzac commemoration, his identity as a returned man, and his memory of the war.

Oral testimony is an invaluable source for uncovering the diverse personal experiences of repatriation. Some of my interviewees were puzzled by questions about this period of their lives. In comparison with the war, the postwar years had rarely been the focus of prior questioning and conversation, or of recognition through representation in books and films. Old diggers usually talked more readily about their war experiences, which they assumed to be of primary historical significance. Yet once my questions had got the men talking about their postwar lives, most opened up about what was a time of great personal upheaval, and in which significant personal experiences — the return, finding a job, joining an ex-servicemen's organisation — were clearly sign-posted in memory. Indeed, because these postwar experiences had not been the subject of regular interest and articulation, stories of that period often seemed more fresh and less influenced by public accounts than veterans' war stories.

Coming home

Australian soldiers waited in Europe for up to a year after the Armistice. Transport boats were in short supply, and Australian civil authorities deliberately spaced out shipping to reduce the social and economic disruption that would be caused by a sudden influx of returned men. Veterans recall a period of relative ease based in London or with the army of occupation. According to James McNair, 'when the war ended it was a big relief':

> Well I was in London for ten months. We did nothing but running around after the ... See, we had to wait that time before we could come home. We didn't have the shipping ... No, the only thing — how soon can I get back home?[2]

Like many soldiers in other units of the British army who were awaiting demobilisation, the Australians were impatient and even unruly in this period, and their indiscipline was a cause of official concern. In England, Non-Military Employment schemes were established ostensibly to assist the soldiers' transition back into an Australian workforce, but they were also intended to mitigate discontent. On the troop-ships overcrowding, bad food and inadequate entertainment sometimes worsened the soldiers' mood. Angela Thirkell recalls that on the trip to Australia with her new AIF officer husband in 1919, the diggers were riotous and virtually uncontrollable throughout the voyage.[3]

Perhaps because of the confusion in these transitional months between army and civilian life, most Great War veterans recall very clearly the exhilaration of home-coming, which was celebrated in emotional family reunions and official 'Welcome Home' ceremonies. James McNair describes 'the nicest view I ever saw':

> I can still see it vividly in my mind. We're coming one evening, in the Indian Ocean, to Fremantle Harbour ... it was Saturday afternoon, because the pubs were open, and there was lots of girls — I couldn't see their faces but there were all their lovely frocks and so on — striped frocks. Australians! Home at last! Finest sight I ever saw.

After a night in Fremantle the boat headed for Melbourne:

> We anchored just off Gellibrand Lighthouse. 'Oh, there's the synagogue over St Kilda. Oh, there's the Exhibition.' Picking out all the sights, you know. Got off the boat, put on my gear. We were going to do a trip around the ... conquering heroes coming home you see. Well, anyway, we got down to the end of the Pier. 'Oh, there's a car.' Me mater, the old man and a friend, and the sister and the brother. 'Oh!' They stopped the car and I got off, and then straight to the discharge place. Went to one man, another, the Doctor. I said, 'How's the ticker?' he says, 'Okay'. The dentist: 'Want any teeth?' I said, 'No, I want to get out!' I was home at 11 o'clock [...]
>
> My brother said to me in the car, he said, 'Have you got your speech ready?' I said, 'What for?' 'Oh,' he said 'you'll see'. Well, when we got home there's a great big marquee [...] My cousins, I didn't know them. They had to introduce themselves. They came from everywhere [...] I dunno what I said. I must have been a bit flustered [...]
>
> In the afternoon went up to Brunswick, the State School there, Albert Street, and voted. The 13th of December 1919. My name was still on the roll after four years.[4]

Rehabilitation and the Repat

Exhilaration was often short-lived, and was overshadowed by the pressing problems of rehabilitation. Men who had served in the trenches were often physically unfit when they arrived home. Some, like Jack Glew, recall that they were 'as good as gold', but most of the men I interviewed (a selective sample who were well enough to survive into old age) had been wounded or gassed during the war. Several of them were in and out of hospitals after they came home, and many suffered recurring medical problems.[5]

Though Australian disability pensions were comparatively liberal, it was not easy to get a pension. The diggers' troubles began at the medical examination upon discharge, as Jack Flannery recalls:

> They asked you how you was. You'd say 'Good-oh'. You'd get discharged 'A-1'. All you thought about, there was 999 out of a 1000 '14–'18 diggers, all they thought about was getting out of the army, back into civilian life.[6]

In the excitement and confusion of return many men did not realise the future significance of that medical code. When an old wound or war-related illness flared up, the 'A-1' on the form made it very difficult to prove that the condition was 'war-related', and the unfortunate veteran often forfeited the right to a disability pension or subsidised medical treatment. The inaccuracy of health records was exacerbated by the failure of the AIF medical service to obtain individual medical history sheets held by the War Office. Many ex-servicemen struggled with 'bloody red tape' over pensions; they complained that their rights had not been explained properly at the outset, and that a veteran had to be missing obvious pieces of his body to receive a disability pension. The government department that administered war pensions was nick-named 'The Cyanide Gang' because of the rejection and 'sudden death' of claims.[7]

Not all aspects of repatriation policy were regarded so negatively. Some men did receive the disability pensions they deserved, and many made good use of their gratuity. It provided the capital for a deposit that enabled them to purchase a house, taking advantage of special terms offered to ex-servicemen for War Service Homes, and thus allowed them to achieve an ideal that otherwise would not have been possible for many years, if at all. Stan D'Altera recalls that his gratuity of £120 'was a fortune then', and that he and his brother used it as a deposit on a house in Yarraville, where they had spent their childhood in rented poverty.[8]

However other returned diggers, who were struggling to get a job, let alone a house, were forced to use the gratuity as credit for a loan, or sold it to contractors who extracted a percentage from the value of the gratuity. They recall a thriving black market for gratuity bonds, which were easily converted into food or drink by unemployed, down-and-out ex-servicemen; about half the bonds were cashed immediately. Out in the bush Sid Norris sold his gratuity just to stay alive: 'that went like anything else. Because there was times there was no work and it was a very rough time up around Cairns, I'm going to tell you, after the war too'.[9]

The Repat became a target for veteran's grievances, at the time and in memory. Perhaps inevitably, many diggers were embittered by the false economy of sacrifice; the scanty rewards of repatriation did not match their wartime sacrifices or the rhetoric about how Anzac heroes would be treated after the war. Bill Bridgeman recalls starkly that 'they were finished with you and you were finished'. Some returned men accepted this state of affairs as inevitable and got on with the job of survival, while others fought to get a better deal. Fred Farrall remembers that there were often violent clashes at the Sydney Repatriation Department offices, and that iron bars were installed to protect the staff. Many veterans joined an organisation of returned servicemen in the hope that it would represent their claims, but some became contemptuous of token gestures of recognition and began to reconsider their attitudes to war service. Stan D'Altera recollects that diggers who complained were called 'hoodlums', and he was so angry that he used his war medals as fishing sinkers.[10]

If a returned digger had some chance of receiving a pension and medical treatment for a war wound, he had almost no chance of getting any official support for psychological damage. There was a staggering incidence of mental disorder among returned diggers. According to Bill Williams, 'we were all rather confused in some way, we didn't exactly know where we were'. Williams was himself stricken by grief for a mate who had died, and tortured by guilt about killing people; Ern Morton was one of many who suffered terrible nightmares for years. The men I interviewed survived the emotional traumas of war and the return to civilian life, yet they all have vivid stories about soldier friends who were not so lucky, who went crazy, became derelict drunks or killed themselves.[11]

The home front

Psychological problems were partly the after-effects of trench warfare, but they were also caused by the difficulty of readjusting to peacetime society, and by the gap between each digger's experiences and the image of the Anzac hero. As soon as they arrived home, ex-servicemen lost the security of army life and the everyday support network of their digger mates. Instead they faced an uncertain future in a society that had been radically transformed, and that did not find it easy to comprehend or cope with the material and emotional needs of ex-servicemen.

Some of the men I interviewed had very positive experiences of their return to domestic life. Alf Stabb brought an English war bride back to Australia, and Ted McKenzie married a woman who had sent him socks during the war. These and other men tried 'to put the war behind them' and focused on raising and supporting a family. Others, like Jack Glew, recall that 'it was like heaven' to be looked after like a lord by parents in the comfort of home. Yet there were many who struggled to cope with the transition from army to domestic life. Young men who had enlisted as teenagers had particular problems. At war, boys had become men and grown independent of family life. That independence was not always so easy to handle without the security of the army, which had provided food, clothing and accommodation, as well as friendship and regular employment. Fred Farrall felt like 'an abandoned pet' after he disembarked in Sydney.[12]

Single ex-servicemen who returned to their parents' homes often found it difficult to settle into family routines and relationships after the itinerant lifestyle of the army. Stan D'Altera recalls that 'many of the younger ones of us, you know, we still had the wanderlust, you know. I used to go off all over the place, you know, New South Wales, South Australia'. The rules and habits of the AIF were vastly different from those of the family at home. Wartime codes of appropriate masculine behaviour — the male comradeship of drinking, swearing, gambling and womanising — did not conform to civilian prescriptions for domestic manhood. Alf Stabb (who lived with his parents and sisters while waiting for his war bride to arrive from England) recalls that 'you had to watch yourself because you didn't care much about what language you used when you were in the army. You had to smarten up'. Stan D'Altera remembers that when he went to visit a digger mate in Adelaide the man's mother would not let him in the house because she feared, correctly, that Stan would encourage her son to miss work and go out on a drinking binge.[13]

Families found it difficult to cope with men who were used to years of rough living in the army. They also found it difficult to understand how and why their men had changed. Censored letters and news reports had done little to inform civilians about the experiences of the men at the front, and when the soldiers came home it was difficult for them to communicate the nature and effects of their experience. Some ex-servicemen enjoyed eager audiences for their stories — Jack Glew recalls that he got 'sick and tired of telling it day after day' — but more often war stories were hard to tell and equally hard to listen to. As A. J. McGillivray remembers:

[...] quite a few people didn't quite believe what we said and from then on, unless we were amongst our own servicemen and that, many things were never related to others because of their attitude [...] And when you come to think of it later on it was quite understandable that they would be hard to ... hard to believe.[14]

Diggers became selective about what they would say about the war and who they would talk to. Jack Flannery found that civilian ideas about war were so different from soldiers' memories that he could only relate 'what good times we had when we was on leave or something'. He recalls that 'a lot of diggers, they won't tell you nothing, they'd have to be terribly drunk to plug the gap'. On the whole, civilians rarely asked Jack about the war, and those who did were sometimes hostile because of jealousy or because they thought the soldiers were 'bloody murderers'. Tensions between ex-servicemen and their families, and with civilians in general, were thus exacerbated by mutual incomprehension, and returned men found it easier to retreat to the familiar camaraderie of their digger mates.[15]

In some cases the diggers' sexual behaviour had also been altered by the freedoms and frustrations of the war, although men were affected in different ways depending on their wartime experience and postwar domestic context. Stan D'Altera blames his bachelorhood on physical disability and awkwardness caused by the war; in contrast E. L. Cuddeford recalls that he 'went for women more' when he came home. Men who had been married before they enlisted coped with — and caused — particular stress. Many couples were happily reunited, but affections were easily blurred by years of separation, and uncertainty or jealousy caused terrible anguish. One married officer on Ern Morton's home-coming ship suicided because — along with one in seven Australians discharged from the army — he had venereal disease. Charles Bowden was the only man I interviewed who was married when he left Australia; the inevitable age bias of my interview sample produced a disproportionate number of young, single recruits. Bowden recalls that he felt very guilty about leaving his wife and young son, and concerned about her prospects as a woman alone. He says that he stayed out of the brothels in France because he was married, but that he was upset because his wife was an infrequent correspondent due to her resentment of his enlistment. Home-coming was not easy for Charles Bowden and his wife, and in our interview he still did not want to discuss that period of his life.[16]

Interviews with old soldiers reveal the awkwardness and pain of the domestic no man's land between army and civilian life. They only hint at the suffering and dogged survival of the women and children who had to live with war-scarred ex-servicemen. Other sources have begun to complete this side of the story. The national divorce rate doubled between the censuses of 1911 and 1921, and women brought the majority of the petitions for divorce for the first time. Local courts were crowded with cases against disturbed returned men who had gone berserk, or veterans being sued for maintenance of pre-war brides they refused to support, or children for whom they denied paternity. Testimony from women provides more personal detail. One Melbourne woman I interviewed in 1982 was a young girl when her father returned from the war. Unable to settle into a job and resentful of the independence his wife had developed in his absence, he retreated to the association of returned cobbers. Her mother forbad the children from entering the lounge room where the men drank, smoked and played cards, and the girls were terrified of the returned men who were 'very sexual'. The eldest sister aborted the child of a soldier when she was twelve, and within a few years the father left his wife and children.[17]

Historian Judith Allen summarises the impact of the return of the soldiers upon Australian women:

> The interpersonal brunt both of the First World War and of the inadequacies of public provision for this population of disturbed young men fell disproportionately on Australian women. Women's bodies and minds absorbed much of the shock, pain and craziness unleashed by the war experience.[18]

There was considerable public concern about desertion and physical abuse of women by veterans, though, as Allen shows, support for women victims declined and returned servicemen tended to receive judicial sympathy because of their wartime status and the acknowledged difficulties of rehabilitation.

A land fit for heroes?

Of greater public concern, and a further source of tension between ex-servicemen and civilian Australians, was veteran unemployment. For the majority of diggers, finding and keeping a job was the most urgent and worry-ing problem of home-coming. Stan D'Altera recalls that 'my main worry was whether I could get a job, because we'd learned that there was much unemployment in Australia, like in, among the working class anyhow'.[19]

One group of diggers who had few problems were those who had man-aged, often because of clerical training, to get a job out of the line during the war. Percy Bird, James McNair and Leslie David had all been posted to AIF clerical work in France or England; railwayman Charles Bowden had served in an AIF train-driving unit. These men had a relatively easy war — as Leslie David recalls, 'it was mainly a pleasant experience for me' — and were generally unscathed by the emotional and physical traumas of the trenches. They benefited from wartime opportunities to travel and broaden their horizons, and gained invaluable work experience; Leslie David and Charles Bowden were both sent to Pitman's College in London. When they returned to Australia they were highly employable, and in some cases the war proved to be an invaluable stepping stone to a successful career.[20]

Other diggers were fortunate to have worked for public service institutions like the railways or the Post Office, which kept their jobs or apprenticeships open for them until they returned. Soldiers from well-off families could sometimes return to their studies or to the family farm or firm, and some enjoyed professional support networks which may have been enhanced by contacts made in the officers' mess. For example, my grandfather Hector Thomson returned to a property within the family estate. My other grandfather, John Rogers, did not find it easy to get work as an industrial chemist when he finished his studies after the war, but he was helped by glowing references from senior AIF staff officers. In contrast, the majority of Australian soldiers who had been workers in primary or secondary industry before the war faced hard times. In a recession that peaked in 1921 and in the Depression of the 1930s, these men were caught between the economic pincers of a rapidly increasing labour force competing for diminishing jobs.[21]

Ex-servicemen were not well-placed in the competition for work. Youths who had enlisted before they could learn a trade, and who had gained no employable skills at war, were perhaps worst off. Ern Morton was in that position. He had to survive on casual work out in the bush when he first got home, and recalls thinking that no trade meant no future. Men who had worked in a trade before the war found they had lost vital skills and promotion opportunities. Other men who got a job found that they just could not settle down, or were not well enough for regular employment. Percy Fogarty went through eleven different jobs when he first came home because he couldn't settle and 'kept getting crook' with pneumonia, and Stan D'Altera remembers that he lost his ambition to be a skilled metal worker, partly because he couldn't get a job in his trade, and partly because he had 'the wanderlust'. Men who had hoped to make a new start after the wasted years of war were often cruelly disappointed as they struggled to find work when they first came home, and then, if they did gain steady employment, lost it again in the 1930s. The unemployed digger became a potent symbol of the failure of the promise that the Anzacs would return to a land fit for heroes.[22]

The various retraining and employment schemes for ex-servicemen were better on paper than in practice. In Commonwealth vocational schemes the government shared the payment of an ex-serviceman's wages with an employer while the man was being trained and settling into work. Albie Linton benefited from this scheme. When he was discharged from hospital he learnt sheet metal work at Footscray Technical College and got a job on the strength of the training. But Stan D'Altera and Ern Morton could not stick at their training course in Geelong, and other diggers recall that the courses were badly organised, that there was often not enough work for the trainees, and that employers were not interested in keeping the returned men on after the government wage-subsidy ended.[23]

The plans to establish ex-servicemen as self-employed businessmen or small-holders were even less successful. Bill Langham 'scrubbed' the idea of establishing his own taxi service because the level of capital support was inadequate. Land settlement schemes were very popular and many veterans were granted small blocks of land on special terms. A. J. McGillivray, who had worked on the land before the war, started a dairy farm on a soldier settler's block in Gippsland, and prospered after 'a dreadful period' of battling to get established between 1921 and 1928. The majority of soldier settlers lost that battle because they lacked experience or were not physically fit enough for farming, and because they were often given poor land and did not have the financial capital or good weather to make it work.[24]

Preference for returned servicemen in employment (by which means a job applicant who was a returned serviceman would be granted preference over other candidates if all other factors were equal) was a central plank of repatriation policy. Commonwealth and State governments legislated for preference in public service employment, and many local councils and some private employers followed suit. Most returned men supported the policy when they first came home. They had made great personal sacrifices during the war and now they desperately needed assistance. Stan D'Altera recalls that preference was necessary because 'the returned soldier had given away opportunities for promotion in their job [and] missed getting money to buy a house and all those sort of things'. The policy undoubtedly benefited some ex-servicemen. Ern Morton, sick of dead-end jobs, applied for a permanent position as a meter reader with Coburg Council. At first unsuccessful, he got the job after the local soldiers' club and the associations of Fathers and

Mothers of Ex-servicemen protested against the appointment of a man who had not served in the forces.[25]

Nevertheless it was notoriously difficult to prove that 'all other factors' were equal, and easy for a potential employer to claim, sometimes with justification, that a soldier applicant could not do the job as well as another candidate. Private employers were particularly ambivalent about taking on ex-servicemen. Some respected their wartime promises about returned heroes and were sympathetic about the special needs of returned men. Others took advantage of the government subsidy for ex-servicemen trainees. But the requirements of business usually came first, and employers were loathe to demote or replace men and women who had proved to be good workers during the war. Bill Langham recalls:

> [...] fellows that I knew, I said, 'Did you get your old job back?' They said 'No [...] Jack Smith's got my job'. [The boss said], 'He's been there since you went away. He didn't go away to war, he stayed there. But he's done such a good job, well I can't very well sack him now'.[26]

Nor were employers keen to take on unfit, ill-trained or undisciplined returned men who did not settle easily into work rules and routines. Stan D'Altera and Percy Fogarty recall removing their 'Returned from Active Service' badge before a job interview so that prospective employers would not think that they were unskilled or unreliable. Employers were also loathe to pay the higher rates given to ex-servicemen, which took into account their years of military service. Stan D'Altera returned to his apprenticeship as a fitter and turner after he finally settled down, but when he turned twenty-one and asked to be paid a full salary the boss refused, saying that he should work as an improver at the lower wage for all the time he had missed from his trade:

> I told him to stick his job, and walked out of the factory immediately [...] What angered me, on his advertising pamphlet, you know, that he sent farmers and that, he had a copy of the Honour Roll there, in colour. That would be to entice them to buy his goods. My name was on the top of it, you know, in coloured lead. That's where his patriotism was — to himself.[27]

The material needs of ex-servicemen and the rhetoric about their special status inevitably collided with the needs of business and of other workers. Whether it was the wartime worker or the ex-soldier who got the job, promotion or pay rise, this competition caused tension between returned men and employers, and between ex-servicemen and other workers.[28]

Hoodlums, revolutionaries and genuine diggers

These nagging tensions, caused by competition for jobs and readjustment to work practices, also prompted ex-servicemen to play an important though ambiguous role in industrial action. Economic pressures, barely contained by insistence on loyalty 'for the duration', exploded in 1919. Trade unionists fought to regain conditions, pay and industrial power which they had lost during the war, whilst ex-servicemen struggled to find work. More strike days were lost in 1919 than in any year until the 1970s. Some diggers took the side of unionists; there was a large increase in union membership as the soldiers came home. Ex-servicemen were prominent in the 1919 Fremantle wharf strike, which was called by the union president (a returned man) to oppose preference for non-union labour. When the police attempted to disperse crowds of strikers, returned men armed with revolvers and home-made bombs used wartime tactical skills to repel them. In other places ex-servicemen were mobilised by employers to serve as strike-breaking workers or to put down strikes by violence. On the Melbourne waterfront, diggers joined the union and the strike-breakers in equal numbers and were pitted against each other in wild street fights. On the Western Australian goldfields, ex-servicemen fought alongside other miners against members of the Returned Soldiers' Association.[29]

In postwar Australia the line between industrial violence and civil unrest was blurred, and the role of returned men in public disorder was a cause of great official concern. Australians struggled to comprehend the contradiction between their Anzac ideal and the uncomfortable presence of ex-servicemen. Whenever soldiers met there was, as historian Terry King records, 'a scent of trouble, a whiff of impending mob violence, a vague sense of things being out of control'. The two most dramatic outbursts of soldier violence were the Red Flag riots in Brisbane in March 1919 and the Melbourne Peace Day riots in July of the same year, which was one of the most violent in White Australian history.[30]

To celebrate the proclamation of peace, 7000 returned soldiers and sailors marched through the city of Melbourne on Saturday, July 19. Though the parade was orderly, in the evening groups of ex-servicemen derailed trams and invaded city theatres. Police made a number of arrests and subsequently battled with soldiers who rushed the Town Hall to release their mates. On Sunday a group of fifty to sixty returned men, trying to reach the headquarters of the civil police, brawled with sentries at the gates of Victoria Barracks. They were bloodily subdued by mounted policemen. On Monday a crowd of

Figure 12 *This article written by Stan D'Altera for* Smith's Weekly *(1 August 1931) conveys the feelings of a veteran about the different ethos of life in the AIF and Australian society.*

mainly discharged ex-servicemen, variously estimated at between 2000 and 10 000, marched to the Treasury Gardens with a list of complaints about police harassment to present to Parliament. In anger and confusion a section of the crowd attacked and wounded the Premier in his office, and it was an hour before order was restored by 300 foot police and twenty mounted troopers. That night 6000 people lined Russell Street opposite the police station where the prisoners of the weekend were being held, until they too were cleared by a baton charge. The Melbourne press was horrified that the heroes of Gallipoli and Pozières could become a 'howling, stone-throwing mob', and blamed the evil influence of 'hoodlums and revolutionaries' while asserting that 'genuine diggers' could not have been involved in the fracas.[31]

In contrast, the ex-servicemen in Brisbane were mobilised before they could get into trouble. When red flags appeared at a local civil liberties march in March 1919, a massive loyalist force, including a large number of ex-servicemen, was organised within hours to attack Brisbane radicals and the expatriate Russian community. The following weeks of 'white terror' saw the formation and drilling of an 'Army to Fight Bolshevism', a prototype of subsequent loyalist paramilitaries which usually included a considerable number of ex-servicemen in their ranks.

This postwar turmoil was precipitated by the ending of wartime restraint and the return of the soldiers, who comprised almost half of the male population in their age group, and who were reshaping Australian domestic, industrial and political relationships. The potent role of ex-servicemen in this turmoil was, in turn, due to the diggers' experience of social dislocation, and their determination to achieve due recompense for sacrifices at war and for the difficulties of return. Some ex-servicemen were motivated by overtly political concerns, but the needs and identities of returned men often cut across those of civilian society and created unexpected, contradictory political positions that confused and worried civilian activists of all persuasions. A new, postwar battle had begun, to win the allegiance of ex-servicemen, to define the 'genuine digger', and to control the legend of Anzac.

CHAPTER 5

THE BATTLE FOR THE ANZAC LEGEND

Loyalists and disloyalists

Although the Anzacs had won their battle honours in Europe, the battle for the Anzac legend was shaped by Australian political culture and by the ideological schism between loyalists' and 'disloyalists' that had emerged during the war. The apparent national unanimity for the war and enlistment that had existed during 1914 and after the Anzac landing had been short-lived. The first to suffer exclusion from the national embrace were interned 'aliens' of German or other suspicious ethnic origin, but as the war progressed the term 'alien' broadened from its ethnic definition to include any individual or group that did not fully support the war effort.

Economic developments ensured that an increasing number of Australians, particularly within the working class, attracted suspicion as they became disenchanted with the war effort. The insecurity of international markets which accompanied the outbreak of the world war caused an initial increase in domestic unemployment. More significantly, prices and rents began to rise before the end of 1914, and were rocketing up by 1915. Federal and State governments imposed a freeze on wages, but the new Labor Prime Minister Billy Hughes reneged on promises to introduce a concomitant prices freeze. While working-class living standards rapidly deteriorated, profits seemed to be unaffected, and this apparent inequality of sacrifice infuriated the Labor movement and precipitated a series of strikes over wages and hours which culminated in a 'general strike' in New South Wales in 1917.[1]

These tensions were intensified by conflicts over Australian participation in the war, and in particular by the conscription referenda initiated by Hughes to boost recruiting. Conducted with great rancour, the conscription 'blood votes' of 1916 and 1917 widened existing social and political divisions. The

Labor movement spearheaded the 'No' campaigns and gained the support of the majority of the working class. Many Irish Catholics, led by the charismatic Archbishop Daniel Mannix, and angered by the British suppression of the Easter Rising in Dublin in 1916, also opposed conscription. Middle-class, Protestant Australians were appalled by this apparent disloyalty, and their powerful representatives in parliament and press fumed about Fenians, traitors and spies.

Opposition to conscription was not necessarily opposition to Australian participation in the war, but the military conduct of the war and the appalling casualty lists fuelled dissent. After the failure of the first conscription referendum in 1916, Billy Hughes was expelled from the Labor Party and took his followers into a coalition with the Liberals. They formed the Nationalist Party and successfully contested the 1917 Federal election on a 'Win the War' platform. The radical remnant of the Labor Party now called for a negotiated peace and supported an anti-war alliance that questioned the sacred icons of nation, empire and patriotic unity, and adopted an increasingly radical anti-capitalist ideology. In turn, the civil authorities used the wide-ranging powers to act for public defence, conferred upon them by the War Precautions Act of 1914, for the surveillance and persecution of militants. As several historians have argued, by 1918 the consensus on the war had broken and 'Australian society [...] had virtually polarised along class lines'.[2]

Some radicals saw in this polarisation an opportunity for political transformation or even revolution. It may be that they misinterpreted the politics of working-class disillusionment and war weariness, and were seduced by the success of the Russian revolution. But whatever the extent or prospects of the radical challenge, there is no doubt that the spectre of disloyalty caused an unparalleled mobilisation of conservative middle-class Australia. This mobilisation was sustained by the constant need to exhort men to volunteer for the forces and Australians generally to support the war effort, and as the news from the front got worse loyalists felt increasingly embattled. After the Russian revolution loyalists perceived a connection between anti-war activity and industrial action, and assumed that they were all motivated by the sinister, foreign influence of Bolshevism.

In the conservative world view 'loyalists' — whole-hearted supporters of the national and imperial war effort at the front and at home — were distinguished from 'disloyalists' who undermined or opposed that cause, and who comprised a conspiracy network of enemy sympathisers, anti-war activists, strikers, shirkers (men who had refused to enlist) and Bolsheviks.

Armed with this potent ideology and spurred by fear and suspicion, conservative State and Federal governments (only Queensland remained Labor), civil and military intelligence agencies, and private loyalist organisations formed a powerful network with, as historian Raymond Evans concludes, 'each one single-mindedly dedicated to the eradication of the radical and alien influences they believed were surrounding them'.[3]

The politics of returned servicemen

The return of the soldiers in 1919 added a potent new ingredient to the cauldron of Australian domestic politics, just as it did in the troubled postwar nations of Europe. Both loyalists and their radical opponents sought to win the allegiance of returned servicemen and the symbolic support of the Anzac legend. Some labor militants hoped that because soldiers were mainly 'members of the working class', the 'economic and industrial interests' of returned men would be 'identical with those of Organised Labor'. Ex-servicemen who rejoined the socialist or trade union movement were given prominent positions on political platforms and were frequently quoted in political leaflets and newspapers. Yet activists were also concerned that returned men might be hostile about labor 'disloyalty' during the war, and might promote the economic interests of veterans ahead of those of other workers. In February 1919, Nettie Palmer warned that if the Labor movement failed to repatriate the diggers into its ranks, they might be 'induced to form dummy unions, employers' unions', and used as 'a wedge driven into the power of Organised Labor'.[4]

Conservative publicists confidently proclaimed that the diggers would surely join the loyalist cause for which they had, after all, been fighting. However their hopes were also shadowed by anxiety about the acknowledged potential of returned servicemen for disorder, and by news of the participation of returned men in European uprisings. The perceived threats of social breakdown and revolution — and the frightening prospect of a radical force with military training — needed to be averted at all costs, and the loyalist network which had been created during the war galvanised into action.

Physical force provided the first line of defence. Police were used to protect strike breakers and control disorder when digger frustration erupted on the streets. Loyalist paramilitaries (like the Army to Fight Bolshevism in Brisbane) were organised by local alliances of military intelligence officers, service leaders and businessmen to put down radical activists and 'disloyal' diggers. This physical force was backed up by zealous official surveillance of what one intelligence report described as the 'alarming number' of ex-servicemen suspected of having links with radical organisations. Officers of naval and military intelligence, and of the Federal government's Counter Espionage Bureau, were active on home-coming ships and at discharge depots. They countered suspected Bolshevik propaganda with their own anti-union and anti-labor leaflets, and supported veterans' organisations with loyalist sympathies. Proven radical diggers were targets for surveillance and physical restraint, and information about the participation of ex-servicemen in industrial disturbances was rigorously suppressed to ensure an impression of their loyalism.[5]

Surveillance and physical force were usually linked to an ideological mobilisation of the 'genuine digger'. 'Genuine diggers' were ex-servicemen whose behaviour and character matched the perceived virtues of the Anzacs, as well as the ideals and cause of the speaker. Despite the attempts of labor activists to define 'genuine diggers' in their own terms as egalitarian democrats, the mainstream press and conservative politicians fused the definition of the digger — as a disciplined and patriotic hero — with the language of loyalism. Aberrant ex-servicemen's behaviour was, by contrast, defined as the fault of 'hoodlums', 'revolutionaries' or 'undesirables'. Through this linguistic distinction the reputation of the diggers was inoculated against unsavoury influences, and the symbolic power of Anzac was enlisted for the loyalist cause. As the Queensland president of the Returned Sailors' and Soldiers' Imperial League of Australia (RSSILA) declared in 1919, the Anzacs would protect Australian society against 'the sore on society that dared to preach disloyalty [...] a microbe that would have to be cut out before it grew into a dangerous cancer'.[6]

Loyalists also sought to harness the diggers' political energy within ex-servicemen's organisations, and in the long term this proved to be a most effective strategy. There were several organisations competing for the membership of Australian veterans. The Returned Sailors' and Soldiers' Imperial League of Australia (the forerunner of the modern Returned and Services League) was formed in 1916 from the merger of several State Returned Soldiers' Associations, and became a national organisation with State branches and suburban and rural sub-branches which served as soldiers' clubs. The RSSILA was active in repatriation campaigns but avoided confrontational tactics. It adopted a national policy against participation in party electoral politics, partly to enhance its effectiveness as a bipartisan pressure group, but also so that it could appeal to diggers of all political persuasions.

However prominent loyalists were usually the driving force behind the establishment of local RSSILA branches, and influential ex-officers from the business and professional class were prominent in the State and national leadership. During the war the RSSILA had actively promoted recruiting and conscription, and it now identified the enemies of returned servicemen as the shirkers who had not gone to war but who had taken the jobs that ex-servicemen deserved, and trade unionists who opposed preference to ex-servicemen in employment. Despite its bipartisan policy, from the outset the League was entrenched in loyalist politics.[7]

Other veterans' organisations had more direct party political links. The Returned Soldiers' National Party in Victoria, and similar organisations in New South Wales and Queensland, were affiliated for a time with the National Political Federation, and sought to combine loyalist and digger causes on a party political platform. In 1919 they claimed to have recruited many members and to be a serious challenge to the RSSILA, but their maverick leaders fell out with Nationalist politicians and the organisations collapsed.[8]

Ex-servicemen in the Labor movement also created veterans' organisations. In response to the perceived attempt by 'Tory politicians [...] to organise returned soldiers to smash trade unionism and the Labor movement', members of the Victorian Returned Soldiers' No Conscription League formed the Returned Sailors' and Soldiers' Australian Democratic League, and similar groups were established in other States. These labor organisations wanted a fair settlement for the diggers, and they identified Tory politicians and 'profiteers' as the enemies of returned servicemen. True to the racist and sexist heritage of the Australian Labor movement they also attacked women and coloured workers for taking 'their' jobs. The 'Democratic' appeal of the title intentionally contrasted with the 'Imperialist' sentiments of the RSSILA, and these radical diggers hoped to inject the egalitarianism of the soldiers into postwar society, perhaps as the leaven for an Australian socialism.[9]

There is little available evidence about the membership of ex-service organisations of the radical left and right, and it seems likely that their success was undermined by digger suspicion of the political conscription of their cause. The labor veterans' groups were also criticised for their association with organisations that were perceived to have opposed the war and thus betrayed the men at the front. But the main factor in their demise, and in the triumph of the RSSILA, was the latter organisation's achievement of State patronage.

Throughout the latter years of the war the RSSILA courted the Hughes government, arguing that it was in the government's interests to grant the League an official role as the sole representative of returned men. The RSSILA national president argued in June 1918 that 'unless you do invest them with some such responsibility I think it will be a sure way to create those divergences that will be an ever-lasting trouble and annoyance to the Powers that be'. As historian Marilyn Lake has shown, a bargain was made and the RSSILA was granted official recognition in return for supporting 'the powers that be'. The League soon proved its value in containing digger militancy. For example, after the Brisbane and Melbourne riots RSSILA sub-branch officers formed special forces of League members to maintain law and order. More practically, the sub-branches offered carefully controlled meeting places as an alternative to the more threatening gatherings of returned men in pubs and on street corners.[10]

In return for its 'law and order' stance, the RSSILA was allotted a place on all the Federal government committees concerned with repatriation, including the Repatriation Committee which had overall responsibility for the national administration of the Repatriation Act. Thereafter the League wielded enormous power in the creation and implementation of repatriation policy. Its efforts were based on the premise that returned men not only needed assistance, but also deserved special treatment because of their wartime achievements and national status. In effect, the League's campaigns for special treatment promoted an image of loyal and disciplined soldiers who had fought and died for the national cause, and the League sub-branches sought to ensure that ex-servicemen lived up to that Anzac reputation.

Not all League members accepted these links between the national leadership and the conservative Federal government, or the methods and policies adopted by the leadership in the political trade-off. Throughout 1919 there were fierce conflicts within most State branches between League leaders and some of the digger members. In Victoria a 'democratic ticket', representing rank and file members with labor sympathies, contested executive elections against a 'centre' ticket of current office bearers and a 'Tory' ticket of employers and officers, 'mostly of the haw-haw brigade'. But the outrage about hooligans and militants that followed the soldier riots of mid-1919 strengthened the hand of the loyalists, and left-wing critics within the RSSILA were silenced or expelled. In this regard the Australian politics of ex-servicemen matched the British pattern in which, according to historian David Englander, 'this potentially disruptive community' was reduced to 'a manageable interest group within a pluralist democracy'.[11]

The RSSILA did not always get its own way, and in the inter-war years there would be many disagreements between the League and Federal governments that could not afford to accede to all the ex-servicemen's demands. However the League was generally successful in its campaigns for special treatment, partly because of its official influence and national status, but also because its preferred policies usually coincided with the ideologies of the conservative Federal governments of the 1920s.

An example of this is the soldier settlement scheme, which appealed to ex-servicemen who wanted to achieve the Australian dream of an independent small-holding (and to publicists like Bean who thought of the Anzacs as natural Bushmen). It was also popular with loyalists concerned about 'cities [...] congested with idle men', and for conservative governments was far preferable to alternative plans for State-supported manufacturing industries and co-operatives. Similarly, conservative governments and councils usually supported preference for ex-servicemen in employment, in principle if not always in practice, and believed it to be preferable to subsidies for their employment. The policy caused friction between the League and the Labor movement, which rightly suspected that preference would be used against trade unionists, and gradually came to oppose it. Yet when Labor gained power at the local, State or federal level, attempts to alter the policy were almost invariably quashed by alliances between the League and other loyalist organisations.[12]

With its vigorous and effective promotion of special treatment for ex-servicemen, it is not surprising that many newly returned diggers joined the RSSILA. League recruitment was also bolstered by government assistance. RSSILA officials were allowed to sign up diggers on the home-coming ships, and to wait at the end of demobilisation queues with badges and membership forms. League membership grew by about 1000 per week throughout 1919, and by the end of the year the organisation could boast 167 000 members, a much higher proportion of ex-servicemen than were recruited by comparable organisations in Britain, the United States and Canada. This success was testimony to the League's political acumen, the intersection of League and loyalist ideologies, and the symbolic power of the digger in Australian society. Alternative veterans' organisations were ideologically and politically out-manoeuvred by the RSSILA, and had virtually disappeared by the end of 1919.[13]

Most of the men I interviewed joined the RSSILA. The interviews help to explain why working-class veterans identified with the League, although they also show that individual motivations for joining and remaining in the RSSILA did not always coincide with the political motivations of the leadership. Ex-servicemen, who shared the extraordinary experiences of war, and the difficulties of home-coming, inevitably turned to each other for support and understanding, and usually wanted to join some sort of soldiers' club. A high proportion of my interviewees joined the League within a few months of their return, in some cases on the home-coming ship. Ern Morton recalls that it was 'natural' for soldiers to want to band together in those first few years. A. J. McGillivray joined because he wanted to be able to talk about the war with people who would listen to him and believe his stories.[14]

The soldiers' club was also a place to get a sympathetic hearing about a repatriation grievance, and to campaign for better conditions for ex-servicemen. In the trenches Fred Farrall had been inspired by trade unionist diggers, who recalled the ill-treatment of Boer War veterans and decided to form an organisation to look after their own interests. Signing the RSSILA membership form in the Sydney Domain on his first day home, he assumed that this was the organisation they had talked about in France. It is significant that very few of my interviewees could recall any alternative to the RSSILA. By the time the majority of diggers were demobilised in 1919 the official status of the League was a *fait accompli*; most ex-servicemen assumed that it was *the* diggers' organisation and automatically joined up.[15]

After the initial success of 1919, League membership plummeted to only 50 000 in 1920, and halved again the following year so that less than ten per cent of returned men were still members. Several of my interviewees left the League in the 1920s and their stories help to explain its decline in that period. Some men were busy with their civilian lives and just forgot to renew their subscription. Alf Stabb and Harold Blake thought that in the early days the clubs were a haven for boozers and 'no hopers', and they preferred to spend more time with their families. Charlie Bowden stayed away from the club because he had few positive memories of his military past; he wanted to 'give the war away' and focus upon his career and family.[16]

Other veterans were alienated by the politics of the RSSILA. In the northern Melbourne suburb of Coburg, Ern Morton found that many Labor supporters despised the reactionary politics of League leaders, and that the local sub-branch wanted to suppress his own militancy; on one occasion he was chastised for proposing a motion that ex-servicemen from all countries, including wartime enemies, should band together to prevent another war. With some difficulty, Ern reconciled League and Labor Party membership, but he eventually resigned from the League after the Second World War when it decided to expel Communist members. Up in north Queensland,

Sid Norris went along to one RSSILA meeting with digger mates from the Australian Workers' Union. After heckling a clergyman preaching king and country politics, Sid decided to give up on the RSSILA and concentrate on union work. Ex-servicemen who had become active in the Labor movement were finding that the League did not represent their ideals or interests, and that there was no alternative ex-service organisation to which they could turn. The League had successfully appropriated the definition of 'the digger' so that 'radical digger' had become a contradiction in terms, and many left-wing veterans shed their identity as returned servicemen and gave their first loyalty to the Labor movement.[17]

Although radical ex-servicemen remained alienated from the RSSILA, often until they died, many other veterans rejoined the League in the late 1920s and during the Depression. Its decline was halted as membership grew to 50 000 in 1932 and 82 000 in 1939, and then received another huge boost during and after the Second World War. Most of the men I interviewed who left the League in the early 1920s, rejoined later in the decade or in the 1930s. The role of the local sub-branch as a meeting place in which veterans could re-establish wartime bonds was increasingly signif-icant as soldier mates dispersed and it became difficult to maintain informal contacts. Unit reunions provided valued opportunities to meet men who had served in the same battalion, but they were often only annual occasions. In contrast, the League club was open all week, every week. Alf Stabb rejoined because he wanted 'somewhere to go' and because he wanted to meet up with old soldiers. A. J. McGillivray recalls that many men rejoined when their families were older and it was easier to leave them at home or to bring them along to club events. Within every suburban and rural community the soldiers' club became an important focus of social life for ex-servicemen and their families, providing dances and other entertainments, sports clubs and, in some States, subsidised bar facilities.[18]

The soldiers' clubs also became necessary as a resource for welfare support. As veterans grew older their war wounds sometimes flared up and required medical treatment, and they often needed increased pensions. Perhaps most importantly, many ex-servicemen, including half of my interviewees, were unemployed at some time in the Depression of the 1930s. RSSILA sub-branches helped unemployed ex-servicemen to find work and provided limited financial and material support for impoverished members and their families. They also campaigned for the retention and more effective implementation of the policy of soldier preference in employment. Many ex-servicemen recall that the League's support at this time was invaluable.

They had special welfare needs — although they were not necessarily worse off than other working-class Australians — and the League was an effective champion of those needs.

In hard times, veterans often used the League's argument that they deserved privileged treatment because of their wartime achievements and sacrifices. Some advertised themselves as ex-servicemen when they were looking for work, and others recall that the sight of the League's 'great big badge' often prompted job offers or favours. The badge was the most obvious way in which ex-servicemen could be identified in civilian life, and was an everyday assertion of difference and status. In effect, veterans used the Anzac legend to gain material benefits, and in doing so they reinforced a legend that represented their experiences in terms of individual honour and national pride.

Some ex-servicemen rejected the digger legacy and its rewards. Radical returned men like Stan D'Altera came to agree with Labor movement criticism of government policies, such as employment preference, that favoured ex-servicemen over other Australians. Stan also resigned from the Footscray sub-branch of the RSSILA during the Depression, because it refused to provide welfare support to lapsed members who could not afford the annual membership fee. As an alternative, he organised an ex-servicemen's section of the local Unemployed Association, which respected the personal importance of a digger's identity but did not place unemployed diggers above other people without work. Not many returned men could afford to take such a principled stand, and during the Depression the League, and the privileged Anzac identity it promoted, gained considerable support among ex-servicemen.[19]

League membership did not necessarily require acceptance of the loyalist attitudes of its leadership. The working-class veterans I interviewed who were active in League welfare work or in the social life of the local club, rarely associated the League with those attitudes, or indeed with any political perspective. It was even possible for members to define their involvement in dissenting terms. According to Jack Flannery, the Footscray sub-branch passed a policy of refusing membership to men who had been military policemen, and when an ex-MP was admitted, Jack and several others resigned their membership and formed their own branch in the adjoining suburb of Yarraville. However the hierarchical structure of the League, and its overall control by conservatives, limited dissent. Local clubs were monitored by State officials — in 1934 a sub-branch in the coal-mining town of Lithgow was disbanded after it published anti-war statements —

and radical members resigned or were expelled. Excluded from the League, radical diggers lost access to the main postwar forum of digger culture and were marginalised in the battle for the Anzac legend.[20]

The many diggers who remained in the League enjoyed economic support, a vibrant social life and a valued opportunity to maintain links with other veterans. The RSSILA also became an influential forum for the articulation of digger identities and memories. The League club provided a refuge from civilian incomprehension, where ex-servicemen could talk about the war, recreating the language, jokes and behaviour of digger culture, and constructing shared ways of remembering their experiences. This reminiscence sometimes provoked intense feelings, but it usually focused upon moments of humour, positive experiences of mateship and good times out of the line. Ern Morton recalled that they would only talk about the 'jovial side' of the war or other pleasant memories such as the sunsets on Gallipoli. More negative experiences were ignored or were reworked in humorous terms. Dead mates were recalled with sadness, but also with an emphasis on their character and comradeship. Through collective remembering veterans thus composed more comfortable memories of the war, and positive, affirming ex-service identities.[21]

The particular ways in which ex-servicemen reminisced about the war were partly influenced by the psychological need to cope with difficult and painful memories. However the remembering of RSSILA club men was also shaped by the language and culture of the League which, in comparison with life in the AIF, provided a rather different forum for the articulation of digger culture and identity. During the war, small groups of rank and file soldiers had been able to articulate their experiences, and the troublesome, disrespectful features of digger identity, with a degree of autonomy, albeit limited by army regulations and mediated through official control of the printed media. After the war, digger culture was primarily sustained within the RSSILA, where it was more tightly contained within, and influenced by, the official culture of Anzac. Within the League it was difficult to sustain a digger identity that rubbed against RSSILA respectability, or to articulate war stories that did not match the League's promotion of brave, disciplined and patriotic soldiers. In short, by bringing the politics of returned servicemen within the embrace of loyalist ideology, the RSSILA not only policed digger behaviour; it also gained symbolic control over the definition of 'the digger', and shaped the ways in which ex-servicemen could remember the war and identify themselves as Anzacs.

The League was also powerfully placed to institutionalise its respectable, loyalist definition of the digger, and of the Anzac legend, through national war commemoration. The monuments and ceremonies of Anzac commemoration would, in turn, provide a further, resonant source of representations about what it meant to be an Anzac.

Commemorating the Anzacs

National war commemoration is a powerful way to disseminate ideas about war and nation because it addresses the intense and widespread emotional need to cope with grief and make sense of loss. Monuments and ceremonies serve as focal points for mourning, where individuals can share their suffering and find solace and meaning through collective affirmation of the significance of death. They represent death at war as a sacrifice for the national good, and help to bind the bereaved into the 'imagined community' of the nation. The participatory nature of public commemoration is the key to its effectiveness. The rituals of commemoration — in the consecration of war memorials or annual memorial ceremonies — facilitate intense involvement in collective practices that are intended to be stirring and inclusive, and are thus potent occasions for identification with ideas about war and national belonging. Furthermore, because commemoration is sanctified as an occasion for mourning, those meanings acquire a sacred significance and criticism is defined as disrespectful and even heretical.[22]

The immense number of casualties in the Great War, and the difficulty of attending personal graves and funerals, generated a great need for public commemoration among the people of the combatant nations. Commemoration was especially important in Australia because it was almost impossible for Australian relatives to mourn at such distant gravesides, and because of the national significance that the war and the war dead had already attained in Australia. The nature and extent of Australian commemorative forms reflected the widespread desire for public commemoration. By the mid-1920s Anzac Day was instituted as a national holiday and almost every country town and city municipality had its own war memorial. Committees in each State capital were producing grand plans for 'national' war memorials, and Charles Bean had outlined his spectacular vision for the one truly national war memorial in Canberra. In terms of the number of memorials in proportion to war dead, and in their physical scale and ambition, Australian war memorials rivalled those of Europe.[23]

Australians wanted commemoration, but they did not always agree on the forms it should take. Some memorials and ceremonies were created in relative harmony, reflecting a common need for mourning and serving as a focus for community integration. Just as often, public interest degenerated into debate over the appropriate form of commemoration, with all sides determined to ensure that it represented and conveyed their perception of the significance of the soldiers' deaths. In effect the debates were about how the war, the achievements of the AIF and the character of the Australians soldiers should be remembered, and about the lessons of the war for peacetime society. Symbolic control of 'the digger' and his Anzac legend would, ultimately, be won through the institutionalisation of a particular version of the war in the resonant rituals and monuments of national commemoration. The following brief account explores the origins of Anzac Day, and the enshrinement of the Anzac legend as an integral part of Australia's war commemoration.[24]

Anzac Day originated during the war. In 1916 there was a ground swell of support for some form of anniversary commemoration of the landing at Gallipoli, though from the outset the anniversary was also linked to recruiting and fund-raising for the war, and was actively promoted by the various State War Councils. State and local commemoration committees, often comprising the same loyalist worthies who dominated war effort committees, were responsible for deciding the form of the day and for supervising its conduct. The Anzac Day Commemoration Committee (ADCC) in Queensland delegated the planning of observances to the Anglican Canon David Garland, who developed many of the practices that were adopted around the country. In 1916 morning church services were conducted throughout the State on Anzac Day. At public meetings in the evening, at which returned servicemen and soldiers' relatives shared pride of place, a uniform resolution about the importance of the Day — to commemorate the heroes who had died preserving liberty and civilisation for their country and empire — was followed by one minute's silence.

The ADCCs sought to ensure united action throughout their regions by subjecting local committees to the advice and regulation of the State committee, and by providing pamphlets with 'Hints for Public Meetings on Anzac Day'. Canon Garland also crusaded by mail in other States and New Zealand, and even in Great Britain, for the wider adoption of Anzac Day rituals. Promoters of the day within commemoration committees and the newly-established RSSILA, concurred that it was necessary 'to educate the people to strictly observe Anzac Day'. Their greatest initial successes were in

State and Catholic schools, to which ADCCs supplied Anzac Day literature, badges and guest speakers to ensure that 'the imperishable tradition' of the landing at Gallipoli would be imbibed by the younger generation.[25]

Although Anzac Day continued to be observed in schools after the war, from 1918 up until the mid-1920s war commemoration was racked by conflict about its purpose and nature. In turn, this dissent hampered efforts to institutionalise Anzac Day through legislation. Some of the bereaved opted for more personal forms of mourning, or preferred simple memorial services without the trappings of national pageant; church services on Anzac Day continued to be well attended in this period. Others, including some ex-servicemen, were 'wearied' by the war and did not want to remember or commemorate it. When Western Front veteran Bill Harney got home from Europe, he felt 'somehow ashamed of the war,' and to 'forget about it' he rode 800 miles into the outback.[26]

As public and municipal interest waned in the early 1920s, Anzac Day virtually disappeared in some parts of the country. For example, in the Melbourne working-class suburb of Brunswick, Anzac Day was celebrated in local schools from 1915 as 'the Anniversary of our boys' heroism', and in 1919 it was joined to Empire Day (May 24) by a public carnival month. Between 1920 and 1922 large gatherings of returned servicemen attended memorial services in a local theatre, but by 1923 the only Anzac Day service held in Brunswick outside of the schools was a brief ceremony at the tramway sheds. This story of inconsistent ceremonies and flagging attendance was echoed in the city centres, and for a time it seemed that Anzac Day would simply die away.[27]

However from the mid 1920s Anzac Day underwent a transformation, and by the end of the decade it was institutionalised as a popular patriotic pageant. The transformation was due to a number of factors. The RSSILA, which had always organised services and marches for ex-servicemen on Anzac Day, became more vociferous in its demands for public commemoration of the 'diggers' day', which would promote national feeling and boost the special status of ex-servicemen. In 1922 the League's National Congress resolved that Anzac Day should be known as Australia's National Day and be observed with a statutory public holiday on April 25, 'to combine the memory of the Fallen with rejoicing at the birth of Australia as a nation', and to 'inculcate into the rising generation the highest national ideals'. With the influential backing of Billy Hughes and 'digger' papers like *Smith's Weekly* and the Melbourne *Herald*, the League galvanised popular support for the National Day. The 1923 State Premiers' Conference was persuaded to recommend that each State take the necessary steps to institute April 25 as a National Day, on which religious and memorial services in the morning would be followed by addresses 'instilling in the minds of the children of Australia the significance of Anzac Day'. There followed further disagreement about the nature of the public holiday, but by 1927 the appropriate legislation had been passed in every State.[28]

The convergence of a number of psychological and ideological factors influenced this revival. From the mid 1920s there was widespread interest in 'spirit-soldiers', as newspapers reported attempts to communicate with the war dead. The new State war memorials provided special places for such spiritual communion, and bereaved family members sometimes chose inscriptions for their soldiers' headstones that reflected this attitude:

I AM NOT DEAD BUT SLEEPING HERE

TO LIVE IN HEARTS WE LEAVE BEHIND IS NOT TO DIE

Perhaps the most famous popular representation of this feeling was a cartoon by Will Dyson entitled 'A voice from Anzac', which was published in the *Herald* on Anzac Day 1927, with an accompanying 'Anzac' poem by C. J. Dennis that urged Australians not to forget the war dead. In Dyson's cartoon a spirit soldier on Gallipoli remarks to his dead mate, 'Funny thing Bill — I keep thinking I hear men marching'. Many Australians wanted to maintain the memory of men who had died at war, and Anzac commemoration helped to fulfil that need.[29]

In the late 1920s there was also an upsurge in the publication of Anzac memoirs. This Australian development matched the trend in European military publishing, and suggests that a decade after the war veterans were finding it easier — or more necessary — to write and read about their experiences. The new memoirs were often nostalgic for wartime excitement, purpose and fraternity, and frequently contrasted the war experience with the dullness of civilian life and the divisiveness of the Depression. German historian George Mosse argues that veterans of the Great War created a 'Myth of the War Experience', which revered fallen soldier mates, emphasised positive memories and played down the terrible aspects of service. In this way, Mosse argues, horror was 'transcended and the meaning which the war had given to individual lives was retained' and transported into civilian society. Many Australian ex-servicemen shared these sentiments. On Anzac Day they were delighted by the chance to meet up with lost mates, and they relished their one day of national esteem at a time when everyday life was often hard and humiliating.[30]

Figure 13   'A voice from Anzac', Will Dyson's drawing for a cartoon in the
Melbourne Herald, 25 April 1927. (1927, brush and ink with pencil heightened
with white, 63.2 x 50.2 cm, Australian War Memorial [19662])

The onset of the Great Depression also had a more material influence.
Returned men now had an urgent reason to highlight the special status
and needs of veterans, and as ex-servicemen rejoined the RSSILA in large
numbers they were encouraged to participate in commemorative activities.
Furthermore, at a time of increasing social and political disorder, the Anzacs
and the AIF were promoted by conservative leaders as exemplars of unity
and discipline. Anzac Day was proclaimed a day of reunion and national
reconciliation, and was enjoyed by some participants for that reason. In

these ways public and private interest in commemoration were mutually reinforcing, and the revival of Anzac Day and enactment of State legislation were due to a combination of political skill and popular demand.

For example, in 1929 the Brunswick RSSILA, which had collapsed in 1924 and been reconstituted four years later to campaign for veterans' welfare support, decided to reinstate local Anzac Day commemoration. An inaugural town hall memorial service was held on the Sunday before April 25. Each year the Brunswick Sunday commemoration became more ambitious. In 1930 the local Citizen Force battalion escorted Brunswick returned men and a motorcade of nurses along Sydney Road to the town hall service, where a thousand people were packed in to hear addresses by representatives of the military, the church, the RSSILA and the Council. The following year Brunswick's Sunday celebration was looked upon as the chief northern suburb Anzac gathering: 'with Sydney Road thronged to watch returned soldiers march to the tune of martial music, and the Town Hall filled for a memorial service, Brunswick citizens commemorated Anzac Day'. At the service, Chaplain Captain Hagenaur expressed his ideal for the day:

> [...] that the wonderful spirit which led, and the inspiration which filled the hearts of the men at Gallipoli, could reach and enter the hearts of every person in the nation [...] there are many in our midst at the present time who deride discipline [...] discipline made men of our Anzacs. May God awaken the Anzac spirit, rekindling it into a new spirit which will animate our nation. If this comes to pass, the dead will not have died in vain.[31]

The Brunswick commemoration was held on the Sunday nearest to Anzac Day so that on the day itself the people of Brunswick could join the celebrations in the city centre. In the city, as in the suburbs, these became more ambitious and better attended each year, with a Dawn Service at the city shrine followed by a march through the city by ex-servicemen, a public memorial service and reunions of AIF units. As General Monash remarked in Melbourne in 1926, Anzac Day had 'grown year by year from small beginnings to a mighty solemnisation [...] on this day a whole people pause to mourn their dead and honour their memory'.[32]

Coinciding with the fall and rise in popular support for Anzac Day in its first decade, was considerable dissent about the form and meaning of the day. This dissent highlights the role of pressure groups with rather different investments in commemoration. The choice of Anzac Day as the national day of remembrance was in itself significant. There were other options. Empire Day was a celebration of both national and imperial loyalty. It achieved a peak of popularity in May 1915, and throughout the war it complemented Anzac Day and was used by loyalists to sustain imperial fervour, especially in schools. Yet Empire Day was unequivocally identified with conservatism and was rapidly eclipsed by Anzac Day, which could be claimed by all ex-servicemen and most Australians as the national day.

Another alternative was Armistice or Remembrance Day, which com-memorated the Armistice on 11 November 1918. The British idea of observ-ing two minutes silence at 11 a.m. on that day each year was adopted in Australia, and in many places memorial services were held on the day or on the nearest Sunday, But whereas in Britain Remembrance Sunday became the national day of remembrance, in Australia it was always of secondary importance. Most Australians preferred to commemorate the war on the anniversary of the Anzacs' baptism of fire, and not on the anniversary of the end of the conflict. This choice highlighted the proud national significance of the war, and reinforced the distinction between ex-servicemen and other civilians.[33]

This preference also meant that Australians would commemorate war on 'the diggers' day', a day which reaffirmed the extraordinary power and status of these men. There were a number of challenges to RSSILA control of Anzac Day, and to the primacy of diggers on the day. One area of disagree-ment, which caused tensions within loyalist circles, was religion. While the churches sought to retain Christian observances, on several occasions the RSSILA tried, with varying success from State to State, to omit Christian references in Anzac Day services and replace them with a secular liturgy emphasising nation, empire and digger. Catholic clergy and believers, who had often been forced to accept the rituals of the Protestant establishment or to hold separate services, supported the RSSILA, but Protestant clergy were horrified, and joined with other military and civilian leaders to denounce the disavowal of religion as the trademark of communism.[34]

There was also disagreement between those who believed Anzac Day should be primarily a citizens' tribute to the dead, and ex-servicemen who wanted the diggers' day to honour all the men who went to war, and to serve the needs of the survivors while also paying tribute to dead comrades. This issue was hotly contested within Anzac Day Commemoration Committees and between the committees and the RSSILA. In Queensland, for example, Canon Garland was determined to maintain the citizen's day against RSSILA wishes, and he was supported by organisations of soldiers' mothers and by a Labor government that was hostile to the RSSILA. Opponents of the diggers' day achieved a measure of success. In school ceremonies the emphasis was on wartime sacrifice and patriotism, and there were few mentions of ex-servicemen. Certain aspects of Anzac Day ritual, such as the memorial service, also emphasised the war dead and gave pride of place to the bereaved.[35]

Yet the centre-piece of the day, which received the bulk of media and public attention, was the march by returned men. Organised and marshalled by the RSSILA, the Anzac Day march through the centre of every capital city, suburb and country town empowered ex-servicemen practically and symbolically. As Ken Inglis argues, Anzac ceremonies and the soldiers' clubs were added to the pub, the sports' game and the races as male citadels. Anzac Day increasingly became an occasion of masculine assertiveness. The RSSILA chided returned men who did not join the march: 'On such a day their place is in the march with their comrades — not on the sidewalks with their wives and families'. Soldiering men dominated on the day and remained central in the public memory of the war; women were confined to a passive role as sidewalk mourners.[36]

Another related issue was whether April 25 should be a day of mourning or celebration. Canon Garland believed at first that it was not 'practicable in any way to mourn the loss of the fallen, and at the same time rejoice at the birth of Australia as a nation'. Some Australians, most notably in the Labor movement, maintained this concern. Yet most Anzac Day promoters perceived no contradiction between mourning and celebration, and the duality was embodied in ritual. The RSSILA and Premiers' Conference recommendations of 1922 and 1923 called for memorial services in the morning and patriotic addresses and carnival in the afternoon; later this duality was embodied in the sequence of a Dawn Service followed by the Anzac Day march and unit reunions. As the day progressed, mourning would thus be ritually transformed into celebration. Although mourning remained a significant motivation for the bereaved and for many ex-servicemen, national celebration became the predominant mood and was emphasised in numerous speeches, such as that made by General Monash in 1924:

> Mourning should not dominate the day; the keynote should be a nation's pride in the accomplishments of its sons. The day should be one of rejoicing.[37]

There were also arguments about whether April 25 should be a public, industrial holiday. This was the most contentious issue when each State tried to implement the Premiers' Conference recommendations of 1923. The RSSILA wanted a public holiday so that no ex-serviceman would be prevented from attending Anzac Day services by work commitments, and so that national commemoration could be properly observed by all Australians. On the Anzac Days before legislation was enacted, RSSILA members sometimes 'waited on' businesses to force them to close on the day, and at the State level the League campaigned vigorously for the holiday and against the 'pusillanimous governments, business greed, and disloyal influences, which have so far succeeded in preventing it'. Many prominent loyalists, including most conservative politicians and officials of the Teachers' Federation, also supported a public holiday as the best way to promote the national significance of the day.[38]

Opponents of the public holiday wanted commemoration to take place on the Sunday nearest to Anzac Day. Employers did not want to lose a day's labour, and campaigned against the public holiday until they were persuaded by other conservatives, with the economic recession and class tensions in mind, of the ideological value of the day. Some church leaders and women's organisations also opposed the holiday, fearing the intemperate dangers of carnival and preferring the more reverent qualities of a sacred Sunday. They too were generally won over when RSSILA leaders in most States proposed that hotels and other places of public entertainment should be closed on the holiday. Labor activists often opposed the public holiday, arguing that workers would be disadvantaged if they lost a day's pay or had to work on the holiday. In Melbourne industrial suburbs like Brunswick and Richmond, even RSSILA sub-branches opposed the League's position on the grounds that wages would be affected.[39]

Underlying labor opposition were ideological reservations about a carnival celebration of war. In the Victorian parliamentary debate over the Anzac Day Bill in 1925, the ex-serviceman Labor politician Pollard argued that he did not want to give 'prominent generals, colonels, and other people of a militaristic turn of mind' the opportunity to 'glorify the spirit of war', or to propagandise about 'the necessity of preparing for more war'. Labor politicians concluded that the best way to remember suffering and sorrow was on the anniversary of the peace, or in the privacy of the home. They were castigated for 'caring nothing for the sanctity of the day' and for 'holding the achievements of Gallipoli up to ridicule'. After protracted debates, the opponents of the public holiday were defeated.[40]

There were further debates about whether public houses and venues for gambling, horse-racing and other forms of public entertainment should be closed on the holiday. The campaign for a 'closed day' was conducted by a temperance lobby that included churchmen and members of the Soldiers' Mothers' Associations. In Sydney, for example, the campaign was led by Dr Mary Booth of the Centre for Soldiers' Wives and Mothers (she was also Honorary Secretary of the Soldiers' Club), who argued that the sacredness of the day would be ruined by drinking, gambling and other sporting or carnival events. Her argument reflected concern about the diggers' larrikin behaviour, and was supported by the civilian authorities who policed Anzac Day. RSSILA leaders and conservative politicians in most States agreed that the day should be 'closed', although many of their members, and ex-servicemen who were not members, resented this policy which did not represent 'true Diggers'. Mr J. T. Moroney was moved to write to the *Daily Telegraph*:

> As a mere Digger [...] I should like to register a kick against petticoat control of war anniversaries [...] To hold solemn grief-reviving memorial services, to close hotels and forbid race meetings, seems a queer way for Australia to celebrate epic deeds. [41]

Labor populists weighed in with support for the 'diggers' against the 'wowsers', claiming that the restriction would 'cut out working-class amusements [hotels and races] while allowing other classes to motor, sail, golf, etc'. However the left was not always consistent on this issue; in 1926 the *Worker* regretted that the sacred day was becoming an opportunity for 'filling the Bookies bags'.[42]

The outcome of the debate varied from State to State. For example, Queensland and Victoria legislated for a closed holiday, but in New South Wales, where the temperance movement was not as strong and Labor populism was ascendant, there were no such restrictions. The debate is significant because it shows that the wartime tension between digger behaviour and the Anzac ideal was reproduced in postwar society, and that attempts were made to use commemoration to reassert the ideal, and to control the behaviour of ex-servicemen.

The tension between the Anzac ideal and digger behaviour also affected the form of Anzac Day ceremonies. In response to fears about unruly ex-servicemen, the RSSILA and civil and military authorities organised and marshalled well-disciplined, orderly Anzac Day parades, in which men marched in ranks in their wartime units. Newspapers commended returned men for living up to the Anzac reputation for discipline, and for providing a good example to all Australians. Threats to this order, such as the attempt by an Unemployed Soldiers' Association to join the march in Melbourne in 1926, were discouraged. On that occasion the unemployed men eventually marched with their old units, and the RSSILA president commented that 'loyalty to the spirit of Anzac prevailed'.[43]

Yet there were limits to this control. Both before the parades, as ex-servicemen gathered with wartime mates, and afterwards at unit reunions or informal gatherings in pubs and on street corners, the diggers were notorious for flouting 'respectable' codes of behaviour by drinking, gambling on two-up games, and even scuffling with police who tried to move them on. Such larrikin behaviour concerned the organisers, the press and middle-class respectability in general, and lay behind the attempts to enforce a 'closed' day. But because of the status and number of the diggers — this was their day and they controlled the centre of every city and town in the country — it was difficult to enforce standards of behaviour. In time an uneasy compromise developed, in which organisers and the press accepted the informality of the march and encouraged the police to turn a blind eye to most digger drinking and gambling. After all, it was only for one day of the year, and the worst excesses could usually be controlled at the march and in RSSILA clubs and at unit reunions.[44]

This concern over the behaviour of returned men demonstrates that the forms and meanings of dominant rituals are never fully imposed, but often exist in an atmosphere of continuous tension, involving behaviour that stretches the boundaries of acceptability. It also shows how the rituals of Anzac Day embodied a range of alternative and even contradictory practices and meanings, which enabled people to value the day and to participate in it for a wide range of reasons. Yet at the same time this openness was contained and channelled within certain established practices that were rigorously defended and were intended to promote particular, predominant meanings about Anzac and Australian society.

This same tension between openness and containment was apparent in efforts to make Anzac Day an egalitarian affair. Officers and other ranks were encouraged by ADCCs to march together and to not wear uniforms or other insignia of rank; when Generals Monash and Chauvel led the Melbourne parade in 1925, Monash remarked that he preferred to march in plain clothes so as not to be prominent among the 'Diggers'. In fact, the egalitarianism of the day was limited and, to some extent, illusory. Protestant services sometimes precluded Catholic participation; senior officers were frequently invited to march at the front of the parade, or at the front of each unit; and the official platform was dominated by vice-regal representatives and civic and service leaders. Yet labor criticism of 'brass hat' control of Anzac Day failed to recognise the appeal of its egalitarian symbolism. For one day of the year many diggers recalled the interdependence of officers and men in the AIF, which was readily contrasted with the tensions of the workplace and of inter-war society. Anzac Day reasserted the equality, dignity and value of the ordinary man, both in the wartime AIF and in postwar Australia, and became a significant, affirming occasion for many working-class ex-servicemen.[45]

The national and military symbolism of Anzac Day also generalised the significance of Anzac in ways which excluded or marginalised alternative understandings of Australian participation in the war. Certain aspects of Anzac Day ritual — in particular the march in civilian clothes and the absence of weapons — were intended to emphasise that the diggers were citizen soldiers, and to play down celebration of the military and fighting. When AIF General Brudenell White was accused of encouraging military ardour in his Anzac Day speeches, he responded that he had seen too much of war to regard it with 'anything but horror'. Yet in many ways the parade was a martial affair, with ex-servicemen marching in their units to the music of military bands, accompanied by regular servicemen in ceremonial dress armed with rifles and sabres. Anzac Day orators usually asserted that Australian participation in the war had been justified, that Australian soldiers had acquitted themselves superbly, and that commemoration of their noble sacrifices was necessary to ensure that the 'rising generation' would be morally prepared to fight in a future war.[46]

In the early postwar years, when Anzac Day and the national war memorials were still to be made, radical Australians had hoped that commemoration might provide rather different lessons. Radical nationalists argued that the digger should be commemorated because his achievements had won international recognition and laid the foundation for independent Australian nationhood. For example, Mary Gilmore, socialist, feminist and poet of the horrors of modern war, campaigned for a memorial 'fit for Goulburn and Australia', which would defy the misapprehension that Australia was 'a back number that can never build like Europe or America'. In 1925 Labor leader Matthew Charlton supported the Bill for a National War Memorial on the grounds that Anzac commemoration might be used to 'train the young minds of the future in the paths of peace'.

However it was not easy to commemorate the war in terms that combined pride in the Australian soldiers and their national achievement with anti-imperialist or even pacifist lessons. During the war the language of

Anzac had become infused with loyalist ideology, and radical Australians had been tarred with the brush of disloyalty. After the war radicals were poorly placed to wrest the symbols of Anzac from loyalist control. Their exclusion from wartime patriotic committees, and from the RSSILA, was usually perpetuated in their exclusion from commemoration committees. Not untypical was the National War Memorial Committee of Victoria in 1921, which comprised five city financiers, two manufacturers and six professional men, two of whom, Monash and Chauvel, had been AIF commanders. Radical Australians did not have the institutional power or ideological coherence to shape commemoration in their own terms.[47]

Towards the end of the 1920s, socialist and pacifist concern about the 'glorification of war' solidified into outright opposition to the official Anzac Day ceremony. For example, in 1927 the Sydney *Labor Daily* agreed that the vital lesson should be 'Never Again!', but complained that:

> In the flamboyant jingo Press the occasion will be recalled in the main as a propaganda stunt associated with the 'glories of Empire'. The fact that the Empire is unrepentant and would have the dose repeated in China or anywhere else tomorrow will be conveniently overlooked.[48]

The prevailing view of the left, derived from international pacifist and socialist literature, was that the war had been an imperialist struggle between European ruling elites for national and economic gain, and that the soldiers of both sides were the unwitting victims of a trade war. Although left-wing critics of Anzac Day sometimes blamed Australian losses on the British and asserted an anti-imperialist Anzac nationalism, they more often criticised any patriotic rhetoric as cant and rejected Anzac Day altogether.[49]

Several of my interviewees — usually men who had become active in the labor and anti-war movements — identified with these criticisms of Anzac Day. Bill Williams felt that the day was used by the organisers to glorify war. Although he marched in tribute to his dead friends, he demonstrated his principled opposition to militarism by not wearing his medals. Ern Morton went to the first few marches to meet up with his mates and remember the good times, but he stopped attending when he realised that the political use of the day was contrary to his own views about the war.[50]

While radicals were alienated by the politics of Anzac Day, some other diggers felt excluded by its particular representation of war and military manhood. Bill Bridgeman gave up on Anzac Day because it was mainly for and about soldiers, who were the 'real' Anzac heroes, and because he felt that the contribution of navy men like himself was not adequately recognised.

Charles Bowden decided that he had no real part in the day because he had been in a train-driving unit rather than the infantry, and was not a real Anzac. Anzac Day was thus an uncomfortable or alienating event for some diggers because it did not recognise or affirm their identities — or their politics — as soldiers and as ex-servicemen.[51]

However for many returned servicemen Anzac Day was a profoundly important occasion. The majority of my interviewees first marched on Anzac Day in the 1920s, and then marched every year until they became too frail or were the only survivors in their units. They stressed that reunion and reminiscence with wartime mates, and remembrance of mates who had died, were the main reasons for their participation. Though some veterans like Bill Williams and Charles Bowden found it difficult to meet and talk about the war, most ex-servicemen actively sought opportunities for reminiscence at the Anzac Day ceremonies and at unit reunions. Stan D'Altera and a group of Yarraville digger mates went to the Dawn Service in the city every year to show their respects to dead comrades, and then returned to Yarraville to share the traditional tot of rum and to reminisce about old friends and old times.[52]

Throughout the year, ex-service RSSILA members remembered and re-constituted their war memories and identities at the League's clubs. Anzac Day focused that process in potent ritual form, and provided generalisations about the war, and about Anzac identity, which ex-servicemen participants used to articulate their own war experience. The nature and sequence of Anzac Day rituals contributed to this memory composure, by evoking a particular sense of the war experience. Digger participants re-enlisted with their unit for the day, shared a drink before the Dawn Service just as they had shared a tot of rum before going over the top, marched together again, re-enacted the funeral rites for dead cobbers, drank at collective wakes for the dead, and then left their soldier mates and returned to their civilian homes and lives. This ritual recreation of the war experience, repeated year after year, was a potent form of collective, participatory remembrance. It facilitated a reconciliation with the past through which positive experiences were emphasised and affirmed, while negative experiences were played down. By reliving their military service in these ways for one day of the year, veterans remembered the camaraderie and excitement of the war, and confirmed its significance as a highlight of their lives.[53]

Participation in Anzac Day did not necessarily require acceptance of the military and patriotic meanings of the day. For most of my interviewees, Anzac Day did not 'glorify war'. Few of them had enjoyed the military aspect

of their experiences or wanted trench war to be romanticised and celebrated. Yet at the same time the dominant messages of Anzac Day — that
the Australians were fine soldiers and men, and that the fighting and dying
had been worth it — addressed their own emotional need for justification,
and provided a public language and sense to articulate the war experience,
and their Anzac identity, in more positive and affirming ways. For men
like Ern Morton or Charles Bowden the gap between that public legend
and their own individual experience was too great. For them, Anzac Day
was a painful or alienating occasion. For many others, Anzac Day helped
to close the gap by providing understandings to help them live with their
wartime past. The battle for the Anzac legend was won by loyalists in the
inter-war years because they achieved control of public commemoration, but
also because the version of the war that they enshrined in commemoration
fulfilled the subjective needs of the majority of Australian ex-servicemen.

Bean's Anzac history

The Anzac legend and the postwar identities of ex-servicemen were also
influenced by the histories and literature of the war. Valuable surveys of
Australian war literature have been produced by Robin Gerster and others.
The focus here, however, will remain on the work of Charles Bean; because
Bean's official history, while informed by his own Anzac ideal and historical
methods, and by the social and political context in which it was produced, was
enormously influential in shaping the ways in which Australians understood
their participation in the war and ex-servicemen identified themselves as
Anzacs.

When Bean was appointed official war correspondent it was also anticipated that he would write Australia's official war history, and from 1915 to
1918 he gathered extensive documentary and oral evidence for this purpose.
When the war finished, Bean completed his historical investigations by
returning to Gallipoli with a group of Australians to solve the military
riddles of 1915, to provide paintings and relics for the proposed national
war museum, and to organise the Anzac cemeteries. In April of 1919 he
embarked from Cairo on a ship bound for Australia.

On the voyage, Bean prepared recommendations for the Australian
government on both the official history and the war museum. Bean's plan for
The Official History of Australia in the War of 1914–1918, which was accepted
with only minor alterations, reflected his ambition for the project. He was
to write six volumes about the Australian infantry at Gallipoli and on the

Western Front, and would be general editor of a photographic history and of a further five volumes on the Light Horse in Palestine, the Australian Flying Corps, the Royal Australian Navy, the New Guinea expeditions, and the home front. The omission of the latter volume from Bean's initial proposal reveals that his main concern was to record and commemorate the achievements of Australian servicemen.[54]

Bean commenced his new work at Victoria Barracks in Melbourne, but from October of 1919 until 1925 the historical team was based at an old homestead at Tuggeranong outside Canberra. Bean preferred the quieter surroundings and was proud to be making the national war history near to the site of the new federal capital. While he began writing, a small and dedicated staff assisted with the immense task of sorting and classifying sources, proofreading manuscripts, and producing indexes, maps and biographical footnotes. Bean worked as hard as he had during the war. He supervised the production of histories of individual AIF units, and corresponded in great detail with the British official historians about the drafts of the respective histories. He wrote a series of journal articles about the making of the history, and a concise, one volume account of the Australians at war, *Anzac to Amiens*. The official history took more than twenty years to complete, and the final volume about the Australian victories of 1918 was not published until 1942. Bean's last publication, in 1957, was a fond remembrance of Generals Bridges and Brudenell White of the AIF, entitled *Two Men I Knew*. By its scale alone, Bean's historical project was and still is one of the most impressive achievements of an Australian historian.[55]

It was also an extremely innovative official military history. Bean was critical of conventional military history, which was mainly concerned with military strategy and which used jargon and generalisation about the experience of battle:

> [...] when reading in military works, that, for example, the commander, 'by thrusting forward his right, forced the enemy to withdraw his left and centre', I had often longed to know just what this meant.[56]

Bean believed that the fate of a battle often rested with the men in the line, and that military history was inadequate if it did not show the interplay between battle plans and the actual experiences and motivations of soldiers. Because of the small size of the AIF, and because Bean's terms of reference were to write about Australian participation in the war rather than the general trend of campaigns, it was relatively easy for him to adapt his journalistic focus on the 'cutting edge' to the writing of 'a new kind of war history':

This could tell how plans, made on the flagged maps in the General Staff office, or perhaps even around polished tables at Downing Street, worked out in the actual experience where Billjim and his beloved Lewis gun lay in the mud of a French crater blazing at German helmets bobbing along a broken-down trench.[57]

Bean also wanted Australians to be able to identify with the men of the AIF. He named about 8000 soldiers in the text and described them in biographical footnotes which showed that the AIF was 'a fair cross section of our people', and which tied 'this national history into the everyday life of our people'. Though recent critics have shown that Bean sometimes neglected other important military factors, such as weaponry, support, training, logistics and leadership, Bean's interest in the experience of frontline soldiers was ahead of most of his contemporaries.[58]

The range and use of sources for the history was also innovative. Apart from Bean's own 300 volumes of wartime diaries and interview notes, which 'provided most of the colour, though by no means all', the main sources were the unit diaries and official records of the AIF (21 500 000 foolscap sheets arranged according to unit and date); soldiers' diaries and letters which were acquired by the War Memorial following public appeals to ex-servicemen and their families; official photographs which were used in the creation of about 2250 maps and sketches of military engagements; and published monographs and histories — including German unit histories — which located the Australians in a wider context.[59]

Bean was particularly concerned about the accuracy of his evidence, and personally read and checked almost all of the source material, in comparison with the British official historians who worked from précis prepared by clerical staff. When he encountered inconsistencies there was 'nothing to do but read more deeply and widely until in most cases one reached the conviction that, of the facts laid bare, there was only one reasonable explanation'. He was also wary of the subjectivity of sources, and was careful to record the origin and authority of each item of evidence, and to assess its veracity: 'Was it first hand evidence? Was it from a man likely to be on his defence? Was it from Captain A or Captain B, to whose trustworthiness one attached different values?'[60] On the whole Bean found that because soldiers were trained to make accurate observations upon which their lives often depended, interview material did accord with other evidence (apart from the second hand reports of commanders, and the confused remarks of wounded men). Indeed, as he remarked to the British military historian, Liddell Hart, the accounts of eyewitnesses were frequently more reliable than official despatches. Bean was thus an early advocate of Australian oral history.[61]

Bean's most impressive achievement was to ensure that the history was not simply the official viewpoint of military authorities or his government employer. His experience of wartime propaganda and censorship, and his determination to write a history that would be trusted and taken to heart by ex-servicemen and other Australians, made him insist that he should be allowed to write an uncensored history. In 1919, a Military Board was asked by the government to approve Bean's proposal for the history. It concluded that unless all page proofs were submitted to a committee for the deletion of libellous references, the books could not be termed an 'Official History' and should instead be called the 'Story of the AIF by the Official War Correspondent'. Bean responded angrily:

> The fact that the public knew that there was any Government authority or body acting on behalf of the Government which was ruling any statement that it considered to be 'libellous' or 'dangerous' out of the book, would entirely destroy the public confidence in it, and rob the history of its value in one blow.

Prime Minister Hughes accepted Bean's reasoning — he was also influenced by a powerful ex-servicemen's lobby that opposed censorship — and the military history was written without government interference [62]

In this regard the official history differed enormously from Bean's wartime correspondence, and represented his own historical judgements rather than the dictates of the War Office. It also bears favourable comparison with the work of the British official historians, who had few qualms about excluding material that reflected badly upon British commanders or soldiers, and who were subject to interventions 'for political expediency' by the Foreign Office and other government departments. The British Cabinet's Sub-Committee for the Control of Official Histories was aghast when Bean's histories included 'uninstructed criticism' of senior British officers, and on several occasions sought to bring 'the histories into line'. In the main, Bean resisted these efforts, and his published history was more critical than other official accounts about British strategy and command on the Western Front.[63]

Bean was also free to write about the terrible conditions and effects of trench warfare. In many frank passages his history revealed the less 'heroic' aspects of Australian behaviour both in and out of the line. For example, Bean wrote about the 'unmanning' effects of life under fire, and quoted an

AIF sergeant's description of the Australian First Division after it had been relieved from the bombardment at Pozières:

> They looked like men who had been in Hell. Almost without exception each man looked drawn and haggard, and so dazed that they appeared to be walking in a dream, and their eyes looked glassy and starey. Quite a few were silly, and these were the only noisy ones in the crowd.

Similarly, at various points in the account of the Gallipoli landing, Bean described weary, half-dazed and confused men running back from the enemy or straggling in the gullies. He acknowledged that there was a proportion of Australian soldiers who avoided action by feigning illness or seeking safe work behind the lines; that Australian soldiers sometimes wounded themselves to get out of the line; that 'at times of strain, or before a great battle [...] a certain section persistently "went absent"'; and that there were at least two AIF mutinies in 1918. When Brigadier Edmonds of the British Historical Section wrote to Bean in 1928 arguing that he should delete references to shell-shock and malingering from the Australian history, Bean simply scrawled in the margin of the letter that 'Edmonds was never in a real bomb [sic]'.[64]

However Bean's uncensored official history was not quite as frank and critical as some subsequent historians and readers have assumed. Bean's published criticism of wartime commanders was less harsh and more qualified than the comments in his wartime diary. In a revealing 1929 correspondence with his friend the AIF general and parliamentarian John Gellibrand, Bean joked about his role as 'the public and official executioner of hard-won reputations', and explained that he was 'pretty cautious about attributing blame' because of the likelihood of extenuating circumstances. Bean was especially generous to Australian commanders, and on occasions even asked the British historians 'to lighten the effect of the criticisms' of senior Australian officers, including Gellibrand. It seems that he was occasionally swayed by loyalty to friends among AIF staff and regimental officers, and by the fact that these colleagues formed a high proportion of his historical informants, both during and after the war. Bean was equally protective of the reputation of Australian soldiers, and in attempts to persuade the British historians to make less of Anzac failures (such as the abortive night-time assault by Monash's Fourth Brigade during the Gallipoli August offensive) he was not averse to concealing damaging information from them.[65]

Yet for the most part Bean had too much integrity to consciously censor or self-censor the historical record. Far more important were his ideal of the character of the Anzacs and his perception of the national significance of their achievement. He had begun to articulate his Anzac ideal during the war, when his preconceptions about soldiering and Australian character were remoulded by experiences among the Anzacs. The postwar crystallisation of that ideal in the history was, in turn, influenced by Bean's methods of historical reconstruction, and by political and historical debates in which Bean was an active participant. Though Bean worked extremely hard to create a comprehensive and accurate history, it was, like all histories, shaped by the circumstances and attitudes of its creator.

For Bean the purpose of the official history was to explore 'a great theme — the reaction of a young, free, democratic people to this great test — slowly working itself out to the climax of the astonishing victories of 1918'. The historical exposition of Australian achievements in the war would also serve as a memorial to the men of the AIF. Bean did not perceive any tension between the historical and commemorative roles of the history — he thought that 'the only memorial which could be worthy of them was the bare and uncoloured story of their part in the war' — but his nationalistic ideal undermined the intention to provide 'the bare and uncoloured story'. Indeed, Bean wanted the Anzacs and the AIF to serve as models for national development, and the themes of his history — the importance of the bush, the digger qualities of decisiveness and discipline, the unity and harmony of the AIF — were intended to promote a particular vision of 'the only country in the world that is still to make'. In turn, postwar developments in Australian social and political life influenced Bean's history-making and caused him to emphasise certain themes.[66]

Take, for example, the ways in which Bean's history constructed a particular version of the Australian experience of battle, and of Australian military manhood. In the final volume of the official history, Bean explained that there was 'overwhelming evidence that the AIF [...] was found to be amongst the most effective military forces in the war'. He understated structural explanations of that effectiveness and argued that the AIF's success was primarily due to the national characteristics of the Australian soldiers. From his pre-war journalistic travels Bean was convinced that the distinctive Anzac qualities of independence, initiative and mateship had been forged in the bush. In postwar Australia Bean was active in the parks and playgrounds movement, which he hoped would maintain 'the digger spirit (or the Australian spirit) in the big cities like Sydney'; in a letter to Gellibrand he described the movement as a 'Society for the Preservation of Anzac Standards'. Bean's history, in which he emphasised the character-forming role of outdoors or rural Australian life, was influenced by his general determination to explain and sustain the formative influence of the bush upon Australian character and society.[67]

In the history, the typical, bush-bred Anzac was also an imaginative and bold fighter. He was the type of man who had so inspired Bean during the war, but this Anzac ideal was very different to the image of soldiers as emasculated victims which was prevalent in European war novels, such as Robert Graves' *Goodbye To All That* and Erich Remarque's *All Quiet on the Western Front*, which began to circulate in Australia in the late 1920s. The RSSILA campaigned against this 'unbalanced' and 'degenerate' fiction, which, it claimed, belittled the dignity and achievements of soldiers on the Western Front. In contrast, the League praised mainstream Australian war writing for depicting military manhood in terms of daring and stoic courage, and claimed that it had 'revitalised [...] the spirit of the AIF'.

Bean was also determined that his history should inspire the younger generation with 'the real nobility in the ordinary unpretentious Australian' soldier. For example, he wrote of the Australian soldier at Pozières that, 'having resolved that any shell-fire must be faced, he went through it characteristically, erect, with careless easy gait [...] in many cases too proud to bend or even turn his head'. The dazed and even 'silly' men who came out of Pozières were, according to Bean, 'utterly different from the Australian soldiers of tradition'. The implication of these passages is that Australian soldiers are characteristically indifferent to fire, and that men who do not cope in this way are not characteristic Australians.[68]

In comparison with some members of the RSSILA, Bean did not wish to portray the Anzacs as 'supermen'. In the history he refined the subtle defin- ition of heroism — that ordinary men who endured life in the line and did their jobs were real heroes — that he had begun to articulate in his war corres- pondence. Yet he believed that the AIF 'contained more than its share of men who were masters of their own minds and decisions', and that the majority of them were motivated by distinctively Australian ideals of duty or of manhood. For example, he used mateship to explain military endurance and discipline:

> [...] to be the sort of man who would give way when his mates were trusting in his firmness [...] that was the prospect which these men could not face. Life was very dear, but life was not worth living unless they could be true to their idea of Australian manhood.

Mateship was the primary relationship and motivating force for most soldiers in all armies during the war. Bean, however, believed it to be uniquely

Australian, and through the history and activities such as the 'Society for the Preservation of Anzac Standards', he hoped to rekindle 'the spirit of stand by your cobber and give it a go' in the divided society of postwar Australia.[69]

Bean also argued in the history that the Australian soldiers were willing to stick at the job because they believed in the justice and necessity of their cause. In the volume about 1918 he wrote that, although the diggers were homesick and tired of the war, they did not want to go home until they had finished the job of showing the Germans that warlike methods did not pay:

> [...] whenever talk of peace crept into the newspapers there was only one opinion in most of the trenches: 'No use going home with the job unfinished, to be done again in ten or twenty years' time by our children'.

The conclusions in this passage about Australian soldiers' dutiful sentiments were drawn from wartime conversations with members of one particular AIF unit, but they were also influenced by Bean's own attitudes, and by postwar political developments. Bean was concerned about the resurgence of pacifism in the 1930s, and by the onset of another war against Germany. He was especially critical of 'the sheltered innocence' of school teachers who believed that 'by sustaining a pride in the military efforts of our countrymen, the history of war encouraged war'. He condemned the 'careless verdict of 'impatient radicals' that both sides were to blame for the origins of the 1914–18 war, and in the history he contrasted the brotherly ideals of the British Empire with the Prussian 'might is right' ideology, and stressed that the cause of humanity had required Britain and Australia to oppose Germany in 1914.[70]

According to the history, national loyalty was another explanation for the Australian soldiers' fortitude: 'men's keenness [...] was for the AIF — for their regiment, battalion, company — and for the credit of Australia'. Bean also argued that, despite some friction between the Australians and the British High Command, the diggers' imperial loyalty, which he described as 'loyal partnership in an enterprise and [...] complete trust', was sustained throughout the war. There was some evidence to support at least the first of these conclusions about national loyalty, but Bean's emphasis upon, and characterisation of, the Anzacs' national and imperial identities was also informed by his own ideological position within Australian postwar politics.[71]

An example of this was Bean's concern that riotous diggers should not be seen as representative of the Anzacs after the 1919 Peace Day riots in

Melbourne. He condemned the 'unworthy demonstration' by 'the inevitable riff-raff', and praised the 'tried leaders' of the AIF who organised returned men to suppress the rioting, and who disclaimed 'genuine digger' involvement. The same emphasis upon Anzac discipline, and the concern to rebut claims about digger indiscipline, is evident in the history. For example, both during and after the war British commentators frequently asserted that, in their initial military engagements, the Australians were often handicapped by their irregular behaviour and indiscipline. Bean refuted such claims at every opportunity, and in the 1934 edition of the first volume of 'The Story of Anzac', he added a preface to that effect.[72]

Along similar lines, Bean was incensed by images of the larrikin digger which were prevalent after the war in soldier memoirs and in the popular press. In 1946 he complained:

> [...] great damage was afterwards done to the Anzac tradition by caricatures, that became popular in Australia, of the indiscipline of her troops in the First World War, portraying the life of the 'dinkum Aussie' as one of drunkenness, thieving and hooliganism [...]

He argued that this 'false legend [...] travestied the First AIF and damaged the Second', and stressed in the final volume of the history that discipline was responsible for the success of the Anzacs, and that the larrikin reputation applied only to a few 'reckless or criminal individuals'. In effect, the history's emphasis on the wartime loyalty and discipline of the diggers, and on the inspirational leadership of the AIF, promoted those same qualities in civilian society, and, in the political context of Australia at that time, reinforced loyalist ideology.[73]

We can see how Bean shaped his history in particular ways and with particular meanings, by considering the process of his history-writing and the strategies used in the creation of his text. In preparation for writing about each major battle, Bean brought together a huge variety of relevant source material into an 'Extract Book'. In each Extract Book, evidence was arranged by theme and by the geography of the battle, usually from the extreme left flank to the extreme right. Guided by the appropriate Extract Book, Bean then produced a handwritten manuscript, which was transcribed by his staff and circulated to senior members of the AIF, and to the British Historical Section, for comments. Although Bean gave due consideration to the comments he received, manuscript drafts that are housed with Bean's other papers at the Australian War Memorial are, in most cases, remarkably similar to the final, published history.[74]

In postwar correspondence with his friend Gellibrand, Bean wrote frankly about the process of researching and writing the history:

> [...] working through the records is like going on a particularly interesting voyage of discovery, with a sort of excited anticipating of what you will find when you get through [...] as I work certain big conclusions, simple ones, seem gradually to stand out of all the matter one assimilates [...] I am finding it easier to see light through a maze of experiences.[75]

Bean concluded that he was less a historian and more a storyteller, narrating the discoveries that emerged from the evidence. Yet in Bean's 'storytelling' the discoveries did not simply emerge from the evidentiary maze. Rather, Bean used a variety of narrative strategies and linguistic techniques to make sense of the evidence in terms of his own preconceptions and historical judgements, and to fashion a compelling literary and historical text.

Take, for example, a key passage in Bean's history of the Anzac landing. Bean wanted to show that the AIF brigadiers who recommended evacuation on the afternoon of the first day had based their recommendation on the evidence of the wounded and confused men on the beach, but were mistaken about the morale of the men at the front:

> But in this first experience of battle few senior officers, even among those immediately in touch with the firing line, had yet realised the character of those whom they commanded. While there were some of weaker fibre who tended to fall back into the gullies, and while here and there even the bravest had been placed under a strain beyond their bearing, there was nearly always present some strong independent will, among either the officers or the men, which would question any order for retirement.

Typically, Bean makes this point by embellishing the history with an account of an incident he had witnessed and recorded in his diary:

> Towards evening some New Zealand and Australian infantry were lying out on Plugge's, when several salvos of shrapnel fell about them. Further to the right the Otago Battalion lost thirty men in a few minutes. A message came shouted from the rear: 'Pass the word to retire!' Lieutenant Evans of the 3rd Battalion, sitting in the open on the edge of the plateau amongst the bullets, caught it up.
>
> 'What's that message?' he asked sharply.

'Word to retire, sir,' said a man lying beside him.

'Who said retire?' Evans asked. 'Pass back and ask who said retire.'

'Yes — who said retire?' called several of the men around him. 'Pass back and ask who said to retire?'

The inquiry could be heard proceeding from mouth to mouth, and the next minute there came back a very different command: 'Advance, and dig in on the forward slope of the hill' The men picked up themselves and their rifles and went forward. Shortly before this a call had come that someone was hit. Two stretcher-bearers of the 3rd Battalion immediately strolled casually across the hilltop, hands in their pocket, pipes in their mouths, past the crouching infantry, exactly as a man would roam round his garden on a Sunday morning. Ten minutes later they wandered back in the same manner. Their attitude was — as no doubt it was intended to be — a sedative to all around them.[76]

In comparison with this passage, the notes that Bean jotted down on the day simply record the change in the order from 'retire' to 'advance'. The diary he worked up for publication in the months after the landing contains an account of the incident which is almost identical to that in the history, except that in the history Bean added more colour (the metaphor about Sunday gardeners), improved the details (only one of the stretcher bearers was smoking a pipe in the diary version) and concluded with the telling generalisation about the intention and effects of the stretcher bearers' actions. The passage in the history demonstrates the way in which Bean used anecdotes about 'typical' diggers to corroborate historical generalisations about the motivations and behaviour of soldiers, and to reinforce his guiding theme of the strong positive characteristics of the Anzacs.[77]

Bean used a variety of other linguistic techniques to marginalise aberrant behaviour, and to deny it any general or typical significance. For example, although Bean had noted in his Gallipoli diary that self-inflicted wounds were 'not uncommon, even among Australians', in the history he mentioned them only in a footnote which stated that there were 'a very few cases' in the AIF. He argued that the few occurrences were inevitable in war, that they occurred mainly among new arrivals who were unable to face the strain, and that a man who shot himself instead of refusing to go up the line was of 'finer fibre' than other malingerers. Similarly, although Bean's history is relatively frank about Australians running away during battles, the incidents are defined as not too serious, not typical of the Australian soldier or the force generally or, at worst, the inevitable consequence of strain. The Australians often 'withdraw' while the Turks 'bolt', and when Australians do 'retire' or 'fall back' it is because of 'a strain beyond their bearing' or the effects of 'murderous fire'. Stragglers usually have the excuse that they are wounded, exhausted or dazed, or else they are men of 'weaker fibre'. Likewise, although Bean accepted that some Australians deserted from the Western Front before major battles, he emphasised that this was 'the very time when the average Australian refused to go sick or, not infrequently, broke away from convalescence to get back to his mates in the line'. In contrast, the deserters who were responsible for giving the Australians a reputation as 'bad boys' were, according to Bean, a small proportion of '"hard cases" and ne'er-do-wells [...] in some cases actual criminals who had enlisted without any intention of serving at the front, and ready to go to any lengths to avoid it':

> A few men — of a character recognised by their comrades as well as by their officers to be worthless to any community — by open refusal to go into the trenches were causing some of the younger as well as some of the more war-worn of their comrades to follow their example.

Bean's deserters are not 'average' or 'typical' Australians. By isolating and stigmatising such malingerers as an alien minority, Bean inoculates the reputation of the AIF and preserves his Anzac ideal. To validate that inoculation and disguise his own role in the textual process, Bean cites the scorn of loyal comrades. Stragglers and deserters themselves have no say in Bean's history.[78]

Bean's war correspondence could be dismissed as propaganda because it excluded incidents that reflected badly on the Australians. In comparison, as a historian Bean did not deny or ignore evidence that contradicted his Anzac ideal, but admitted and then reworked it so that it was no longer contradictory, but instead reinforced his glowing characterisation of the typical Anzac.

Bean's Anzac legend was also effective because he ensured that his history was widely read. In comparison, the British official history was not written or marketed 'to be a popular success'. Members of the British Historical Section hoped that their publications would attract 'sufficient interest' from the 'well informed' section of society to 'influence public opinion when questions of imperial policy are being debated', and thereby contribute 'to the national education', but their work was primarily intended 'for the education and instruction of officers'. Not surprisingly; there was little popular interest in the heavily censored and turgid volumes of the British official history, and sales were comparatively low.[79]

Bean was criticised by Australian army staff for not writing a history on the British model, but he made no apology for writing 'a national history and not a military one', which would provide the people of Australia with a commemorative record of the achievements of their men at war. Bean and his colleagues at the Australian War Memorial devised a variety of imaginative schemes to 'spread wide the true knowledge of the AIF'. They used 'national grounds' to persuade the government to subsidise a retail price of one pound and one shilling per volume, and commercial logic to commit the publishers at Angus and Robertson to circulate ex-servicemen or next-of-kin with details of a subscription discount scheme. The set that I inherited from my grandfather, John Rogers, was purchased by subscription and still has the original dockets for pre-payment tucked into each book (the photo volume was a gift from an ex-serviceman friend, who enclosed a note saying that 'from the point of view of history, as it affects you and your children [...] I forward this volume with pleasure').[80]

In the early 1920s the first few volumes of the Australian official history sold well, but sales decreased towards the end of the decade and the Commonwealth expressed concerned about the burden of the history upon government resources hit by the Depression. With Bean's support, the Australian War Memorial took on greater responsibility for financing the history, and devised a series of 'special selling campaigns'. Mail-order and lay-by schemes were set up to 'bring the history within the reach of the ex-soldier wage earner'. Sales representatives were employed on commission to sell the books at workplaces and servicemen's clubs, and RSSILA sub-branches agreed to sell sets of the history in return for a percentage of the takings. Most effective of all was a scheme for Commonwealth public servants, who were able to have the purchase price docked from their fortnightly pay. By 1934, Bean and the director of the War Memorial, John Treloar, were celebrating the 'astonishing success' of these schemes and a 'boom in sales of the Official History'. By 1942 most volumes had been reprinted many times over and total sales exceeded 150 000 copies. Vigorous marketing, and a renewed interest in the war which was also reflected in the sales of war fiction and in Anzac Day attendances, ensured that Bean's history was indeed a popular, national success.[81]

The official history was also well received by the critics. Although reviewers in Britain and New Zealand noted that Bean overstated Australian achievements and understated those of other Allied forces, in Australia the volumes were celebrated as our *Iliad* and *Odyssey*, and reviewers highlighted the key passages about the 'mettle' of Australian men and the national significance of their achievements. There was some concern that the volumes about the Western Front were too soft on both British and Australian commanders, but reviewers generally praised Bean's vivid and realistic depiction of the 'barbarous business of war', and compared it favourably with the censored and rather dull reports that Bean had produced during the war. Some reviewers were concerned that the scale and detail of the history would deter readers (Bean responded that the detail *was* the main story), but others applauded the innovative focus on front-line soldiers, and the use of personal anecdotes which brought the history to life for Australian readers.[82]

The history was read and used in many different ways, and it had an indirect influence as a major source for public history and remembrance. Bean's historical tradition was perpetuated in the unit histories, which he often supervised, by the Australian official historians of later wars, and by popular war histories which freely used Bean's evidence and conclusions. Excerpts from the official history were reproduced in school readers and then quoted in student essays, and Anzac Day writers and speakers borrowed passages from Bean to explain the significance of the day. Furthermore, as Michael McKernan has recently recorded, Bean was a key figure in the creation of the Australian War Memorial, and thus enshrined his version of the war in a building which served the multiple purposes of memorial, museum and archive, and which would become one of the country's most popular tourist attractions. For the most part, representations of the Anzacs that drew upon Bean's history were informed by his generalisations about the national meaning of the war and the positive qualities of the Australian soldiers. Very rarely did they use or discuss the more complex and even contradictory evidence which is hidden away in the 10 000 pages and four million words of the history.[83]

For readers, Bean's history was a resonant and appealing narrative. In 1938, Tas Heyes at the War Memorial wrote to Bean about 'the reception accorded it by all classes [...] That it stands high in the regard of Australians is amply indicated by the hundreds of unsolicited letters of appreciation which we hold'. Among those letters were some from civilians applauding Bean's contribution to Australian understanding of the war and its significance. Mr M. E. Marshall of South Australia railways claimed that by contrast with the usual 'droll and uninteresting' school history books, Bean's history evoked 'the courage and life and interesting features which every young Australian, who looks with admiration at the name of ANZAC, will be sure to enjoy and derive from them a wider view on the immortal history of our Australian soldiers'. Other readers were delighted with personal references to ex-servicemen relatives, and some remarked that family members who had died at war lived on in Bean's prose. My brothers and I used to look up 'J. D. Rogers' in the index and then read about his exploits, discovering, just as Bean had intended, a proud, personal connection with the AIF. We also thumbed through the photographic history (which included some of 'Papa's' photos) until it became the most battered volume, as it often is on library shelves.[84] Ex-servicemen also recorded their approval of the history, and in doing so revealed the range of ways in which they read and used its account of their war experiences, and what the history meant for their identities as ex-servicemen. R. L. Leane, an ex-AIF officer who became Chief Commissioner of Police in South Australia, wrote to Bean 'to appreciate as no doubt do Diggers as a whole, the opportunity to go over old actions, see where mistakes were made and successes occurred';

> [...] as usual you have given me [the] most generous lookout. I feel proud of the fact because you were there and therefore speak from personal experience [...] When finished the history will provide a tradition for future generations, and if they live up to it, we need have no fear for the future.

Harold Gieske, who had served in the 26th Battalion and after the war worked in north Queensland, claimed that to be mentioned in a volume of Bean's history was 'a greater honour than any war decoration', and Gellibrand was only half-joking when he told Bean, 'I propose to have my headstone marked merely; "see Bean Vol iv"'. From Western Australia, G. H. Nicholson wrote of his postwar economic tribulations in the mining industry, and claimed that 'your history more than compensates':

> I never thought Passchendaele was ranked so high in the war efforts [...] and to think that you appreciated the part played by my men is just thrilling. It has made me realise that after all my life wasn't wasted.[85]

Bean's official history seems to have been equally popular among working-class ex-servicemen. A couple of the men I interviewed, who had not wanted any contact with the war because it was such a painful memory, had not been interested in the history. But I was amazed to find that almost half of my interviewees had bought copies, and that some had acquired the full set (many also had copies of their unit history). They had bought it as a source of information about events and names, to read about the actions of their own unit, and so that they could get a wider picture of the war. A few were concerned that Bean had not been able to cover particular events, and that no writing could convey the real experience of the trenches, but most admired the accuracy and frankness of Bean's account and enjoyed it as 'real good reading'.[86]

Radical diggers who had been alienated by the loyalist politics of the RSSILA and Anzac Day were critical of the official status of the history, but even they were impressed by the absence of any obvious political censorship or bias. Although wary at first, Fred Farrall and Stan D'Altera both became Bean enthusiasts. The history, even more than Anzac Day, could be read and enjoyed for many different purposes, and was thus appealing to a wide range of diggers. Above all, it was popular among ex-servicemen because it focused on their experiences in the line — including actual battles they had fought in — recognised what they had been through, and admired their achievement.[87]

Bean's official history thus provided positive ways in which a veteran could make sense of his experience. However, by producing an account that generalised the Anzac experience according to the themes of national identity and achievement, and that marginalised war experiences that did not fit those themes, Bean's official history played an influential role in shaping the ways in which ex-servicemen articulated their own war experiences and identified themselves as Anzacs. In the next chapter we return to the memory biographies of the three diggers to discover their individual experiences of repatriation, and to explore the relationships between their war memories and identities, and the postwar organisations, rituals and histories of Anzac.

CHAPTER 6

TALK AND TABOO IN POSTWAR MEMORIES

Percy Bird

Percy Bird arrived back in Melbourne on 30 December 1917. He went to Sydney for a fortnight's paid leave with his fiancée ('she had a single room on the ladies' side and I had a single room on the gents' side'), and then reported back to the Caulfield Repatriation Hospital in Melbourne for an observation week. At the end of the week he went before a Repatriation Board:

> Course, we didn't know anything about pensions. I said, 'I feel all right'. 'Oh, according to your papers so-and-so and so-and-so.' I think that was a hint they gave me but I didn't take it. I said, 'I've had sixty days on the boat and I'm all right'.

He wrote on a form that he was in good health, not really knowing why it mattered, and as a result never received a war disability pension. The way in which Percy told this story suggests that he felt that the soldiers were cheated by this procedure, and he repeated the common complaint about the treatment of veterans: 'But of course, we were told at the time, oh, government was going to do everything for us, but they didn't do anything'. Yet in practice he seems to have been comparatively unaffected and unconcerned about this situation, and in the interview he projected his own feelings upon other veterans who 'just took it like I took it'.[1]

The lack of bitterness in Percy's remembering is explained by the nature of his transition back to civilian life. In February of 1918 his old boss at the railways asked him if he wanted to return to his job, and by the middle of the month Percy was back at work. The boss treated Percy like a son (he had marital aspirations for Percy and his daughter), and was sympathetic

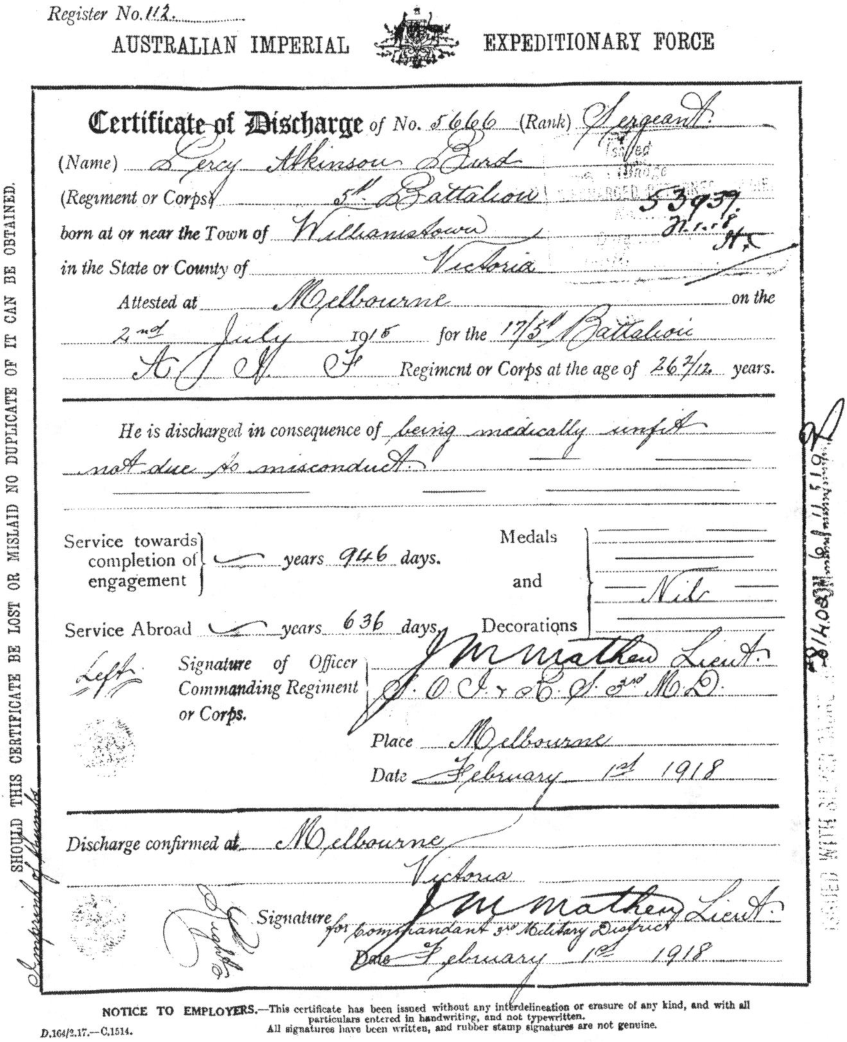

Register No. 112

AUSTRALIAN IMPERIAL EXPEDITIONARY FORCE

Certificate of Discharge of No. 5666 (Rank) Sergeant

(Name) Percy Atkinson Bird

(Regiment or Corps) 5th Battalion 53937

born at or near the Town of Williamstown

in the State or County of Victoria

Attested at Melbourne on the 2nd July 1915 for the 17/5th Battalion A I F Regiment or Corps at the age of 26 7/12 years.

He is discharged in consequence of being medically unfit not due to misconduct.

Service towards completion of engagement — years 946 days.

Service Abroad — years 636 days.

Medals and Decorations Nil

Signature of Officer Commanding Regiment or Corps. J M Mathew Lieut. S.O.I. & E.S. 3rd M.D.

Place Melbourne

Date February 1st 1918

Discharge confirmed at Melbourne Victoria

Signature J M Mathew Lieut. for Commandant 3rd Military District

Date February 1st 1918

SHOULD THIS CERTIFICATE BE LOST OR MISLAID NO DUPLICATE OF IT CAN BE OBTAINED.

NOTICE TO EMPLOYERS.—This certificate has been issued without any interdelineation or erasure of any kind, and with all particulars entered in handwriting, and not typewritten. All signatures have been written, and rubber stamp signatures are not genuine.

D.164/2.17.—C.1514.

Figure 14 *Percy Bird's Certificate of Discharge from the AIF, 1917.* (Kath Hunter)

to the needs of an ex-serviceman, allowing Percy to take a walk whenever he needed a break. The clerical work was easy for Percy because 'I was doing so much of it in the army, you see, it didn't affect me'; in this respect his experience was very different from that of soldiers who had remained in the line. He stayed with the railways for the rest of his working life, keeping his job through the Depression and the Second World War, and finishing up as an Audit Inspector. As ever, he took a great deal of pride in his work and in his career success, and he was relatively satisfied with his job.[2]

Percy also recalled that he had had no problems readjusting to domestic life. He could not remember any tensions caused by his separation from family and fiancée, and it may be that the continuity of the relationship with his fiancée smoothed his transition from soldier to civilian; they married in July of 1919. Though neither his wife nor his children were a significant part of his remembering with me, it seems that he enjoyed a stable family life. For much of his working life Percy was away during the week on the railway audit circuit, and Eva Bird ran the household while also being active in local voluntary organisations.

The relative ease of Percy's return to civilian and working life meant that he had no grievances that might have made generalisations about the neglect of ex-servicemen more personally resonant. Although in our interview he occasionally mentioned postwar meetings with soldier mates who were having a hard time, the emphasis in these stories was always upon the pleasure of reunion rather than the man's difficulties. When I asked Percy about other ex-servicemen's problems he remarked: 'No, no I can't place any of that [...] I didn't feel any effects from it or anything'.[3]

The absence of bitterness or disillusionment in Percy's postwar experience was also significant for his memory of the war and his identity as a soldier and ex-serviceman. In comparison with other soldiers who had difficulties after the war, Percy was not provoked into rethinking whether it had been worthwhile to enlist and fight. Indeed, his attitude to the war and his identity as an Australian soldier were affirmed through various public practices of remembrance.

The war was still very much on Percy's mind after he came home. In those first years he dreamt about the war in ways which suggest that anxieties from the trenches were still present in his subconscious; in one dream 'we had a surprise with the Germans and ourselves'. Like many ex-servicemen he did not like to talk about the war with his family or civilian friends and only talked about it easily with other veterans. He started to attend meetings of ex-servicemen even before the war ended, and in 1919 helped to form a Williamstown branch of the RSSILA. Percy did not have time to be very active in the branch because of his family commitments, and as a result he only attended the social 'turnouts'. But even this social membership provided a ready-made support network of ex-servicemen, as well as an affirming public acknowledgement that Percy belonged to the prestigious elite of men who had been 'over there'.[4]

More important for Percy were contacts with men from the 5th Battalion. As the 5th was a Melbourne battalion, he often met up with his old pals. He relished the opportunity to 'talk about the old times to the chaps, different things happening', and clearly remembered such meetings as a highlight of his postwar years. Percy also attended the annual battalion reunion because it brought the battalion together again and was an opportunity to 'meet my old pals. That's the most important. We thought the world of each other'. In chance meetings and organised reunions, stories about the war were told and retold:

> Thinking of our old pals and what we had to put up with, of the fun we used to have at different times. I will admit every day wasn't sad but all the same we used to enjoy, have little concerts and things like that.

Reunions provided an important social forum for the continuation of the wartime process in which some aspects and meanings of the war experience were actively highlighted while others were silenced. As both storyteller and audience, Percy gradually refined his set of war stories.[5]

In some ways the relationships between the men, and the stories they told, were different in the postwar reunions. Wartime stories had been immediate articulations of everyday incidents and issues; after the war the stories became more generalised and nostalgic. For example, Percy's reminiscences came to include stories about the successes of the AIF in 1918 (when he was already back in Melbourne), which he heard from 5th Battalion veterans and which reaffirmed his impression of the Australian soldier. Crucially for Percy, distinctions that had been troubling for his own identity during the war — between men who served in or out of the line and between larrikin diggers and conscientious NCOs and officers — were now eclipsed by the pleasures of reunion and the common, remembered experiences and identities of the battalion and the AIF. Above all, the shared Anzac identity was reinforced by the much greater distinction between Australians who did or did not go to war.

The pleasure of reunion, and reassertion of the special status of the Anzacs (to 'let all the people see all the old fellows marching'), were also the main reasons for Percy's participation in Anzac Day. Anzac Day confirmed the general theme of Percy's stories about the distinctive qualities of the Australian soldiers. It also reinforced the underlying assumption of Percy's remembering: that his personal experience of the war, and that of the Australian soldiers in general, had been worthwhile.

Anzac Day was a pleasant event for Percy that consolidated the ways in which he had already begun to compose his memory of the war. This was because the themes of the day closely matched Percy's own attitude to the war; because his postwar experiences did not force him to question the gap between the legend of the Anzacs and their treatment as ex-servicemen; and because collective participation in the ritual played down distinctions within the AIF.[6]

Percy also shaped and affirmed his memory of the war through reading. He never read Bean's history but he bought and 'thoroughly enjoyed' the 5th Battalion history, A. W. Keown's *Forward With the Fifth*, which 'brought back memories to me about what we did at lots of times'. The purpose of Keown's history was to strengthen the ties binding old comrades and to record and perpetuate the memory of 'deeds nobly done, of gallant men; of their bravery, their endurance, their cheerfulness; of Death bravely met and sufferings bravely endured'. Percy valued the book because it recounted the stories of his unit in great detail and validated his experience as a soldier.[7]

Although Percy claimed that reading the history did not affect his memory because he already knew the story (he even corrected a couple of minor errors), the history was influential because it highlighted certain events and reinforced particular ways of remembering them. The focus on the battalion confirmed Percy's own tendency to remember the war in terms of the battalion experience. The book provided a clear chronological outline of the period in which Percy was in D Company, and helped him to recall and fix in his mind the exact dates of his movements with the company. It was a compendium of stories, meanings and even words that Percy adapted and used as his own; some stories in the history are almost exactly repeated in Percy's remembering.

Although the history is more descriptive than Percy about conditions in the line, the themes and tone of the book — celebration of the 5th Battalion and of the independent Australian soldier; exclusion of tensions within the battalion; and emphasis upon the humorous ways in which the men endured life in the line — were used by Percy to articulate his war experience in terms of the Anzac legend, and were powerfully affirming for Percy's positive identity as a veteran of the 5th Battalion and the AIF.[8]

By the time Percy Bird lost the battalion history when he lent it to a mate in the 1950s, he had, through reunions, commemoration and reading, composed a way of remembering his overseas service which enabled him to handle the personal issues that had troubled him during the war.

Percy Bird and his fiancée Eva Linklater, photographed in the backyard of his family home in Williamstown soon after Percy's return from the war. (Kath Hunter)

Percy Bird and his fiancée Eva (at right), photographed with friends in a Melbourne park soon after Percy's return from the war. (Kath Hunter)

Bill Langham after the war, wearing his RSSILA badge. (Bill Langham)

*Bill Langham (seated) with his cousin George Macumber (wearing an RSSILA badge),
photographed in Bendigo, 1919.* (Bill Langham)

Fred Farrall pictured outside DaSilva's upholstery factory in Marrackville, Sydney, in 1926, showing signs of his physical and nervous collapse. (Fred Farrall)

Fred Farrall, the confident trade unionist, pictured outside Trades' Hall, Melbourne, May Day 1940. (Fred Farrall)

Percy Bird in his Williamstown home after out interview in 1983. (Alistair Thomson)

Bill Langham at his home in Yarraville after our interview in 1983. (Alistair Thomson)

Fred Farrall at his home in Prahran after our interview in 1983. Fred and Dot are shown in the pictures above Fred's head, with Fred wearing the robes of the Mayor of Prahran. (Alistair Thomson)

Bill Langham

The ways in which Bill Langham remembered the war were also fundamentally shaped by his postwar experiences. Although during the course of the war Bill had made sense of his experience in ways that matched Anzac and digger images, when he came home the theme of disillusionment might easily have become predominant.

Bill vividly recalled the day in January 1919 when he was driven from the home-coming ship to Melbourne's Sturt Street barracks and saw his mother and two brothers waiting for him at the gate:

> That was one of my lovely memories. After being away and through that. There she was standing at the gate, and my little brother standing alongside her. Always that scene, I always remember that. Yeah.

After a two-hour wait for a medical inspection, Bill told the army doctor that he 'didn't give a continental' whether or not he was marked in A1 health, just as long as he could get out to meet his family. It was a lapse he would regret.[9]

The chronology of Bill's life in the following years is not clear from the interview, because of the confusion of the times, and because his remembering highlighted steady progress rather than the uncertainties and frustrations which in fact characterised his postwar years and those of many other ex-servicemen. It seems that in the first couple of years after his return Bill lived off the deferred pay of four shillings and sixpence per day of service which his mother had saved for him, and that he moved between Melbourne and the Victorian bush playing football and doing casual labouring work. He made his base in Yarraville, where his mother and siblings had moved after a wartime separation from his father who makes no more appearances in Bill's story.

Bill was offered his pre-war job at the Caulfield stables, but declined because he had put on weight during the war. Determined to find independent self-employment after years of obeying orders — 'I thought what a lovely thing it would be to be your own boss' — he applied to the Repatriation Department for a grant so that he could start a taxi business. They would only give him enough to buy one taxi, but he decided that a one-car business would be too risky, and 'scrubbed it'. Next he enrolled in a Repatriation upholstery course, 'but the fellows that were instructors and things there, they didn't give a continental, so I scrubbed that'. His money was running low and after a period of unemployment he was forced to cash his gratuity.

Eventually he got a job in a barber's shop, learnt the trade and then set up on his own. Heavy smoking and indoor work contributed to ill-health, and on a doctor's advice he left the trade to work outside, where he recalled that he made a success as a quarry worker and eventually earnt seven pounds a week as a powder monkey and foreman.[10]

In our interview Bill was ambiguous about the end of that career. At one point in the interview he explained that he hurt his back and had to leave the job, but he also recalled the day 'the Depression came on' and all the workers at the quarry were put off. By this time he had married and had a young son, and for three tough years during which he was unemployed they lived off a small government sustenance payment, which was supplemented by occasional quarry work and mushroom picking, and by the local Unemployed Self Help Association. Bill repaid help from the latter by giving free haircuts to the unemployed. Though, like many who were unemployed, Bill remembered the vital role of initiative and community support during this time, the story of the Depression was relegated to a minor position in a life history that emphasised progress and success; in one telling, he shifted from the quarry job to council work after the 1939–45 war, neatly ignoring the difficult intervening years.

The next main signpost in Bill's memory of work was the Second World War: 'we all came under control again then. You lost your freedom'. He was assigned by 'the Manpower' to a sugar works in Yarraville where he lumped sugar bags on night shifts for the duration. After the war he got a job with the Melbourne City Council, starting off as a cleaner and retiring twenty-three years later from a job 'on my backside in the Town Clerk's private office [...] that's what I wanted and I got it, see'. The fact that work provided the main signposts for Bill's postwar life shows how career success was vital to his identity and pride, at the time and in memory.[11]

These postwar experiences affected Bill's understanding of the war, and his identity as an ex-serviceman, in a number of ways. Firstly, the desire for work independence both grew out of and confirmed his dislike of military regulation ('I'd had the khaki'). Yet the postwar frustration of that desire made him bitter about repatriation. Apart from the disappointments of the taxi business and the upholstery course, Bill was also angered by the Repatriation Department when it reclaimed a pair of special glasses on the grounds that his eye problems were not due to a war injury. Worst of all, in Bill's view, was the rejection of his war disability pension claim. When Bill applied for the pension after the war he was knocked back on the basis of an X-ray, which he claims was faulty, and because he had marked himself 'A1' at discharge. These grievances made Bill receptive to the prevailing postwar criticisms of the treatment of ex-servicemen, which in turn encouraged him to articulate his own experiences as 'the same old story':

> When we want you to go away and fight we'll give you the world, but when you come back we'll take it off you again.[12]

Bill's bitterness about the plight of ex-servicemen provoked different ways of reinterpreting the war and identifying himself as a soldier and ex-serviceman. Characteristically, he did not consider that his postwar misfortunes may have been in any way his own responsibility. He blamed the ex-servicemen's lot on the politicians, businessmen and other 'big nobs', who had sent the soldiers to fight their war but neglected the veterans when they came home; in this light he contrasted his own small gratuity with the massive profits of arms manufacturers. Yet he also perceived ex-servicemen's problems to be the fault of an ignorant and uncaring civilian society, and asserted that ex-servicemen were special and deserved privileged treatment because of their wartime sacrifices. Each of these contrasting attitudes — one confronting the Anzac legend and the other reinforcing it — was present in Bill's ex-service identity, and either one might have become predominant. The ultimate pre-eminence of the latter attitude can be explained in terms of the postwar public contexts in which Bill made sense of his war and postwar experiences.[13]

During the interview Bill recalled that when he was alone in the years after the war, memories of his time in Europe— both pleasurable and unpleasant — were often jostling beneath the surface of his consciousness. At the time he wanted to talk about these memories but found that few civilians were able to listen or understand. Some were jealous, while others thought he had just gone 'for a good trip', and Bill equated their lack of understanding and respect with that of the government and Repat. Like Percy Bird, mates from the war provided Bill with essential relationships for social support and collective remembering. He lost contact with some of the men from his unit because they returned to different States, but sustained significant relationships with others through occasional interstate visits and regular Melbourne smoke-nights and Anzac Day reunions. Bill recalled that he never missed the Melbourne Anzac Day reunion and march: 'Well, you met your mates. It was lovely to meet them, and have a drink and things with them afterwards, have a yarn, over the good times, I'll say, not the bad ones'.

For Bill, the reunions provided an intense and pleasurable affirmation of the camaraderie of the war experience and of his identity as a soldier and ex-serviceman. This affirmation was also selective:

> [...] people used to say to me, 'What do you talk about when you go to [reunions] ... Do you talk about this battle you was in and that battle you was in?' I said, 'Oh, don't be funny mate, we talk about all the funny incidents that happened. You're taboo if you start talking about war if you go to a smoke night. You pick out all your funny incidents'. It's like, you want to forget it, see. You want to forget the bad parts, which we all do.

This social remembering provided collective validation of the pasts that were easiest to live with. Bill's difficult and dissenting memories — of driving over the dead Germans, or wartime disillusionment — were not erased, but nor did they become favoured public stories. In contrast, positive anecdotes about humorous experiences, or about the nature of the diggers and the AIF, were favoured, and the war experience became characterised in those terms.[14]

Bill valued participation in Anzac Day because, besides the pleasures of reunion, it provided public recognition of his special Anzac status. The reception that the marchers received from 'terrific crowds' showed that 'somebody remembers anyway'. Bill also joined the RSSILA soon after his return, and attended its Footscray club because it seemed to be a good way to keep in touch with other ex-servicemen, and because of the practical benefits of membership. The League's 'great big badge' often came in handy as evidence of veteran status; Bill related with pleasure the story of an ex-service railway ticket collector who let Bill and another unemployed old digger travel for free because he recognised the badge. He was also grateful for help from the Footscray branch of the League, which eventually secured his war disability pension. The League kept the Anzacs' special status in the public eye and ensured that status brought material benefits. However, Bill's loyalties were not confined to the League. During the Depression he was active in the local Unemployed Self Help Association, and not in the RSSILA sub-branch's Unemployed Section, which refused to help non-League members. This suggests that community, working-class loyalties sometimes conflicted with ex-servicemen loyalties, and that Bill used the identity that was most useful at a particular time.[15]

During the war, the diggers' anti-authoritarian ethos had acted, to some extent, as a counter to official and media Anzac identities, and had sustained

Bill's ambivalence about the war. It may be that Bill's involvement in the local unemployed organisation sustained something of his anti-authority and anti-war tendencies. Yet the ex-serviceman identity that Bill generally adopted was, on the whole, closely linked to the official legend and identity. In reunions, Anzac Day marches and League clubs, the proud and deserving status of veterans was stressed and wartime dissidence was played down. There was some space for deviance. Bill liked to get on the grog with digger mates on their 'one day of the year'. He was also concerned that rather than glorifying the 'whole flaming business' of war, Anzac Day should tell people that they didn't want war, and he believed that it did just that. But because he wanted Anzac Day to ensure that the soldiers' sacrifices were not forgotten, he also argued, using the rhetoric of Anzac Day speeches, that those sacrifices had been necessary in order to prevent aggression and maintain democracy. It is significant that in our interview Bill's criticisms of the war were never made during discussions of Anzac Day and unit reunions; in his memory those occasions were linked with entirely positive accounts of the war.[16]

In contrast, Bill's dissenting war memories were not articulated into a coherent political critique of the cause or conduct of the war. Bill was never active in the Labor movement — he was too much the individualist — and his enthusiastic reading of war stories did not include pacifist or socialist literature. He did not forget his wartime disillusionment, and his memory still contained sparks of criticism that could be lit by a particular prompt or question. Yet the postwar interface between his own emotional and practical needs, and public remembering among ex-servicemen and in Anzac commemoration, ensured that he highlighted positive memories of his wartime role as a mate, a digger and an Australian soldier.

Fred Farrall

Fred Farrall was in poor shape when he was finally discharged from hospital and the army in January 1920. He could not go back to work on the family farm because of the trench foot condition, and as he had always been attracted by city life he stayed in Sydney. He had little relevant work experience, however, and had never learnt to fend for himself as an independent adult. On the farm and in the army his life had been organised and provided for, but now he felt like a pet that had been thrown away by its owners:

> And then when I got into civilian life, well this was something new, and to some extent it was, it was terrifying. You're out in the cold hard world. Nobody to look after you now. You've got to get your

own accommodation, your own meals. In short, you've got to fend for yourself.[17]

Fred's predicament was worsened by his poor emotional state; he was not 'the full quid'. He recalled that he very nearly shared the fate of several diggers he knew whose 'lives had been changed so much, shattered so much [...] they seemed to have lost all control over themselves' and drank themselves to death or committed suicide. Fred was lucky. After a chance meeting, his cousin and her digger husband gave him a room in their house and helped him to get back on his feet. He lived with them until the end of 1921, when he moved out to board with another digger mate. In 1923 he married and used his deferred wartime pay and veteran's gratuity to buy a War Services Commission house at Brighton-Le-Sands.[18]

Employment was more of a problem. Fred's legs barred him from labouring jobs, and because he had not been discharged until 1920 he missed out on the best retraining courses. Eventually he found a place on an upholstery training course run by the Repat, but the course was a 'farce' because the training school was badly managed and could find few employers willing to hire the men; those that did were not keen to continue the veterans' employment when the government stopped subsidising wages. Fred and his digger mate Roy O'Donnell searched for work for almost two years: 'We walked God knows how far, how many times and of course got sick and tired of it in the end, and, well, what's the use'. In March 1922, Fred finally got a job as a motor car upholsterer at the Meadowbank Manufacturing Company, and then, after another period of unemployment in 1924, he got into the furniture trade.[19]

Although the Repatriation Department had attempted to retrain Fred for civilian employment and had helped him to buy a low-cost house, in those first few years of intermittent employment Fred became very bitter about the government's treatment of ex-servicemen:

> Well, it'd be hard to explain other than that first of all we, of course, had been disillusioned. What we'd been told that the war was all about, didn't work out that way. What we'd been told that the government would do when the war was over, for what we'd done, didn't work out either.
>
> *In what ways?*
>
> Well, you see, the pensions in the 1920s, unless you had an arm off or a leg off or a hand off or something like that, it was almost as hard to get a pension as it would be to win Tatts. There was no recognition

of neurosis and other disabilities [...] they treated the diggers as they interviewed them and examined them as though they were tenth rate citizens. Something like we look upon the Aboriginals. There was great hostility between the diggers on one hand and the Repatriation officials on the other [...][20]

Fred's recollection may well have been articulated in these terms through his subsequent political critique of State patriotism. Yet his personal experience of mistreatment was itself a catalyst for reflection about the worth of the war and of his contribution as a soldier, and fuelled the radical politics that he adopted later in the decade. Fred could not get a war pension for his physical or nervous complaints, and shared the anger that many diggers directed at Repatriation officials, even though they were often also ex-servicemen:

They'd be referred to by the diggers outside as a pack of bastards. That's how they viewed them. So the AIF that was all for one and one for all during the war, no matter whether they were right or wrong, didn't exist any more when they got into civilian life. It was survival of the fittest, and those that had got into jobs were the fittest.

Unwilling to be treated as an undeserving malingerer, Fred decided not to have anything more to do with the Repat. He maintained this resolve until, in 1926, the emotional after-effects of the war and the difficulties of his repatriation caused a nervous breakdown: 'you just get that way where you don't want to do anything. You don't appear to have any energy or any, you know, desire to stir yourself'. Forced to give up his job, Fred 'had no option, at any rate, but to eat humble pie and go back to the Repat'.[21]

In the years between his discharge in 1920 and the nervous breakdown in 1926, Fred Farrall's identity as a soldier and veteran was confused and traumatic, and characterised by a striking contrast between disturbing private memories and public silence and alienation. On the one hand the war retained a haunting and debilitating emotional presence. The feelings of vulnerability and terror that had been induced by repeated shelling in the trenches were relived in harrowing dreams:

Oh well, the dreams I had were dreams of being shelled, you know, lying in a trench, being in a trench or lying in a shell hole, and being shot at with shells. And being frightened, scared stiff [...] you don't know when the next shell that is coming is going to blow you to pieces or leave you crippled in such a way that it'd be better if you had been blown to pieces [...] You'd be going through this experience and you'd be scared stiff, you'd be frightened. You'd be frightened, and wakened up, probably, by the experience.[22]

The nature and power of these dreams suggest that unresolved memories and feelings from the war were a contributing factor to Fred's debilitating nervous condition and eventual breakdown. In the interview he explained how and why his condition worsened in the 1920s:

I didn't realise it at the time, but I long since realised it. But I had neurosis, that was not recognised in those days, so we just had it. You put up with it. And that developed an inferiority complex, plus, really, and I mean extremely bad [...] Well, I had reached a stage with it where, when I wanted to speak I'd get that way that I couldn't talk. I would stammer and stutter and it seemed that inside me everything had got into a knot, and that went on for years and years and years.[23]

Although Fred was not able to untangle these emotional knots for many years, he did develop ways to cope with other aspects of his war memory. He chose to marry on the anniversary of a war wound, he named his house after the places where his two best mates were buried, and he remembered in exact detail the places and dates of his friends' deaths. These private forms of commemoration, which transformed grotesque experience into relatively safe lists and rituals, were Fred's way of coping with the past. He explained to me that everyone had different ways of coping, and that his was to remember dates: 'Anything like that is planted indelibly on my mind'.[24]

Yet in those first years, Fred's personal remembrance never gained the public affirmation that might have helped him to develop a more positive identity as an ex-serviceman, and to resolve the causes of his nervous condition. Digger mates were a vital source of postwar friendship and support, and Fred shared lodgings, his social life and the search for work with men he had known at the war. However, in contrast with Percy Bird and Bill Langham, the war was a taboo subject amongst Fred and his friends: 'we'd talk about racehorses and all sorts of things, but I can't remember us ever sitting down and having a talk about the bloody war'.[25] In old age Fred explained this silence as a result of the soldiers' bitterness and the inability or unwillingness of the civilian population to comprehend or even listen to their experience.

Although Fred remembered his own silence as representative of ex-servicemen, the contrast with Percy Bird and Bill Langham suggests that Fred's response was specific to veterans with particularly negative experiences of the war and return. They were men who wanted to block out their wartime life because recollection too easily revived painful memories and feelings. They were also men who felt that they had been badly treated upon their return, and whose postwar disillusionment made them feel even more negative about the war. Perhaps most importantly, and as a consequence of these factors, they were men who could not or would not participate in the various forms of public affirmation that were available to the Anzacs in the 1920s. Fred refused to wear his war medals because he didn't value them (he cited the common story of ex-servicemen who threw their medals into the sea in disgust). He shut away his beautifully embossed discharge certificate in a dusty drawer, and declined to attend battalion reunions or Anzac Day parades.[26]

The reasons for Fred's disengagement from Anzac commemoration show how some diggers came to be alienated from cultural practices that other men, like Percy Bird and Bill Langham, found so useful and affirming. One contributing factor to Fred's lack of participation was the fact that he was in no fit state for social events. He joined the RSSILA on his first day back in Australia, but was never an active member because he was so 'tongue-tied' and insecure. For the same reason battalion reunions and Anzac Day parades were embarrassing occasions that Fred found easier to avoid. He also stayed away from Anzac Day because he considered it a drunken binge. The difference between his own sobriety and the more larrikin behaviour of some of the diggers, which had troubled him during the war, now contributed to his exclusion from one of the most effective Anzac affirmation rituals. Fred and more radical digger mates also refused to march on Anzac Day because the patriotic rhetoric did not square with their doubts about the worth of Australian involvement in the war, or with the bitterness they felt about the mistreatment of ex-servicemen. But the main reason for Fred's non-participation was the extreme confusion and distress that he still felt about the war. The public celebration of Anzac heroes was a painful reminder of his own feelings of inadequacy as a soldier and as a man. In turn, that self-imposed exclusion from the rituals of collective affirmation reinforced his sense of masculine inadequacy.[27]

Alienation from Anzac commemoration did, however, make Fred more open to alternative ways of comprehending the war and his military exper-iences. In his remembering he related his entry into labor politics as the second and most significant stage of his life story of conversion. He recalled that although he felt disillusioned after the war he was still politically confused:

> I didn't know where I was. I was disillusioned with Hughes and his party and I wasn't, in the beginning, attracted to the Labor Party either because they'd been painted as red-raggers, that at that time I wasn't very sympathetic to.

Despite his doubts, when Fred started work at the Meadowbank Manufacturing Company in 1922, he was persuaded to join the Coach Makers' Union by an old worker who explained that decent wages were due to the union. He discovered that the so-called 'red-raggers' were 'the most honest and reliable' men in the trade union movement: 'That sowed the seeds for my socialism that I developed a few years after and have had all my life'.[28]

Fred became active in the union and in 1926 joined the Labor Party and was prominent in his local branch. Through this activism Fred found supportive comrades and a sense of purpose. He gradually regained his social skills and self-confidence, though he did not overcome his stammer until 1940 when a Melbourne psychologist, who shared Fred's interest in the Soviet Union, treated him for free and taught him relaxation techniques to reduce physical and emotional tension. Despite the stammer, Fred became a Justice of the Peace and was nominated by his Party branch for a position on the New South Wales Legislative Council. He declined the nomination because he couldn't afford to take up the position, and because he was beginning to have doubts about the Labor Party. In March of 1930 he became unemployed again and, disillusioned by the Federal Labor government's imposition of wage cuts and its band-aid treatment for unemployment, joined the Communist Party in the same year. As an organiser for the Communist Party led Unemployed Workers' Movement, he addressed street meetings and produced broadsheets to promote the message that the capitalist system, and subservience to the Bank of England, were the causes of unemployment in Australia.

At about this time Fred's marriage broke up. He had married into a conservative family and the political differences were now too great for a successful marriage (that was Fred's version — it may be that his nervous condition and increasing political workload also undermined the marriage). In 1932 Fred 'teamed up' with Dot Palmer, who was active in the Clothing Trades Union in New South Wales, and also a member of the Communist Party. According to Fred he was politically ignorant in comparison with Dot, who had grown up in a radical socialist tradition. Although they never married because they couldn't afford or be bothered with the ceremonies, they lived together and shared their political struggles until Dot's death in 1979.

The new and empathetic peer group of the Labor movement, which included radical ex-servicemen, helped Fred to articulate his wartime and postwar disillusionment. Fred was lent his first political books by the union representative at Meadowbank. They included *War and Armageddon*, Harry Holland's account of the ill-treatment of New Zealand conscientious objectors, and *John Bull's Other Island*, George Bernard Shaw's critique of British imperialism in Ireland. Shaw's book prompted Fred to reconsider the cruel treatment meted out to the Egyptians during the war by the British forces: 'Well I suppose that was the first concrete thoughts, politically speaking, that occurred to me'. He also read and approved the anti-war books that came out of Europe in the late 1920s, including Erich Remarque's *All Quiet on the Western Front*. During the inter-war years Fred had no desire to read mainstream Anzac literature. When a man from the War Memorial tried to sell sets of Bean's official history at Fred's workplace in the 1940s, he was 'so disinterested in it that I didn't, I hardly gave him a hearing'.[29]

Through reading and talking, Fred adopted the left's version of the war as a folly in which working-class soldiers were the victims of imperialist economic and political rivalries. His political views encouraged Fred to highlight certain war stories and led to particular understandings of his own experience as a soldier. Thus it seems likely that at this point Fred came to emphasise the story of Bill Fraser's warning against fighting the rich man's war, represented himself as an unwitting victim of an imperialist war, and began to place the war as a key event in the political development of himself and other ex-servicemen. A new, Marxist analysis of class prompted him to compare the relationship between officers and men in the AIF with that between employers and workers in peacetime Australia, to emphasise the role of trade unionists in the AIF, and to stress that the diggers were often rebellious towards authority.

As a proponent of this radical Anzac tradition Fred also articulated his disillusionment about repatriation, and deduced that Anzac Day was 'a clever manoevre' intended to stifle veterans' anger about their mistreatment:

> Well I would say that if it wasn't for Anzac Day, the First World War would have probably been — met the same fate as the Eureka Stockade [the armed rebellion of gold miners in 1854]. That is, it wouldn't be recognised. It wouldn't be recognised. And whoever thought up celebrating Anzac Day, which was a — had nothing to recommend it in a way, first of all we were invading another country, Turkey ...

Secondly, it finished in a defeat. So what was there to celebrate, looking at it from that angle? So they celebrated it for another reason. That was to cultivate a spirit of war in the community. Of admiration or respect, or honour or something for war. And that's all Anzac Day really does. But they had to do it in a certain way, and it was done in a way whereby they could get them together on a social basis. First of all they marched and paraded and showed themselves to the public. And then when that was over they got into their clubs or their pubs or whatever, and did what they wanted to do.[30]

According to Fred, this 'clever manoeuvre' brought many diggers into official commemoration. Yet he no longer felt excluded by Anzac Day. Instead he contested the politics of the day and, with other unemployed activists, opposed the building of grand civic memorials and supported alternative, utilitarian memorials like veteran's hospitals.

Fred also became critical of the RSSILA. In the early 1920s his physical and emotional handicaps had kept him away from League meetings. Now he had no time for an organisation that was not acting 'in the best interests of the ordinary digger [...] it was a political organisation of the extreme right wing'. By the end of the 1920s Fred Farrall had aligned himself against the RSSILA and was fighting with members of the Unemployed Workers' Movement in street battles against RSSILA club men and the proto-fascist New Guard movement. He became a confident opponent of the official Anzac tradition and its RSSILA organisers, and in 1937 was arrested at an Anzac Day parade for distributing the pacifist leaflets of the Communist Party-organised Movement Against War and Fascism.[31]

Fred and other socialists and pacifists hoped to subvert the nationalist and militarist aspects of Anzac Day, and sometimes tried to assert an oppositional Anzac tradition. Fred wore his service badge on special occasions — it is in his lapel in a May Day 1940 photo (see the second section of photographs in this book) — because he wanted to show that his criticisms of war and jingoism were based on experience. Yet it was difficult for him to integrate his identities as a radical and as an ex-serviceman, because by this stage the Anzac tradition had been intertwined with loyalist ideology. Socialist ex-servicemen were targets for venomous intimidation from RSSILA members in street fights because they were regarded as traitors. Unable to forge a positive identity as a 'radical digger', Fred, for the most part, shed his ex-service identity and adopted the more affirming identity of a 'soldier of the labor movement'.[32]

Within the Labor movement Fred was able to develop a critical analysis of the war and to compose his own war experiences into the story of a victim of imperialist rivalry. Though he sometimes used this story in his public life, it did not help him to resolve memories of terror, guilt or inadequacy, and it did not provide a positive affirmation of his wartime manhood. At the height of his power in the Labor movement Fred was still deeply troubled by the fact that he had been so frightened during the war. In the interview he explained that he was able to talk a little about his war experiences and feelings with Dot, but that on the whole he still didn't discuss that part of his life, even among socialist colleagues. The most positive public affirmations of Anzac manhood were available in the institutions and rituals of official commemoration, but they were off limits to Fred. Alienated from mainstream recognition, and still unable to resolve the most intimate emotional traumas of his war, for many years Fred would not talk about his military past and was haunted by painful memories that he could not resolve.[33]

The postwar memory biographies of Percy Bird, Bill Langham and Fred Farrall demonstrate that the various experiences soldiers had of repatriation shaped quite different attitudes to the war, and different relationships with the inter-war institutions of Anzac. The memory biographies also show that the ways in which veterans articulated the war experience, and their identities as ex-servicemen and Anzacs, depended on the availability and appropriateness of public narratives about Australians at war.

For Percy Bird, the Anzac legend of mateship, good times and national achievement, which was retold at Anzac reunions, ceremonies and in histories, made less of his wartime feelings of inadequacy and difference, and instead provided positive ways to articulate his war story and an affirming Anzac identity. Bill Langham had at least two, contradictory ways of comprehending his war. Although collective remembering among digger mates, and the rituals and rhetoric of Anzac commemoration, highlighted positive memories of digger culture and Anzac achievements, they provided no narrative to make general sense of experiences of horror and disillusionment. In contrast, Fred Farrall was unable to relate to war commemoration or to the digger culture of the RSSILA, because the ways in which they constituted the war experience were so different from his own memories. Instead, in the narratives of the Labor movement Fred found more appropriate ways to articulate his years in the AIF, and to identify himself as a victim of war.

The inter-war years were not the last stage in the development of these men's Anzac memories and identities. For each of them, the meaning of the war, and their identity as soldiers and veterans, also changed as they became old diggers, and as Anzac took on new cultural forms and meanings in Australian society during the 1970s and 1980s.

Part III

Anzac comes of age

CHAPTER 7

OLD DIGGERS

Another war

After the Depression — in which half of my interviewees lost their jobs and several lost their homes — the next major event in the lives of the generation that went to war in 1914–18 was the war of 1939–45. Some men of that generation, including both my grandfathers, re-enlisted in the Second AIF, but most of the diggers from the ranks whom I interviewed were wary the second time around. Bill Bridgeman and Doug Guthrie had compulsory call-ups because they were members of the fleet reserve. In 1918 Doug Guthrie had joined up too late for active service, and he didn't want to miss out again, but Bill Bridgeman was not so pleased about leaving his growing family:

> It was a drag to drag yourself away from your wife and young kiddies
> [...] but I thought to myself, better go quietly. I didn't think they would
> send me back overseas again. I told them I wanted ... generally given
> home service, see. 'Oh no', they say to me, 'you got to, on your papers
> here, you've got a high gunnery rate'. That finished the subject. So I
> went to sea.

A. J. McGillivray joined up again so that his younger brother — who had a wife and children — could remain on the family farm in a reserved occupation:

> I couldn't bear to think of my brother going away and leaving little
> children and that, and actually that's one of the things that did ...
> 'cause I hated really, it wasn't because I had any liking for war.

McGillivray's wars make for a poignant life story. In his first war he had been wounded and then almost drowned when the hospital ship was torpedoed. This time he was captured in Singapore and spent most of his military service in Japanese prisoner of war camps, along with another brother who died on the Burma Railway.[1]

Ted McKenzie recalls that he would have enlisted if he had not been married, and that as an alternative he joined the 'Dad's Army' unit set up by the local RSL branch. Percy Fogarty was too old to go to war, but he enlisted for home duties in different parts of the State. More often, 1914–18 diggers had had enough of war and had no desire to be soldiers again. Alf Stabb recalls that 'there was no argument with me. NO! [laughs …] I'd finished with the army. No way! Not after that turn out, no. Anybody that went for a second lot was … he was hungry'. Charles Bowden and his wife had not forgotten the pain of their own separation in 1914, and now tried to prevent their son Kevin from enlisting:

> He really wanted to go to the war. He would have walked to the war
> if he could have […] We didn't want to let him go […] Oh well, he
> was only nineteen years of age. He didn't know his own mind. What
> war was like. I did [laughs]. His mother wouldn't let him go. Course
> I couldn't let him go without her having some say in it. Anyway, he
> pestered the life out of us that much that we eventually said, 'All right'.[2]

Although many of the men I interviewed did not want to go to war again, most of them were active on the home front. Stan D'Altera decided that this was a 'just war' (despite being a communist he felt uncomfortable about Stalin's initial treaty with Hitler), and as the Secretary of the Yarraville Citizens' Club collected money for the war effort and organised amusements for servicemen. Ern Morton was now a Shire Secretary in rural Ararat, where he took on responsibility for organising plans to evacuate people to the country from the Bellarine Peninsula: 'I had no time to take an active part in the Peace Movement. I thought I was doing something that was very beneficial at the time'. Bill Langham was less happy to be assigned to factory work 'by the Manpower', but at least this lifted him out of the unemployment that had plagued his inter-war years.[3]

The Second World War had a major impact on the lives of some Great War veterans, like A. J. McGillivray and Bill Bridgeman, and it became a signpost in the life stories of all my interviewees. Yet in most cases the Second World War did not have the emotional impact or life-changing significance of the First, and it did not take on the same, central role in memory. By 1939, working-class diggers who had survived the Great War and the traumas of repatriation and the Depression were mostly settled into family, home, work and neighbourhood. The 1939–45 war was usually a temporary blip in the mid-life decades which were characterised by continuity in employment, residence and domestic life. The next, dramatic life change followed retirement in the 1960s.

Growing old in Australia

Old age was the third major context in which World War One diggers remembered and composed their life and war stories, and it was the context in which I came to be told these stories. The experience of old age was not an explicit focus of my interviews, and even in the second set of popular memory interviews I was only vaguely aware of the relationships between old age and remembering. For me this awareness was heightened by subsequent participation in reminiscence work in England. Reminiscence work usually takes place in care settings with groups of elderly people. Whereas in oral history the primary aim of remembering is to contribute to historical understanding, in reminiscence work remembering is primarily intended to enhance the emotional and social well-being of participants. Reminiscence workers use a range of approaches to help make remembering a positive experience which reaffirms an individual's identity and sense of worth. Informed by my own experiences in reminiscence work, I returned to my notes and memories of each interview in order to write about growing old in Australia and remembering in later life, and about the distinctive experiences and remembering of old Anzacs.[4]

By the time I conducted my interviews in the mid-1980s, all of the interviewees had been retired for well over a decade. There is a gap in my own knowledge about the initial years of their retirement, but certain themes in their experience of growing old in Australia were apparent when we met, and were recorded in my interview summaries.

Considering that these old diggers were well into their eighties, and had often suffered wounds and other war-related illnesses, they were in relatively good health. I was walked around well-kept back gardens and along well-trod local paths, and ninety-year-old Ern Morton drove me to Maryborough railway station in his brand new car! My interview sample of survivors was inevitably somewhat biased in this regard. What was more surprising and impressive was the spirit with which most of these men coped with their physical disabilities and rejected the stereotypes of old age. Jack Flannery had suffered a stroke in the previous year and was now less active in his garden or on interstate hunting trips, but he still went rabbiting with a set of ferrets he kept out the back. Stan D'Altera had recently returned from seven weeks in the Heidelberg Repatriation Hospital after being knocked over by a car and, though he was very fragile, he was teaching himself to walk again and joking with local columnists about not using the pedestrian crossing. The following comment from one of my interview summaries was a typical impression:

> Charles Bowden [...] has lived alone since his wife died a few years ago. Mr Bowden is old — 94 — and frail, with crook legs, gout, bad skin cancer and very poor hearing. Yet, as he says, he is not 'a geriatric'. His mind is keen and alert, and though his speaking voice is slow and feeble, his memory for dates, people and events is quite stunning.

What I was realising as these men introduced me to their lives and life stories was that old age was not necessarily a negative experience, and that a spirit for life was often more durable than a worn-out body.

I was also realising that this spirit, and the enthusiasm and capacity for remembering, was affected by the particular material circumstances of each man. None of the men I interviewed was destitute or suffering extreme poverty. Only a couple of them were wealthy, but most lived in relative comfort. They often had savings from a long working life to supplement the State pension, and as war veterans some of them also had a service pension. In comparison with their complaints about the Repat between the wars, most spoke favourably about the good deal they now received from Veterans' Affairs, including high quality free medical treatment and 'gold' travel passes for public transport.

Housing conditions provided the most obvious sign of this material well-being. Almost all of the men I interviewed owned their own home, usually a free-standing bungalow or semi-detached house built between the wars in a residential neighbourhood of the western suburbs. Most had lived in the same house, or at least in the same neighbourhood, for much of their adult lives. A few were slightly less well-placed. Stan D'Altera and his brother had pooled their gratuities to buy a Yarraville house, but as a lifelong bachelor Stan had moved out when his brother started a family, and then lived in various rented accommodation, including the caretaker's hut at Yarraville Football Club. When I met Stan he had moved into a tiny pensioner's flat on the sixth floor of a Housing Commission block in Footscray. Another man who had lost his own house in the Depression now lived in a crowded house with his son's family.

Despite examples of difficult individual situations, what stands out among the digger interviewees is that none of them were institutionalised. It's not surprising that institutionalised veterans were not included in my sample; they would have had less contact with the organisations that introduced me to the veterans. The men I did interview had not suffered many of the ill-effects of institutionalisation: the loss of familiar places and people, and of privacy, independence and self-esteem. In their own homes and their own lifelong neighbourhoods, these old diggers had in most cases sustained a sense of place and self-esteem.

What had changed in old age, in some cases with dramatic consequences, were their primary relationships of care and support. Just under half the men I interviewed were still living with their wives in what appeared to be traditional marital relationships, with the female partner responsible for the bulk of domestic care while the male partner was more active as handyman or gardener, or in local clubs and associations. In most of these relationships the wife was the younger, more active partner and played a vital role in maintaining the material well-being of a less capable husband. Some elderly couples were also supported by their neighbours and by adult children, but on the whole husbands were supported by wives, and their different experiences of growing old were profoundly shaped by gender roles.

Ten of my interviewees were widowers (McGillivray and D'Altera were bachelors) and their situation in old age was markedly different from that of the married men. These widowers had learnt to fend for themselves as best they could. Fred Farrall told me that he had learnt to cook after Dot died, and he treated me to his new-found culinary skills; Fred was also looking after an old socialist friend whom he had taken in as a housemate. More often, widowers were dependent on a variety of new carers, including grown-up children, neighbours and social services staff. Bill Williams still lived in the large eastern suburbs house he had bought in 1920, but as a widower he now coped with the support of a cleaner, Meals on Wheels volunteers, a helpful woman neighbour and visits from his four sons. When I arrived to interview Albie Linton and Bill Bridgeman, I was met in each case by a woman neighbour who wanted to keep an eye on the visiting stranger, and whose support had helped Albie and Bill to remain in their own homes. Stan D'Altera was virtually trapped in his sixth floor flat after his accident, but even while I was with him a string of teenagers from the same housing estate looked in to see if Stan was all right and to do his shopping. When I first interviewed Percy Bird in his sprawling Williamstown home he did seem isolated and lonely, but not long afterwards his daughter persuaded him to move to a home for veterans and war widows which was in the same street as her house on the other side of town. Once he got used to the change, he was delighted by the new company and care.

I don't want to romanticise the experience of growing old. One of the old diggers — who refused to move into an old people's home — declined badly and wandered lost through the cold streets in the middle of the night. Others were angry about their failing bodies or a changing world which they could not understand or accept, and for some the interview was a rare opportunity to spend an empty afternoon talking about their younger, vigorous days. But for most of these men an impressive 'community' of carers made old age a relatively secure experience.

Alongside these care relationships — and often an important part of the care — were a variety of networks and associations that offered friendship and activity, and thus helped to sustain emotional well-being. Daughters, sons and other family members had often become more important in old age, and in most cases were regular visitors. Although some of the men had lost touch with workmates after retirement, and long-time friends and neighbours were dying off, they usually still had local friends (despite suspicion or wariness of postwar immigrant neighbours). Albie Linton attended local football matches and, like several other western suburbs Anzacs, made occasional visits to his RSL club. After a lifetime of local public service, Stan D'Altera was regarded as the 'uncrowned King of Yarraville', and in 1977 the City of Footscray named him their Citizen of the Year. Right up until the car accident he maintained an active interest in the Footscray Historical Society and the Yarraville football, RSL and Citizens's clubs. According to a relative who used to invite Stan for Christmas lunch with their family in another suburb, he was like 'a fish out of water' when he left the familiar places and people of Yarraville.

In old age these men also developed new activities and contacts; this seemed to be particularly true for widowers who needed to fill the gap left by the loss of a partner. Bill Williams became active in the Gallipoli Legion and would make regular trips into the office for 'work' which kept him 'alive and ticking in retirement'. Percy Fogarty took up pensioner politics and became secretary of the Footscray Pensioners' Association. After he retired from local and State politics, Ern Morton initiated a campaigning pacifist group in rural Maryborough.

Old age was a period of both loss and gain for these old diggers. While they had often lost partners, workmates and neighbours, and though their bodies were faltering, they had gained new friends and support networks, and new activities and identities. It may well be that my sample of old diggers was especially well-placed to make the most of old age. Living in their own homes and neighbourhoods these men had often retained a positive self-esteem and identity, and could usually rely on familiar local carers. With the material privileges enjoyed by Great War veterans they were perhaps marginally better off than other Australians of their generation and class, and their emotional well-being was often bolstered by the public veneration for old Anzacs.

I am not arguing that the remembering and identity of these Great War veterans can be simply and solely attributed to the common factor of old age; as the memory biographies of Percy Bird, Bill Langham and Fred Farrall show, different life courses made for very different war memories and identities. Rather, I want to stress that the specific material and emotional circumstances of growing old affected the ways in which, in later life, each man remembered his life, and that being an old Anzac was an influential part of that process.

Remembering in later life

Although we compose our memories at every stage of our lives, there are aspects of remembering that are specific to the later years of life, and these were apparent in my interviews with old diggers. The remembering of a couple of my interviewees was confused and disconnected, perhaps due to physical decline or as a result of isolation and neglect. More often, in the particular set of men I interviewed, remembering was vibrant and clear, and it was influenced more by the social experience of old age than by physical or emotional deterioration.

It was clear that for many of these men the rapid technological and social changes of recent decades were difficult to comprehend and to accept. To some extent this lack of understanding was due to limited access to new ways and younger people, but it was also born of genuine fears about displacement, loss and danger. City centre gatherings of teenagers with shocking new clothes, language and manners, or the extended families of ethnic communities that had moved into western suburbs' streets with their different lifestyles and languages; these were regular subjects of conversation over a cup of tea after an interview, reflecting the social isolation and fears of this particular set of white, working-class men. Some men — especially those with friends or family members who helped make change familiar and positive — were open to new ways, but most found them threatening and difficult.

A common response of older adults to the discomforting present is to compare it with a more comfortable and familiar past or, rather, to render the past in ways that emphasise familiar, acceptable and appropriate behaviour. We can see this attitude in the ways in which, for example, Percy Bird portrayed his childhood in idyllic terms, or Bill Langham contrasted his own practical, disciplined youth with that of 'young people today'. A rather different response, less apparent among the men I interviewed, is to retreat into confusion and painful silence because the modern world makes no sense and no connection with a person's past and identity.[5]

Different again, and another significant aspect of remembering in later life, is the attempt to articulate and make sense of the life journey as it nears its conclusion. This process of 'life review' is often especially significant, indeed urgent, in later life, as we are faced with our mortality and try to explain or justify our life. Most of the men I interviewed clearly appreciated the chance to relate and relive their life story, and life review was often a process that had become important in their old age. As Fred Farrall commented about his own renewed enthusiasm for relating the past, 'I suppose as you get older you have some sort of feeling for what happened long ago'. While reflecting upon the effects of age upon memory, Ern Morton concluded that his own memory was 'much more vivid than twenty years ago'. As I met these old men I was often impressed by their efforts — sometimes difficult, troubling efforts — to make sense of the life they had led.[6]

One further motivation for life review is the desire to ensure that a memory of the life, and of the lessons learnt along the way, lives on after death. The life stories I heard were often told in the interview, and in other contexts, so that they would be heard by children, grandchildren and future generations. In some cases — especially when a man felt that significant aspects of his life had been neglected by history — the life story was told to represent a forgotten history and to ensure that it survived after its bearers had died. For example, that attitude was a driving force in the remembering of radical diggers like Ern Morton and Fred Farrall. After discussing the need to remember men and women who had opposed war, Ern Morton rounded off our final interview as follows: 'Yes, oh well it's nice to meet you. I'll have a few moments now I suppose of recollections, but still, if it's made public and does some good, it's worthwhile'.[7]

Life review is also driven by the emotional need to come to terms with unresolved issues and experiences, to compose a past we can live with and a life history with particular emphases and silences. At the end of my interview with Charles Bowden I asked, 'how important an effect do you think your going off to war had upon your life?':

> Oh, I don't know that they had any great … impression on my life at all, to get it out of the routine or anything like that. I've always been a hard worker, and my job always, since I've been married, is to look after my wife and kiddies. That's my main object in life. I've done all I can for them.

Unlike almost every one of my other interviewees, he shrugged off the impact of war, and instead emphasised his successful identity as a man who supported his family. Enlistment against the wishes of his wife had caused Charles Bowden to neglect his family role for the duration. To make up for that failing, and to appease his own sense of guilt, he worked tirelessly after the war and highlighted his identity as a breadwinner in his remembering.[8]

In this process of life review, the personal histories we create are also shaped by the ways in which other people represent our lives. This was true for all the digger interviewees, but was most obvious for successful civic men like Ern Morton, Fred Farrall and Stan D'Altera, whose later years were capped with glowing public tributes. Ern Morton's ninetieth birthday was attended by notable local and State labor politicians, and recorded with a celebratory life history in the Maryborough newspaper. Stan D'Altera received similar newspaper tributes after his car accident, and had in fact been living with a life story set in stone since 1962, when the Yarraville Club honoured its retiring secretary with a plaque that read: 'The Yarraville Club honours in his life-time Stanley V. D'Altera, a gentleman, an Anzac at 15 years of age, a Worthy Australian Always — Secretary of this club for 34 years'.[9]

This focus on life review reminds us that, in old age, remembering is an important part of the process of personal and public affirmation of the worth of a life. In remembering and being remembered we can reaffirm that our lives have been worthwhile, that we are valued for our achievements, and that we are heard and respected as the bearers of family or community history. Indeed, the spirit and enthusiasm for life amongst most of the men I interviewed was strong precisely because these men still felt that their lives, and their life stories, were of interest, not just to this oral historian, but also in their everyday social relationships. And, crucially, this self-respect was reinforced by their local status as ex-servicemen, and by positive social regard for the military contribution of their generation.

At every stage of life our identities (and life histories) are defined and affirmed within particular public relationships. In old age, as I have already noted, we depend upon new social networks which in turn shape our identities and remembering in new ways. For the men I interviewed the social and support networks of extended families, neighbours and local associations were also the particular publics with whom they now related and articulated their life stories, and through which they gained self-esteem for their privileged knowledge of the past. The role and influence of new particular publics is especially obvious in relation to Anzac memories.

In the mid-1980s some Great War veterans were still active in the subculture of ex-servicemen. Many of the western suburbs men I interviewed had known each other through the local RSL clubs (and in social or political organisations), and a few of them still tried to get out to the major club or Anzac Day events. Some of the stories that were told to me had been told many times among these mates, and were thus shaped by the culture of ex-service organisations. Bill Williams was in regular contact with many Gallipoli veterans through his voluntary work with the Gallipoli Legion, and Doug Guthrie was a prime mover in the Minesweepers' Association. Stan D'Altera continued a life of writing by contributing snippets to the 7th Battalion Association journal, *Despatches*, and by winning a prize in a Veterans' Affairs short story competition.

But for most old diggers these service contacts had declined as their own mobility was reduced, and as wartime mates died. Percy Fogarty had been active in both the Pioneers' Unit Association and in the Sailors', Soldiers' and Airmen Fathers' Association of Victoria: 'I finished up State President [of the latter]. As far as I know I'm still State President but where's the others? [...] I think they're all dead, that was in it'. Ted McKenzie decided that Anzac Day 1983 would be his last march because only three members of his battalion had turned up and he 'didn't see it through' anyway. In future years he might watch the parade from the roadside or on television, but he would miss the company and reminiscence of wartime cobbers.[10]

The most familiar faces from the war were now in photos, and I was struck by the number and prominence of such photos and other wartime memorabilia in old veterans' homes. Doug Guthrie brought out a vast collection of newspaper clippings about his two wars and ex-servicemen's affairs; Ern Morton told me that he had tried, unsuccessfully, to have his AIF number reproduced on the licence plates for his new car. On the walls of Percy Fogarty's living room were two almost life-size AIF photographs of Percy and a brother who died in the First World War — enlarged and framed by their mother — and the testimonial they received from Footscray Council.

The physical presence and prominence of wartime memorabilia indicated the continuing emotional significance of memories of war, and of the identities they recalled and reaffirmed. Anzac was not the only identity on display. Albie Linton's living room was full of sporting trophies because — as a prominent sportsman who played football for North Melbourne and Victoria — sporting prowess was an important feature of his identity as an old man. But the prominence of wartime memorabilia, and its resurrection for display and family reminiscence, highlighted the renewed significance of the war in old age.

Throughout history a soldiering youth has been important in most veteran's memories, but in the 1980s an Anzac past attracted new interest and new audiences, and became especially significant for its bearers. Several of my interviewees commented that their grandchildren were becoming more interested in grandfather's Anzac past, and that the interview tape or transcript was received with great interest by family members. Alf Stabb told his one surviving sister that I was coming to interview him about the war:

> 'Oh', she said, 'it's about time'. 'No, it's not about time at all. None of you ever asked me,' I said. 'I wouldn't volunteer it.' We went through it but we didn't want to remember it that much. I said, 'I wouldn't talk war to you'. 'Oh, we wondered why you never spoke.' 'Well', I said, 'you never asked me. If you'd have asked me I might have told you something, but now', I said, 'this interview is going to come out', I said. 'I believe I'll have tape of it and then you can have a borrow of it.' 'Oh, fine', she said.[11]

The resurgence of interest in First World War Anzacs went beyond the family circle. Some veterans, like Percy Bird, were invited to local schools to talk about their experiences of war; others attended special events for First World War men at the local RSL club, the League's Victorian headquarters and even at the Australian War Memorial. Several were interviewed for other oral history projects, and some became Anzac media celebrities in their own right. Bill Williams was cast as one of the 'Children of Federation' in a television documentary that traced the lives of that generation and highlighted the impact of war upon them. Ern Morton was disappointed that the local RSL disapproved of his pacifist politics and would not have him speak at Anzac Day, but he found a national audience in an Anzac Day special for ABC Radio's Social History Unit.

The declining membership of Australia's First World War contingent undoubtedly contributed to this renewed media interest. In turn, the revival of public interest in the Anzacs gave new life and meaning to the wartime reminiscence of old diggers. The remembering of Anzac in later life was shaped in part by the particular personal and social situation of each veteran, and by the need for life review in old age, but it was also influenced by growing public interest in Anzac, to which I now turn.

CHAPTER 8

THE ANZAC REVIVAL (1939–1990)*

The Anzac mystique under fire ... and re-emergent

Australian military experiences in the 1939–45 war boosted the Anzac legend but did not alter its fundamental meanings and significance. In Britain, the RAF pilot had replaced the infantry man as the popular military hero, but in Australia the soldier retained his pre-eminent position and, as Robin Gerster argues, writers 'big-noted' the men of the Second AIF in terms similar to those that had been used to portray their First AIF predecessors.

There were some changes in the Anzac character, or at least in the way in which it was depicted. The Great War and the Depression had tempered king and country patriotism — poor recruiting figures in 1939 confirmed this change — and writers were now less restrained in their representation of digger larrikinism and virility. Australian soldiers and their legend also fared differently in different war zones. The initial victories in north Africa were the stuff of classical legend; the grisly, jungle war in the Pacific required new ways of writing about war and an emphasis on the Anzac qualities of humour and resourcefulness; the experiences of prisoners of war were, until very recently, the most difficult to represent in positive terms.

For the most part, members of the Second AIF perceived themselves to be living up to the legend of their 1914–18 predecessors, and Australian publicists represented them in those terms. Writing in 1943 of the 'War Aims of a Plain Australian', Charles Bean commented that the new generation of Australian soldiers had re-established the 'Anzac spirit of brotherhood and initiative', and offered renewed hope and vision for the postwar nation. Australians at war also won new admirers. George Johnston recalled that

---

*   New edition note: I have not changed this chapter, which was written in the early 1990s. It provides my perspective, at the time, on the Anzac revival, and words such as 'recent', 'modern' and 'today' refer to the 1980s and early 1990s.

between the wars he had been cynical about the Anzac story, but that experiences as a reporter during the Second World War had opened his eyes to 'the remarkable new breed of men' — cynical, carefree and masculine, and with their own codes of loyalty, patriotism and comradeship — who comprised the First and Second AIFs.[1]

Despite the boost provided by the Second World War, which was reflected in improved Anzac Day attendance figures and a growth in RSL membership during the immediate postwar years, subsequent decades saw a decline of interest in the Australian experience of the Great War. The historians who pioneered the study of Australian history in schools and universities after the Second World War focused on more 'suitable' social and political topics. Liberal academics found it distasteful to write about war, and left-wing historians were disconcerted by the politics of the Anzac legend; in 1958, Russell Ward barely explored the obvious similarities between the soldiers' legend and the 'Australian Legend' of the convicts and bushmen. It may also be that the scale and stature of Bean's official history stunted re-evaluation of the Australian military experience; certainly his historical work was neglected in the 1950s and 1960s.[2]

In these decades the RSL was identified by many Australians with social and political conservatism, and Anzac Day gained a reputation in some quarters as a boozy veterans' reunion that had little relevance for other Australians. Dramatists articulated the mood of a new era and generation. Ric Throssell's play, *For Valour,* was based on the life of his father, a Light Horse man who had come home from the war with a Victoria Cross for bravery, but who was driven to suicide by personal and economic failure. First performed in Canberra in 1960, the play contrasted Anzac rhetoric with veterans' confusion and pain, and suggested that the legend was a cause of suffering as well as pride. Alan Seymour's play *The One Day of the Year* sparked controversy when it was published in 1962 and performed at the Adelaide Festival. Although the subtext of Seymour's play reveals great respect and sympathy for the original Anzacs, as represented by the character Alf, the message that spoke most clearly for the times, and which was pounced on by RSL stalwarts, was that of young Hughie, with his contempt for the drunken rituals of Anzac Day and for the glorification of war and soldiers:

> All that old eyewash about national character's a thing of the past. Australians are this, Australians are that, Australians make the greatest soldiers, the best fighters. It's all rubbish.[3]

The chastening experience of the Vietnam War added to the widespread disillusionment with Australia's military past and present. Opponents of Australia's involvement in that war scorned attempts to create a 'New Anzac Legend' which praised the men of the Royal Australian Regiment as proud bearers of their forefathers' military traditions, and which justified 'fighting the Vietcong in defence of the [Vietnamese] "people"'. Recent novelists and historians have sought to restate these positive themes, and to reaffirm national respect for our Vietnam veterans; writing in 1986, Lex McAuley characterised the Battle of Long Tan as 'the legend of Anzac upheld'. But at the time of the Vietnam War and moratorium marches against conscription, many Australians were sceptical of the nation's military tradition, and of its relevance in a changing world.[4]

In the 1960s and early 1970s, defenders of the Anzac faith wrote with great concern about its critics, and about the apparent decline of interest in the first Anzacs. In 1963, Peter Coleman mused in the *Bulletin* about 'the tendency among some Australian historians to play down the place of Anzac Day in Australian history', and noted RSL fears that Anzac Day was losing its popularity. Two years later, George Johnston wrote an article for *Walkabout* which assessed the state of the Anzac legend on its fiftieth anniversary, at a time when 'it seems to have come to a point where it could be debased or twisted or even lost altogether in ambiguities of social misunderstanding' as exemplified by Alan Seymour's play. Johnston argued that Gallipoli still offered valuable lessons about the universal truths of the human spirit, and about the 'legendary and undoubted' qualities of Australian soldiers. Writing for *Advance Australia* in 1973, Jack Woodward recalled previous decades of misplaced criticism: 'Anzac Day was accused of jingoism, cant and the glorification of war — whatever that means [...] an image of bawdy, boozy, authoritarian camaraderie was seen by some to pass unworthily for patriotism, and the Anzac mystique came under fire'. For Woodward, Anzac was still 'a credible legend of stoicism and fraternalism', and patriotism, 'bound up with national honour and freedom', and was far preferable to the 'non-patriotism of today':

> It has an alien ring to it and an unruly look about it. It seems to be promoted by ideological vagrants, spiritually footloose and with no visible moral means of support[5]

And yet, within a decade, the conservative columnist Gerard Henderson was claiming that Anzac Day 1982 took place 'at a time of unprecedented revival of interest in the Australian involvement in the Great War and, more particularly, in the Dardanelles campaign'.[6] This revival was not the work of traditional Anzac guardians in the RSL or among conservative patriots, but was led by historians and film-makers responding to a burgeoning popular interest in Australian history and national identity.

Historians can be credited with some responsibility for the reawakening of interest in Anzac. In 1965, Australians celebrated the fiftieth anniversary of the landing at Gallipoli and Ken Inglis wrote an article about 'The Anzac Tradition', in which he criticised historians' neglect of the tradition and its founder, Charles Bean, and speculated about the nature of Anzac and its lessons for Australia. The article was rejected by the academically prestigious *Historical Studies*, but was published in *Meanjin Quarterly*, which then provided a forum for several prominent Australian historians to debate issues about the diggers and their legend. Inglis embarked upon a major study of the origins and history of Australian heroes and national identity, which resulted in many innovative publications about the significance of the Great War in Australian political culture and popular memory.[7]

Other historians begin to research and write about the war. In Melbourne, Lloyd Robson explored the origins and character of the First AIF, and introduced a generation of school and university students — including me — to debates about the Anzac experience. In 1974, the impressive results of Bill Gammage's doctoral research using soldiers' letters and diaries was published as *The Broken Years: Australian Soldiers in the Great War*, and in 1978 Patsy Adam-Smith combined oral testimony with soldiers' writings in her best-selling account, *The Anzacs*. In their different ways, members of this generation of historians were influenced by the new social history, with its emphasis on ordinary people's historical experiences. In British military history this new approach generated important new works, such as Martin Middlebrook's *The First Day on the Somme*, about the experiences of soldiers in the ranks. In Australia, it led to the recovery of Charles Bean's historical tradition of writing about war at the cutting edge. The new wave of Australian war historians acknowledged their debt to Bean and paved the way for a renaissance of Bean's historical work, which to date has included publication of his Gallipoli diary, a major biography, a new imprint of the official history, and numerous newspaper and journal articles. Recognising a growing demand for writings about war, Australian publishers developed extensive military history lists in the 1980s, which ranged from academic studies to a Time-Life Books' series telling 'the whole bloody history of Australians at war', and offering a 'specially commissioned cassette of catchy songs from the years of Australians at war'.[8]

The Australian War Memorial played a significant role in this resurgence of military history and publishing. In 1980, new legislation provided official sanction for the War Memorial's growing commitment to fostering military history. Its huge archive of primary source material about Australians at war was made more accessible to researchers, and a Research and Publications Section was established to coordinate a research bursary scheme and to organise a journal and national conferences dedicated to the history of Australian experiences of war. In the 1980s, it was probably easier to get funding to research and write military history than to be supported for work in most other fields of Australian studies. Many young scholars, intrigued by the issues raised by study of Australians at war, and suffering from the squeeze on university funding, took advantage of such opportunities.[9]

Underlying popular and academic interest in the history of the Great War was the blossoming of an Australian nationalism with particular interest in the national past. The enthusiastic independence of the 1972–75 Whitlam Labor government was one spur for the new nationalism, but it was subsequently fuelled by the economic and political insecurity of the mid-1970s, and promoted by both Coalition and Labor governments in the 1980s. Cultural nationalism directed intellectual and financial stimulus into an Australian film industry which, in turn, produced influential representations of national identity and Australian history.

In 1981, Peter Weir's film *Gallipoli* sought to breathe life and meaning back into the Anzac story. Screenwriter David Williamson claimed to be motivated by the quest for national identity: 'as for myth, a country, for its own psychological well-being needs to generate its own myths, otherwise it doesn't feel whole'. At the film, Phillip Adams 'watched the faces of the young audience and saw them yearning for a world in which they could believe in the transcendental power of an abstract idea [...] where once more people could have the sort of faith in values, institutions and themselves that would give them such a rush of conviction and courage'. British audiences were less convinced about such faith, as Jill Tweedie wrote from London a few months before the Falklands War, and at a time when European citizens were becoming more vocal about the threat of nuclear war:

> We, here, can no longer afford to admire war games, however heroic they may once have been. We, here, can no longer afford to confuse ignorance and a boyish wish for adventure (clearly depicted in the film) with nobility, patriotism or courage.

Australian letter-writers attacked Tweedie for 'degrading the memory of their compatriots in the first world war', and in Australia *Gallipoli* was a huge popular success. On the one hand it matched contemporary attitudes about the tragedy of war, and suited the Australian desire to blame the British for the loss of Anzac lives (in the film, the man who gave the command for the Light Horse men to continue their suicidal attack at The Nek is portrayed as a British officer, even though he was an Australian). At the same time, the film also evoked for Australian audiences strong feelings of sympathy for the Anzacs, pleasure in their Australian characteristics, and pride in their courage and their achievements.[10]

The success of *Gallipoli* in 1981 was echoed by a general interest 'in all things military and Australia's war history'. The War Memorial recorded that in September 1981 it had had 83 570 visitors, fifty per cent more than in the same month the previous year, and that the Memorial was now Canberra's most popular tourist attraction. RSL membership was surging in all States and, after a recruiting drive lasting twelve months, the Army Reserve boosted its numbers by about 8000 to just under 30 000 members. At schools and universities more students were taking courses that examined Australian participation in wars, particularly the Great War, and publishers were doing particularly well with war history. A Penguin spokesman reported that sales of Bill Gammage's *The Broken Years* had increased in 1981, and that 17 000 copies had been sold since 1975. *The Anzacs* had sold 30 000 copies in hardback since its publication in 1978, and in the first two months of the paperback release in August 1981, 15 000 copies were delivered to bookshops which were already placing orders for additional stock.[11]

Anzac Day also benefited from this extraordinary revival. With the added poignancy of the decreasing numbers of 'original' First World War Anzacs, April 25 became the subject of intense media interest, and Anzac Day crowds and marches became larger every year. Key anniversaries were important catalysts for media attention and public participation. In 1965, a return to Anzac Cove by a party of veterans reignited interest in their fifty year-old story; and in 1990 the visit of a second party to Gallipoli for a seventy-fifth anniversary ceremony generated extraordinary national interest. This fascination with Australia's military history has made a significant contribution to the obsession with the national past in Australian political culture, and to recent debates about national identity and purpose.

Anzac histories and Australian popular memory

Within this resurgence of interest in the Anzac legend there has been both continuity and change in the meanings attributed to the Australian experience of the Great War. Charles Bean's official history has played a pivotal role in the revival of interest among historians in Australian military participation in the Great War, as a source of evidence, as a historical model, and as the focus of debate. Almost invariably, modern Australian historians of the Great War acknowledge their debt to Bean, and very often they draw upon his work for anecdotes and for explanations of the Anzac experience. John Vader's 1972 account of *Anzac* is scattered with stories lifted straight out of 'Dr Bean, the historian who wrote the Diggers' story so accurately'. In Jane Ross's 1985 analysis of *The Myth of the Digger*, over a third of the references to the First World War are from Bean's writings.[12]

There are, however, a number of differences in emphasis and characterisation between Bean's history and the mainstream of recent, popular Anzac histories represented by the work of Bill Gammage and Patsy Adam-Smith. In the new histories there is a greater emphasis on the personal tragedy and horror of the war, and a suggestion that not only was the war a waste of young lives, but also that they were wasted without good cause or reason. Adam-Smith describes the Great War as 'the greatest tragedy the world has known', and implies that Australians died 'for no cause at all'. Time-Life's 'Australians at War' series is promoted by the slogan, 'Our Men ... Other People's Wars', and the overall impression conveyed by these histories is that Australians were fighting a British or European war, and that they were sacrificed by British politicians and generals. Whereas Bean was at pains to emphasise that Australian soldiers sustained their imperial loyalty throughout the war, most modern historians, whose Australian identity does not generally include an imperial sentiment, highlight the wartime decline in the soldiers' bonding with 'the mother country' and empire.[13]

This interpretation of Australian participation in the Great War has been contested. During the 1980s and 1990s, a number of conservative commentators have challenged what Gerard Henderson describes as an attempt by the left (he includes Gammage, Adam-Smith and Peter Weir) 'to win back lost ground' and 'redirect the Anzac Legend'. Henderson outlines the 'newly emerging mythology':

The current Anzac revival [...] attempts to distinguish between the Anzacs as individuals and the cause for which they fought. The former are to be glorified but the latter is condemned since it led to the death or disfigurement of so many Australian sons.

Henderson makes a case against such 'pacifist platitudes', arguing that, in 1914, Germany was the aggressor, that Australian interests were threatened by that aggression, that there was no viable alternative to defeating Germany in the field, and that this was the understanding and motivation of most Australian soldiers. Other historians have added academic weight to this response; in 1991 in *Anzac and Empire: The Tragedy and Glory of Gallipoli*, John Robertson argued that 'these men sacrificed much for what they rightly considered a noble cause'. Yet the direction of most Anzac histories, fiction and film suggests that the 'futile waste' and 'other peoples' wars' theses have been incorporated into Australian popular memory of the Great War.[14]

In modern Anzac histories there are also changes in the characterisation of the digger, who is generally less chaste and virtuous than Bean's ideal Anzac. Patsy Adam-Smith apologises for any distress she might cause to Anzac survivors or their families, but argues that it is necessary to include, for example, details of the extent of venereal disease in the First AIF. Bill Gammage uses the testimony of diaries and letters to evoke the horror of modern battle, and to show that in such circumstances Australians could be both the victims and perpetrators of brutality. Yet in the final analysis these histories come to the same conclusion as Bean, that through their endurance of terrible circumstances the Anzacs were truly heroic. Adam-Smith cautions that although 'War is hell', in our attempt to denigrate and outlaw it, 'we must remember not to castigate the victims of war — and every man who fights is a victim'. She argues that details about digger boozing, womanising and disrespect for authority do not in any way 'lessen the immortality that the men's endurance made legendary'.[15]

Like Bean, today's historians emphasise that it was the positive national characteristics of the Anzacs which made them such stoic and effective soldiers. This is Bill Gammage's summary of the Australian achievement at the Gallipoli landing:

> They were not experienced soldiers, they were too precipitate and they made too many errors to be that. They were ardent, eager, brave men, naive about military strategy, but proud of their heritage and confident of their supremacy. Despite their mistakes, they did what few could have done.

In other chapters Gammage notes that 'mateship was a particularly Australian virtue'; that trench raids suited the Australian temperament, 'for its chief weapons were stealth, individual initiative, patience and skilled bushcraft'; and that 'Australian success in battle was largely attributable to that same unrelenting independence which so regularly offended law and authority'. This is vintage Bean, both in its use of national character to explain Australian war experiences and achievements, and in the particular nature of the Anzac characterisation.[16]

Not all recent histories have used the lens of national character and achievement to explain the Anzac experience. Military historians, like Jeffrey Grey, have focused instead on the influence of factors such as command, training, fire-power, logistics and supply. Other historians have confronted the nationalist Anzac history head-on. The work of Carmel Shute, Robin Gerster, Richard White, Marilyn Lake, Lloyd Robson and David Kent, among others, has highlighted the neglected or 'dark' side of the Anzac experience, and has explored the processes by which a selective ideal or legend was installed, and aberrant experiences were marginalised or excluded. Tony Gough summarises this approach in his conclusion that the Anzacs were 'not so bronzed, not so democratic, not so courageous, not so physically superior nor so well behaved as has been popularly imagined'.[17]

Yet the limited impact of these critical reinterpretations of the Anzac legend, and the predominance of nationalist accounts, is reflected in the representation of Anzac histories in recent Australian films. Films are the pre-eminent myth-makers of our time, and in the 1980s a string of films and television mini-series about Australian soldiers in the Great War — including *Gallipoli*, '1915', 'Anzacs' and *The Lighthorsemen* — have had a major influence upon Australian popular memory of Anzac. In particular, *Gallipoli* and 'Anzacs' were major commercial successes, and provoked extensive media attention.

The makers of these films acknowledge their debt to Charles Bean's official history. John Dixon, one of the producers of 'Anzacs', recalls that his interest in the diggers was 'ignited' by Bean's history, and he encouraged his co-producer to read Bean's 'masterpiece of fact and observation'. Although Bean was the starting point, the history underlying these films is the modern reworking of Bean's themes as represented by Patsy Adam-Smith and Bill Gammage. Gammage served as the historical adviser on *Gallipoli*, and Adam-Smith was brought in as a story consultant for 'Anzacs' because of her 'fine work' using 'the reminiscences and insights of old soldiers', and because her 'feminine perspective' would curb any tendency for the series to become a 'macho romp'.[18]

These films convey the main themes of the new Anzac legend. There is no homage to the imperial alliance, and every Anzac film blames arrogant British generals for Australian losses, and exonerates Australian leaders. Australia's egalitarian soldiers and society are contrasted with their class-ridden British counterparts, and in Australian war films the grazier's son and the working-class lad invariably find their national brotherhood in the trenches. The films celebrate digger larrikinism and disrespect for military regulation, and direct it at British officers (or Australian officers with British accents). Paul Hogan of *Crocodile Dundee* fame plays the shamelessly larrikin Pat Cleary in 'Anzacs' because, according to John Dixon, 'men like Hoges are scattered through Bean's volumes and digger diaries', and Hogan's public image embodied the qualities 'that made Australian soldiers different and successful'. The films show trench warfare to be miserable, bloody and terrifying, yet the Australians endure, and emerge with pride in their manhood, their military achievements and their new national identity.[19]

There are elements of these films that gesture towards more complex and even critical tellings of the Australian experience of war. In 'Anzacs', for example, some Australian soldiers are deserters and others suffer from cowardice and nervous collapse. Civilian voices condemn participation in the war; servicemen express bitterness and disillusionment; Australian military nurses have more than walk-on parts; and we see something of the impact of the war upon women at home.

However, as critic James Wieland argues, 'these threads are not picked up', and the films give in to the demands of fiction and romance, and of the legend of Australian character triumphant. As in Bean's history, contradictory experiences are not ignored; rather, they are worked into the films so that the legend is reaffirmed. For example, of the two deserters in 'Anzacs', 'Pudden' is represented as a dim-witted and confused soldier who is brought back to his senses by his mates (in the book and original telescript he redeems himself through heroic death). The other deserter, caricatured as 'Dingo', is an evil, selfish man who cares nothing for his mates and only wants to stir up trouble. Just like one of Bean's 'undesirables', he is a trouble-maker, and his murder by one of the 'genuine diggers' is thus exonerated by the film.[20]

By framing the Australian experience of the Great War in terms of national identity and achievement, and by using cinematic strategies of characterisation and simplification which draw the audience's sympathies towards particular characters and highlight particular meanings about the war, films and series like *Gallipoli* and 'Anzacs' offer a narrow but appealing representation of Anzac.

Anzac Day, 1987

The popular resonance of this revitalised Anzac legend is reflected in impressive film and television viewing figures, but also in the growth of public participation in Anzac Day commemoration. When I returned to Australia in 1987 to conduct my second set of interviews, I went along to the Anzac Day ceremonies in Melbourne in order to assess continuity and change in the ways in which the Australian experience of war is represented, and received, on April 25.[21]

At the start of the day, a mixture of veterans and young onlookers waited in the grounds around the Shrine of Remembrance for the Dawn Service and the moving bugle notes of the Last Post, and then filed through the Shrine and received paper poppies from the attendants. Although the ceremony emphasised the sacred nature of commemoration and suggested certain ways of remembering the war and the war dead, the Dawn Service was a relatively unstructured occasion for personal reflection and reminiscence.

At the parade later in the morning, the First World War veterans numbered fifty-eight, thirteen less than in 1986; the number of Second World War veterans was also reduced, and the strongest contingents of the 15 000 marchers were from the post-1945 wars. The streets were thickly lined with onlookers, waving flags and clapping as each unit marched past. Municipal and military bands marched between the service units, playing an assortment of tunes which echoed and overlapped in ways that reminded me of a carnival or festival.

One striking feature of the parade was the diversity of groups included in Anzac Day. Young people were particularly prominent in the crowd, but they also marched alongside service relatives, sometimes wearing the medals of veteran relatives who could no longer attend. The newspapers emphasised this aspect of the parade, one commenting that 'World War 1 veteran Charlie Stevens and Timothy Doughty, 8, symbolise the two faces of the Anzac Day march'. At the official ceremony the Governor, Davis McCaughey, highlighted the presence of young and old at the parade, and one of the themes of his speech was that Anzac Day is an occasion to recollect past experiences and to renew commitment to visions of the future: 'The old dream their dreams. The young see their visions. Each needs the other'.

One Second World War veteran told a *Sun* journalist, 'The warmongering accusations and attitudes of 15 to 20 years ago have gone. Youngsters seem to respect us a lot more now'.

Other groups were marching for the first time. Members of the Women's Land Army, previously excluded by the RSL because they were not part of the defence force, were clearly delighted to join the parade, and newspapers featured their forgotten wartime experiences. Eighty veterans of the South Vietnamese army — 'boat people' refugees whose war 'has not ended' — were also new faces in the 1987 parade, marching between members of Australia's ethnic communities who served in allied forces: Greeks in national costume waving Greek and Australian flags, Poles, Serbs (only the Chetniks, as the pro-Tito partisans would not march alongside them) and Italians. Ethnic community involvement in Anzac Day — albeit only from a selective sample of those communities — seems to counter the concern expressed in the 1960s and 1970s that the Anzac tradition would be irrelevant to postwar migrants. It shows how Anzac Day has come to embrace and espouse a broader definition of Australian racial and national identity.

Perhaps the loudest cheers of the day were for Australia's Vietnam veterans. This Anzac Day was perceived by many commentators to be a precursor to the 'Welcome Home' events that were held throughout Australia in October 1987, and that both symbolised and reinforced the new public mood of recognition and reconciliation towards the Vietnam veterans. In Sydney, Vietnam veterans were given pride of place at the head of the main body of Anzac Day marchers, and newspapers like the *Canberra Times* headlined the theme of 'Vietnam veterans "home"': 'in emotional scenes after the parade, Sydney's Vietnam veterans said they felt they were finally being accepted by the Australia community'. Certainly more veterans from that war were marching than in previous years. Melbourne Herald journalist, Bernard Clancy, recalled that after returning from service in Vietnam he rejected his father's request to march on Anzac Day 1969.

> He didn't understand. I couldn't explain. It was like that, the Vietnam war [...] you began to try to teach yourself to forget. The bad dream is over.

After 'eighteen years in hiding', he realised that people were interested in the Vietnam war, and he decided that he could face his memories in public and join the parade. Hearing the clapping from the crowd, 'that's when you stop marching and start floating':

That's when your throat constricts and your chest swells. That's when you realise there *are* people who care. There are people saying 'thanks, mate'. 'Good on you boys!' There it is again ... and again ... and again. And you notice something else. All the way along Swanston St there was chiaking in the ranks. Now there is absolute silence. 'Eyes right!' for the red hats — and suddenly it's all over. *Gee it's good to be back home.* It *wasn't* like that, in Vietnam.

Not all Australians felt included on Anzac Day 1987. For the two previous Anzac Days, members of the National Aboriginal and Islander Ex-Service Association had staged their own march in the suburb of Thornbury, and in 1987 they erected a cross for members of their Victorian communities who had died fighting for the Australian forces in twentieth-century wars. Representatives of the Association were angered by the attitudes of the RSL's Victorian President, Bruce Ruxton, and by the League's refusal to allow Association members to march as an Aboriginal and Islander contingent in the official parade. RSL officials responded that the march would be destroyed if 'thousands of splinter groups' marched under their own banners rather than under the banners of their military units.

Ruxton's 1987 attacks on the visiting South African Archbishop, Desmond Tutu, and on Asian immigrants (four years previously the State RSL conference had passed a motion in favour of a greater percentage of Anglo-Saxon and European migrants), caused anguish in other circles, and demonstrated that Anzac commemoration is still a political, and politicised, occasion. In an attempt to disassociate his church from what he saw as 'Mr Ruxton's statements of [...] discrimination', a vicar in the Melbourne suburb of Prahran refused to hold an Anzac commemoration service and received the backing of the Anglican Church. The President of the Prahran branch of the RSL responded that 'you have denied us the opportunity to pray for and honour our dead [...] I have always been under the impression that it was a Christian concept to pray for the dead and I do not recall anyone sponsored by Moscow, your Bishop Tutu included, doing so'.

Throughout the 1980s, feminist groups organised protests on Anzac Day, and in 1987 an Anti-Anzac Day Collective — an umbrella organisation including Women Against Rape and Women for Survival — staged a protest march from the United States Consulate and along St Kilda Road to the RSL's city headquarters. The motivations of feminist participants, as expressed in their broadsheets and by spokeswomen, ranged from protest against male violence and rape through to more specific critiques of Anzac

Day and its reinforcement of 'militarism, male glorification of war and institutionalised mourning':

> [...] no other day of the year embodies a celebration of manhood, military thinking and all things associated with it that continue to oppress us.

The women protestors received short shrift from the official marchers and the mainstream press. The *Sun* recorded that 'Anti-Anzacs face cold war', and quoted one Second World War digger's remark that 'I reckon they should line them up and shoot them. It spoils your bloody day [...] To think they're trying to get rid of Anzac Day'. As the women protestors marched in one direction chanting 'One, two, three, four, Anzac glorifies the war', some of the younger ex-service marchers responded, 'Kick all dykes to the floor'. Yet by marching away from the Shrine the feminists avoided the violent conflict that had marked Anzac Day protests in previous years; they also rejected another tactic from earlier years, of seeking to join the official march and lay a wreath for women victims of war. Press coverage in 1987 was, correspondingly, much reduced in scale and indignation.

Ken Inglis has shrewdly noted that the radical feminists of the 1980s had a tactical dilemma 'similar to that of the communists half a century ago: to attack a popular tradition head on might alienate sympathy; to seek incorporation in it might make the radical critiques invisible'. Although the position of women within Anzac Day has clearly shifted in recent years — witness the inclusion of a greater range and number of women participants — Anzac Day is still, pre-eminently, an ex-service blokes' day. Feminist protests highlight this fact, and the ways in which the Anzac legend sustains a particular, gendered construction of the Australian experience of war, and of Australian national identity.[22]

On the Monday after Anzac Day 1987, John Lahey wrote in the *Age* that Anzac Day 'is one of the most spectacular things we do': 'the public own this ceremony, in a way that it does not own Moomba [Melbourne's annual peoples' festival], which is structured for it anew each year, or even the Melbourne Cup, which has its areas of social privilege'. Anzac Day is certainly a popular occasion, and whether they are in the march or on the roadsides, Melbourne people do play an active, participatory role, with thoughtful respect but also with humour, fun and occasional unruliness.

Yet Anzac Day is structured and institutionalised in particular ways, and it does emphasise certain attitudes and understandings about Australians and war. The organisers control who is and who isn't allowed to march, and they ensure that the primary allegiance of participants is to military units rather than to other identities of gender, race or even sexuality (gay ex-servicemen have been refused permission to march as a separate contingent). Anzac Day may be a popular pageant, but it is also a martial affair with military music and ritual that uncritically endorses the role of the military services in Australian history and society. The two-up games which are played by ex-servicemen throughout the day, and which for the press symbolise the unofficial and larrikin element of the event and of Australian military manhood, come across as an institutionalised ritual that offers little real threat to the order of the day.

The rhetoric of Anzac Day also shows that while the event is open to reinterpretation, it reasserts certain inalienable values and understandings. As a State Governor with liberal sentiments, Davis McCaughey was able to make an official address in 1987 that explored issues not usually associated with Anzac Day. In the International Year of the Homeless, he claimed that the men who died on the first Anzac Day would have been horrified to know that 100 million people, including Australians, were living without shelter in 1987. Fatalism was 'not the Anzac spirit', and he urged us to renew our commitment to the service of others and to banish selfishness and greed. McCaughey's speech shows how the idea of 'the Anzac spirit' can be like an empty box, and that it is possible to fill that box with new and even radical meanings. In this context it is significant that Bruce Ruxton, Victoria's most vocal Anzac traditionalist, was on a pilgrimage to Gallipoli during April 1987, and thus missed Melbourne's Anzac Day and his usual opportunity to fill 'the Anzac spirit' box with his more conventional ideas about racial purity, the importance of the monarchy and Australian defence preparedness.

Yet even without Bruce Ruxton, much of the rhetoric of Anzac Day 1987 reaffirmed traditional Anzac meanings. There were hymns about 'knightly virtue proved', and Sir Eric Pierce's lyrical oration of the Anzac Requiem explained that Australians fought to 'defend the free world and Commonwealth against a common enemy', and died so that 'the lights of freedom and humanity might continue to shine'. After televising the march and service, Channel Nine broadcast an Anzac Day special about 'Our Magnificent Defeat' at Gallipoli which 'brought Australians together as a nation for the first time'. Features in the press and electronic media highlighted the distinctive qualities of Australian soldiers — 'resolute, brave and resourceful fighters' — who forged a legend.

A very few news features — such as Michael Cathcart's article for the *Age* about 'The Dark Side of the Diggers' who were involved in right wing secret armies between the wars — suggested dissonant views. But for the most part, Anzac Day 1987 evoked the same messages as contemporary Anzac histories and films: that Australians are good fighters and good mates; that the Australian armed forces have been successful military units; that through their sacrifices and military achievements Australian service men and women have established a nation in spirit; and that Australians today can and should learn lessons from the tradition of Australians at war. Modern Anzac Days do, however, differ from recent histories and film in one respect. In order to explain and justify the sacrifices of service men and women, the rituals and rhetoric of Anzac Day affirm that Australian participation in war has always been justified, and imply that Australians have always been unanimous in their support for such participation.

The voices and views of old diggers still contribute to the Anzac Day representation of Australians at war. In 1987 most news coverage by the print and electronic media included interviews with veterans. The theme that sounded most clearly from their published testimony was mateship. For veterans, the primary purpose and value of Anzac Day is the opportunity it provides to remember wartime mates who died during or after the war, and to meet up with other survivors and recall common experiences. As one Second World War veteran remarked: 'Anzac Day is a day you remember your mates who are not there. It's not about glory, it's about mateship. You can let go everything you've bottled up over the year'.

Sometimes the interview excerpts which are cited in media coverage of Anzac Day are used to distil key themes about the Australian experience of war, and to show how memories of war are interwoven with popular memories. Bill Owens, one hundred years of age, a shopkeeper before 1914 and a 58th Battalion veteran, recalled in 1987 that 'the war made me a man'. Cyril Feathers, an intelligence officer with the 23rd Battalion on the Western Front, and only a month away from his own one-hundredth birthday, explained that 'General Monash showed the British how to fight a battle'. Veterans highlight themes in their remembering that have been reaffirmed by Australian popular memory of the war. The media uses such quotes because they are clear, effective and acceptable 'sound bites' about the war, and because this testimony can be represented as the authentic voice of Anzac.

Yet a striking feature of Anzac Day media interviews with veterans is the fact that so many of the men who are asked about the war find it difficult to remember or relate their experiences, and can only do so in particular contexts and using specific narrative forms. One man, an officer in the British army who won many medals during the Normandy landings, was unable or unwilling to tell his war stories on Anzac Day 1987, despite the urgings of his family. He deleted the details of his war experience from a statement provided by his family, and would 'only say that there was a tragic story behind each medal'. I hate war. I lost too many friends.'

Similarly, Australian Second World War veteran, Jack Nolan, had 'no desire to talk about the enormity of what he saw, beyond jocular exchanges with old soldier mates. Not even his family is privy to what still gives him nightmares'. For Jack Nolan the legacies of war are 'nerves', an arthritic condition and friendships. Each year he camps overnight in his car near the Shrine of Remembrance so that he is ready to meet his mates for the Dawn Service: 'it is moving. I can't say much; I think a lot'. The only way in which he can articulate his experiences of war is through the 'teasing jocularity' shared with wartime mates.

Other ex-servicemen interviewed on Anzac Day are explicit about the process of using public narratives and images to help remembering. Gallipoli veteran Roy Grant explained: 'These days I need help to remember, photographs, books and medals. They help […] They make the memories more vivid'. As he picked up various books and memorabilia, they prompted anecdotes about mates who 'bought it', about battles — 'so many of them' — and about 'lovely days' on leave in Paris:

> Some of the memories haven't been very good to me though. Not sad so much because they were brave mates who were killed and I helped to bury them.

> There are times I try not to think about today. But I got my medals out yesterday and gave 'em a bit of a polish. Always do. I'll be there all right … good suit, shoes, neat, medals tucked into the top pocket. As it should be. Been going since 1919. Don't know if I'll be around for the next one though […] Saw a young naval officer there last year. Went up to him. Saluted, cutting away sharply like we were taught to. Hope he's there this year.

These interview excerpts suggest that Anzac Day is still a special and affirming occasion for old diggers. They show how public remembrance prompts individual remembering which can be both difficult and pleasurable, and that representations of Anzac in ceremony, history or film support the recollection of certain memories while silencing others. Thus the remembering of old diggers is shaped in relation to the particular forms and meanings of the public legend, and according to the particular publics of remembering. These are the themes of the final memory biography chapter, which explores the experiences and remembering of Percy Bird, Bill Langham and Fred Farrall as old Anzacs in the 1980s.

CHAPTER 9

LIVING WITH THE LEGEND

Percy Bird

After completing his career as an Audit Inspector with the Victorian railways, Percy Bird retired to his comfortable weather board house on the Esplanade at Williamstown. He cared for his wife when she suffered a long-term illness in her seventies, and lived by himself after her death. Not long before his one-hundredth birthday, he moved to a Vasey Home for aged ex-servicemen and war widows in the eastern suburb of Sandringham, where he enjoyed the company of other residents and visits from family members until his death in 1990.

In old age Percy continued to take an active interest in all things Anzac. The new Anzac films and books of the 1970s and 1980s reaffirmed many of the meanings and identities of Percy's war stories. For example, in 1975 the AIF Victoria Cross winner W. D. Joynt, whom Percy had known as a member of the same brigade, wrote a book about the crucial role of the Australians in the arrest of the British retreat of 1918. In *Saving the Channel Ports*, Joynt relied heavily on Bean and Monash for his depictions of the independent, effective Australian soldier and the classless AIF. Percy was very impressed by Joynt's account, which matched and confirmed his own understanding of the war, and a number of the stories that he related in the interview as evidence of the quality of the Australians were taken directly from this book.[1]

Percy was, at first, equally impressed by the television series, 'Anzacs', and he began our second interview by telling me with great excitement of a recent viewing. The episode that excited Percy portrayed a brigade concert at which he had been booked to sing until prevented by illness:

> And I said [to the residents of the Vasey Home] little did they think
> that an old man just on ninety, nearly ninety-eight years of age, was

booked to sing at the concert [laughs]. If they'd a known that they probably would have got me to sing and it would have been on the film.[2]

For Percy the pleasures of 'Anzacs' were the recognition — at least in his mind and his telling — of his wartime role as a performer, and the affirmation that his identity as a performer was still valued in his old age.

However 1980s' representations of Australians in the Great War also contained elements that threatened the ways in which Percy had composed his war memory in a fixed repertoire of anecdotes. 'Anzacs' touched off disturbing memories of Percy's war, and he decided not to watch all of the series because he 'wasn't very much impressed with it'. He regarded it as 'the Hogan film', and didn't like the way it used the Paul Hogan character to show the diggers having a wonderful time in the *estaminets*: 'Most of the time we were looking around to see if we could get something to eat'. 'Anzacs' reminded Percy of the larrikin aspect of digger life about which he had felt uneasy during the war, and which he had sought to exclude from his own remembering and from his Anzac identity.

Percy also criticised 'Anzacs' for not accurately depicting life in the trenches and 'what we had to put up with, in lots of cases'. Yet, paradoxically, the main reason for Percy's decision not to watch the whole series was that its representation of wartime death and trauma recalled experiences that were not part of his composed memory:

> I turned it off because I ... was, it brought back too many sad memories to me [...] about all my pals getting killed [...] I like to forget all those things nowadays.[3]

Percy's response to the television series reveals a tension between a desire for recognition of his experiences and the need to maintain composure in his remembering. It shows how, even in old age, Percy had to negotiate between shifting public versions of the war and his own memories and identities, and that even though he tried to filter out uncomfortable reminders, they could still be painful and troubling.

Another example of the ways in which changes in the public account of the war require renegotiation of individual memory and identity, is Percy's explanation of enlistment. The awkwardness of that section of testimony reveals Percy's ambivalence about volunteering in 1915, but it also hints that in old age he was still uncertain about whether, in the long run, the war and his decision to enlist had been worthwhile, and suggests that he was affected by recent questioning of Australian involvement. In the second interview Percy came close to articulating this unease, in a story about a friend who visited Germany in the 1970s and was asked why the Australian soldiers had come from so far away to fight in Europe. I asked Percy for his response to that story (he had not offered one): 'Oh I just laughed and I said, "Now". I didn't say any more. But I was impressed with what he said'. Percy quickly shifted back to one of his standard stories, not wanting to fully consider the disturbing possibility that he had been fighting someone else's war.[4]

Although this process of negotiation with public accounts had been part of Percy's war remembering throughout his life, old age was a very different social and emotional context for remembering. By 1983, most of Percy's 5th Battalion mates had died, and few of them were well enough to attend reunions or to march on Anzac Day. Percy now watched the parade on television:

> Oh well, it gives me a thrill [...] Well it just goes through me, goes through me body and I can feel that there's the thrill there, thinking of my old pals. What we did and what we put up with.[5]

Anzac Day was clearly still a resonant and affirming occasion for Percy Bird, but with the passing of battalion veterans and their shared remembering of the war, a more general identity of the Anzac elite — frail and few but increasingly revered — was being provided by the media. This new, general identity worked well for Percy because it did not probe the tensions within the AIF or in his own experience as a soldier. It was also tremendously rewarding for Percy, in his old age, to be valued by 'the nation' in this way.

Percy also received affirmation of his Anzac identity from new audiences for the performance of his memories. He had always enjoyed performance, but as an old widower, with few other positive features in his life, it had become especially important for his emotional well-being. He loved to sing and perform and was able to win the acclaim of a variety of audiences — his family, school children, elderly residents at the home ('they think the world of me'), RSL club men and historians — and thus feel good about himself. Furthermore, because there was little about his old age that Percy felt worthy of performance, the past was the mainstay of his shows. More specifically, he focused on his childhood and wartime pasts. Stories of those periods provided the most positive images of youth, and of masculine and national importance. They were also the most well-received stories; few audiences wanted to hear about the infighting of the railways' Auditing Department, even though it was significant for Percy at the time.[6]

Percy's performance of his memories in old age demonstrates a number of general points about remembering. It highlights the emotional value of remembering, which helps people feel good about themselves because they are listened to, and because they are able to represent themselves in positive ways. It also shows how the particular public for the performance is influential, on the one hand applauding and validating certain memories and identities, and in turn causing the performer to shape remembering in response to that validation. Percy told the old ladies in the Vasey Home about his childhood in Williamstown because they didn't want to hear about the war, and he related his selective war stories about humorous events behind the lines to school children and to younger members of his family who were becoming interested in their Anzac grandfather. The new publics of old age continued the process of shaping Percy's remembering, by causing him to emphasise certain aspects of his past and to 'forget' other experiences and meanings. They reaffirmed his safe Anzac identity as a singer, performer and raconteur.

Percy had also developed a particular way of remembering for historians, of which I was not the first to come his way. He liked talking to oral historians because he believed he had important historical experiences to record, and because our interest made him feel that his life had been noteworthy. As with all the audiences for his remembering, Percy liked to please me and to gain my approval. In our interviews he often stopped at the end of a story and asked 'Is that all right for you?', or 'Are you impressed with what you're getting?' He then waited, expectant, to answer the next question. But then, often before I could ask that question, he would launch into another story from his repertoire. Despite his desire to tell me stories I wanted to hear, Percy's composed memory of the war was so strong, and so clearly defined, that he often ignored or reworked questions so that he could recount his standard stories about concert parties and life behind the lines. If my question did not directly refer to one of those stories, he usually found the cue he needed in a word or phrase that I had used.[7]

By the time I interviewed Percy Bird in the 1980s, his account of the Great War was well worked out. Over the years he had composed a story of his war that he could live with, and that gained affirmation from the public legend of Anzac. Although depictions of the Great War in recent histories and films sometimes troubled Percy's Anzac identity, for the most part he was able to exclude challenging reinterpretations by drawing upon modern accounts that matched his own Anzac stories and identities, and by performing those stories to eager and responsive audiences. Percy Bird's remembering thus worked to exclude the aspects of the war that had been most difficult for him, and to highlight positive experiences and the proud, collective identity of Anzac.

Bill Langham

After finishing his working life in the Town Clerk's office at Melbourne City Council, Bill Langham enjoyed an active retirement in Yarraville, where he and his wife participated in a number of local clubs and societies, often with music as a common interest. Bill was not a public storyteller like Percy Bird, and he had not been sought out to perform his Anzac memories for various audiences. Nevertheless he was interested in recent representations of the Great War, and was pleased to discuss those representations and his own remembering with me.

Like Percy Bird, in old age Bill Langham no longer felt strong enough to march on Anzac Day, and there were too few survivors from his unit for a reunion. Books and films provided the main influences upon, and affirmations of, Bill's war memories. He enjoyed reading war stories, including Bean's histories, because 'they give you a bit of a kick'. He liked the positive representation of his own experience and the validation of his youth, though he was not an uncritical reader. He argued that Bean must have got some things wrong because he could not have known about all aspects of the Australian experience of war, and he explained that books cannot recreate war as it really was.

This latter criticism was his main concern about recent Anzac films. For Bill, parts of the 'Anzacs' television series were 'bloody terrible' because they did not show how, for example, men were sucked under the mud at Passchendaele until they died of suffocation. Like Percy Bird, Bill wanted films to accurately represent his own experience of the trenches, even though this recognition brought back painful memories. Despite his reservations, Bill felt that the series was 'mostly pretty good' because at least it attempted to convey the horrors of trench warfare to a civilian audience.[8]

Bill claimed that he was 'immune' to the emotional effects of 'Anzacs', which just came 'matter of fact'. This wording is revealing, and suggests how Bill came to terms with his war memories. In a reflective passage, he described movingly how he handled the risks of remembering:

> Sometimes if you sit on your own and [...] you start to think, then that's the time they all come back to you. Then you wish there was somebody there, with you to speak to so's you could forget those things.

I often lay awake at night [...] where I go back over the old trails there. Right back, I suppose it's a funny thing, you never forget them. They're always there. Although, as the years go on they get, they get milder and milder, they're not as bad as they were when you were, like in the early years and things like that. Like everything is, you get used to them. But you always try to remember the, as I said we try to remember all those funny incidents and things that had happened. Then occasionally, as you're remembering a few of them, one of the other ones slips in between somewhere. Yeah. Then you got to bring yourself [...] back to reality then, and as I say one of the greatest things in the world to bring you back to reality is that music.[9]

This passage reveals the unpredictable nature of memory, and how disturbing memories can slip out unasked and unexpected. When this happened to Bill he tried to pull out of memory lane. Yet Bill's memories did not become milder (in Bill's words: 'now they're memories and I've got used to them') just because 'time heals'. Rather, 'getting used to memories' so that they are less disturbing is an active process, through which certain memories are emphasised while others are played down or ignored, or are worked through in satisfactory ways.

For Bill, like all of us, this composure was a social activity in which memories were reconstituted through particular and general public relationships. It was also a dynamic process. As the public contexts for Bill's remembering changed over time, or between different social situations (even within the oral history interview with its different cues for remembering), Bill recounted different memories and different meanings about the past. For example, within Bill's memory there were certain stories that he used when he wanted to explain the disillusionment of soldiers and ex-servicemen. Yet there were other stories that he recounted — often spurred by criticism of the Anzacs — when he wanted to argue that his war service, and Australian participation in the war, was worthwhile.

In this regard, Bill Langham differed from Percy Bird and Fred Farrall, each of whom had honed their war memories into a relatively fixed repertoire of significant stories. The flexibility of Bill's remembering may have been due to the fact that, in contrast with Percy and Fred, he neither sought nor attained the public role of Anzac storyteller and historical source, and had not constructed his remembering into a fixed public performance. That flexibility may also reflect the fact that Bill's war experiences were less traumatic and troubling than those of some other veterans, so that he had less need to compose a safe but rigid Anzac identity. For the oral historian the many threads of Bill Langham's remembering are both confusing and enlightening. They show how within one person's memory there may be different and even contradictory understandings of the past, and that a person can have a range of different identities which are adopted in appropriate circumstances. It may well be that, for Bill, this ability to adapt his Anzac identity served as a source of social survival and emotional strength.

Bill Langham's remembering also shows how the Anzac legend worked for a man whose experience of the war and postwar periods could have easily facilitated an oppositional stance. In comparison with Percy Bird, Bill Langham's experiences at the war, and when he came home, were provocative and challenging. Yet he did not become alienated from the war and its memory. He actively participated in the wartime life of the diggers and in the postwar rituals of remembrance, which addressed his practical and emotional needs and affirmed certain ways of understanding of his war. Bitterness and criticism were not erased from his remembering; indeed, the distinctive quality of Bill's memory was that it included many different and even contradictory renderings of his experiences. Yet public affirmation of the aspects of Bill's memory that accorded with the Anzac legend had, on the whole, caused him to highlight such aspects and to play down more dissenting memories. Bill Langham's remembering thus reveals the interplay between individual subjectivity and public myth, which is the key to the resonance and effectiveness of the Anzac legend for war veterans.

Fred Farrall

In 1938, Fred Farrall and his de facto wife Dot Palmer left Sydney to look for work in Melbourne. In the years during and after the Second World War, Fred was an influential figure in the Melbourne Labor movement. He became a leader of the Federated Clerk's union, staunchly opposing a right-wing coup that occurred within the union, and retained his pro-Soviet communism during the schisms on the left following the crises of 1956 and 1968. He was also active in the Campaign for International Cooperation and Development, a peace movement organisation with close links to the communist parties of Eastern Europe. After his retirement from work and trade-union activities, Fred became active in local politics and was elected to Prahran City Council. His work for old people won him great popularity, and with the support of the Combined Pensioners' Association he was appointed Mayor of Prahran. Fred retained an active interest in local and national politics up until the final months of his life.

During retirement, Fred Farrall experienced another major shift in his relation to the war and to his own Anzac past. In the 1960s and early 1970s he started to read and talk about his war outside of the Labor movement. He went to Sydney to attend the Anzac Day ceremony and reunion of his old battalion, he pinned his war service badge back in his lapel, and he retrieved the discharge certificate that had been hidden away for many years and hung it up on his living room wall. After years of silence he now talked eagerly and at length about the war to students, film-makers and oral history interviewers.

In our interview Fred explained the change in a number of ways. It was partly an old man's renewed interest in his youth: 'I suppose as you get older you have some sort of feeling for what happened long ago'. But Fred was only able to have positive feelings about his wartime youth — which he had shut away in a drawer of his mind for many years — because of changes in the way Australian society responded to the Anzacs and remembered the war. In his old age Fred was able to enjoy the respect, even veneration, that the few remaining Great War diggers received from people in the street who noticed an AIF badge, and from Veterans' Affairs officials who described it as a 'badge of honour' and who gave veterans free passes on public transport and paid their increasing medical costs.

> Well, there was a time when it just didn't fit into that picture at all [...]
> Well, we've never had much over the years of value from that sort of
> thing so if there is anything now, even to the extent of getting some
> respect, well I think it's worth doing.[10]

Renewed public veneration of the Anzacs was affirming for Fred in his old age. More specific shifts in the Australian popular memory of the war recognised aspects of Fred's experience that were once neglected, and thus encouraged him to recall his wartime past and to identify himself as a digger. He was particularly impressed by the recent British and Australian histories of the Great War, based on soldiers' accounts, that attempted to convey the effects of conditions on the Western Front. When we conducted our second interview in 1987, Fred was reading Peter Charlton's new book *Pozières: Australians on the Somme 1916*:

> I was there later and I know all that country and all those places that
> are mentioned [...] in your mind you possibly go back all those years

[...] It seems to ... when he's writing about Albert, the town of Albert, you know ... you know it revives, it certainly revives memories if you've been, if you've been to that place. Whatever it is. And this is the grip, this is the grip the book has on me.[11]

Pozières was compulsive reading for Fred because it was about his own experience, and because it portrayed trench warfare in terms that connected with his own sense of what it had been like. Fred was equally appreciative of the ways in which the 'Anzacs' television series depicted life on the Western Front (he was less happy with its representation of digger larrikinism). Certain themes of the new social histories and films — innocent patriots or adventurers sacrificed by politicians and generals and transformed into weary or disillusioned soldiers — had always been part of Fred's war story, but his story was now recognised and affirmed by the popular public narratives. Perhaps most importantly, the representation of the war in *Pozières* and 'Anzacs' showed Fred that his wartime fears and feelings of inadequacy did not reflect badly on his manhood, but were in fact common results of trench warfare. When Fred described his postwar nightmares about being shelled, he added that, 'Here, to now, I didn't know that there were so many others like me until I read this book on Pozières'.[12]

One practical consequence of this recognition was that it opened up public platforms outside of the Labor movement that now welcomed Fred's participation in war remembrance. Fred took to these new opportunities with great relish, and in turn benefited from direct social affirmation of his memories. For example, during the Vietnam War and in the anti-nuclear campaign the peace movement became a prominent force in Australian society. As an old soldier who was also a pacifist Fred was uniquely placed to contribute to this movement. In the 1980s he wore his war medals at the Melbourne Palm Sunday peace rallies and, sought out by reporters, held up the medals as an ironic symbol of the folly of war and remarked, "This is why I'm here. I don't want to die, but there's a lot of young people who want to die a lot less'. Oral historians, including college students, film-makers and academics, also sought Fred's account of the war, and Fred used each interview to make similar points about the folly of the war and the terrible lot of the soldier.[13]

Fred's new attitude to the war and public war memories was affirmed by an incident that occurred when he visited the Australian War Memorial in Canberra in 1985:

Nearly got a job there. I was there about eighteen months ago, you know, and oh gee, look here, I got the surprise of my life [...] I was treated like a long lost cousin [and was asked to talk about the Western Front to other visitors]. 'Well', I said, 'I wouldn't mind doing that, but', I said, 'I'm a worker for peace and not for war.' 'Oh', the bloke said, 'you know this place was built as a Peace Memorial and so you're at liberty to express your opinions along those lines as you see fit.' [...] So up I went. Well I was there for two or three days really. It looked as though I was going to have, at eighty odd, as though I was going to get a permanent job.[14]

Fred was very impressed with the War Memorial. He told his audience that the dioramas made as good an attempt as possible to portray trench life, and then added, from his experience, the sounds, smells and feelings that they could not convey. Fred brought the old models to life for an eager audience, and felt satisfied that at last his story of the war was being told. He believed that by describing the conditions of his war he was making his message of peace.[15]

As these examples show, in the last two decades of his life Fred made a profoundly important reconciliation with his wartime past, and between his own memory and the public narratives of the Anzacs. This affirmation of Fred's military past, and the opportunities to tell his story of the war to younger Australians, were immensely fulfilling for Fred in his old age. As public representations of Australians at war changed, Fred Farrall came to live with the legend.

In old age Fred's memory still had a radical edge. In social contexts that reaffirmed his identity as a radical digger, such as the Palm Sunday rallies or our interviews, he used the new interest in the Anzacs to make criticisms of war and of Australian society which rubbed against the Anzac tradition. Fred was still critical of the guardians of that tradition. Although Fred enjoyed meeting up with old mates at the reunion of his battalion in Sydney, he still refused to march on April 25. In 1985 Fred invited me to join him for Anzac Day. He wore his medals and we went to watch the parade, but Fred had no desire to join the march. He agreed that the day provided an important opportunity for veterans to meet old friends, and that some form of commemoration of Australians who had died at war was necessary. But he was critical of the patriotic nature of the day, and remarked that, by contrast, his mates on the Somme had been 'the least patriotic blokes you could think of'. Anzac Day remained alienating for Fred because it explained and justified wartime sacrifice in terms of national achievement and pride, which he did not accept.[16]

Yet the partial recognition of Fred's war experience which was provided by the new Anzac legend also caused the displacement of certain aspects of Fred's radical analysis of the war. Anzac histories and films showed that for the poor bloody infantry 'war is hell', yet between the lines they usually promoted the digger hero and a national legend. Fred was so pleased with the new recognition that he did not always see that other aspects of his experience were still ignored or denied by the legend. He did not consider the absence, in most modern books and films about the war, of depictions of tensions between officers and other ranks in the AIF, of the postwar disillusionment of many diggers, or of an analysis of the war as a business, all important themes in his discussions with me. Fred assumed that any museum depicting the horror of the Western Front must be a 'peace memorial', but did not recognise the ambiguities of a museum that is also a memorial celebrating the nation's military achievements. When I asked Fred if he had considered these issues when he was at the War Memorial he was confused, and he responded, 'No well ... that, no I never, I never got down to considering that really'.

The Anzac legend of recent years, with its representations of the horror experienced by soldiers and its blame for politicians and generals, is appealing and inclusive for old diggers, perhaps especially for radical diggers who feel that at last their war story is being heard. The new Anzac narratives offer fresh ways for veterans to articulate their military past and to reconstitute their Anzac identities. Yet the stories and meanings that do not fit today's public narrative are still silenced or marginalised, and at best only resurface within a sympathetic particular public, such as a gathering of fellow radicals or an oral history interview.

This process of assimilation to the dominant narrative is similar to that undergone by diggers like Percy Bird and Bill Langham when they joined the RSSILA and marched on Anzac Day back in the 1920s. Through participation in collective remembrance, ex-servicemen enjoyed recognition and affirmation of a particular, positive Anzac identity, and they articulated their memories of the war using the public narratives of the legend. This memory composure was essential for individual peace of mind but, in the process, memories that were not recognised by the legend were displaced and marginalised. In the 1980s Fred Farrall learnt to live with the new Anzac legend and as a result of that process the way he remembered the war, and his identity as an Anzac, changed.

A past we can live with?*

The concluding sections of the memory biographies of Percy Bird, Bill Langham and Fred Farrall show how their Anzac memories and identities were affected in different ways by popular memories of Anzac in the 1980s. For each man the influence of new Anzac representations depended on his original experience of war, on the ways in which he had previously composed his war remembering, and on the social and emotional context of old age. But for all of them the Anzac narratives of the 1980s facilitated new ways of remembering — sometimes troubling, sometimes positive and affirming — and different Anzac identities.

This book has explored the creation of individual and collective memories of the Australian soldiers' experience in the Great War. Oral testimony indicates the variety of Anzac experiences: of enlistment; of battle and life in the trenches; of digger culture and life out of the line; of rehabilitation and repatriation; of the culture and politics of ex-servicemen; and of becoming old diggers. The diverse and even contradictory experiences of Australians at war have been narrated in an Anzac legend constructed in terms of the preconceptions and ideals of its narrators, according to the requirements and constraints of different media, and in relation to the social and political demands of the AIF and Australian society. In this process, the sharp edges of the Anzac experience have often been rubbed smooth, as legend-makers have fashioned a compelling narrative and a homogeneous Anzac identity defined in terms of masculine and national ideals. While the specific contents of the Anzac narrative and archetype have changed over time and in different circumstances, the legend has always worked to construct a 'typical Anzac' or a 'genuine digger' and, in turn, to render aberrant experiences and identities as alien, atypical and un-Australian.

At the individual level, Australians who served in the Great War have struggled to compose memories of their war. Through participation in Anzac remembrance, and in the culture of soldiers and ex-servicemen, they have drawn upon public narratives of Anzac that have provided interpretative categories to help them to articulate experience in particular ways. In turn, the public narratives and identities of Anzac have recognised key aspects of the diggers' experience, such as comradeship, endurance, personal worth and national identity, and have provided a positive affirmation of that experience.

---

*   New edition note: this conclusion to the first edition was written in the early 1990s
     and has not been changed. It provides my perspective at the time.

The Anzac legend has thus helped many veterans to compose a past that they can live with.

Within this process of memory composure, experiences and understandings that are not recognised and that cannot be articulated through the public narratives are displaced or marginalised within individual memory. For some men the gap between personal experience and public sense is too great to allow for reconciliation and, unable to make acceptable sense of their war, they have been forced into alienation or silence. Memory composure, however, is not a static process. As personal circumstances and needs shift, and as the public narratives and meanings of Anzac change over time, the possibilities for remembering, and for fashioning new identities, also change. Old diggers like Fred Farrall are thus able to forge a new relationship with the legend of their lives.

A historical approach that emphasises the relationship between individual and public memories has implications for Anzac history-making and politics. It provides a starting point for critical analysis of Anzac narratives — from Bean's writings to modern-day films — which assert a single, homogeneous Anzac identity, and which assume that the legend is the accurate and uncomplicated account of an essential Anzac experience. This critical analysis informs recent studies that have explored neglected aspects of the Australian experience of the Great War and the processes by which a selective Anzac legend has been created and installed.

Such revisionist Anzac histories provoke outrage and determined refutation from Anzac traditionalists. For example, in an article entitled 'History as a Kangaroo Court', which appeared in a 1988 *IPA Review* ('Australia's journal of free enterprise opinion'), Tim Duncan attacks the rewriting of Australian history by young, radical historians who are critical rather than celebratory about Australia, and who blame their forebears for its faults:

> Some of these people may not like their country and its history. But they cannot stop from enjoying its comforts, the reality of which every day contradicts the vile stuff they write.

Duncan is particularly concerned that histories using this approach, such as the *People's History of Australia*, are cheap and well distributed, and may be 'heavily used in schools'. In his criticism of that book, he savages my own chapter about the Anzacs:

> As for the Anzac legend, according to Alistair Thompson [*sic*] it 'forgets the black Australians who fought against the invasion of their country

[...] ignores the inequalities and conflicts of class and status, sex and race'. The Anzacs were simply too brutal, they killed Germans like rabbits, and were neither more resourceful than any other soldiers, nor any more protective of their mates.[17]

Some academic historians are equally concerned about challenges to the Anzac legend. In a 1988 *Australian Historical Studies* article entitled, 'No Straw Man: C. E. W. Bean and Some Critics', John Barrett assesses recent studies by myself, David Kent and Lloyd Robson. We had tried to show how Bean constructed a particular version of the Anzac experience in his war correspondence and history, but Barrett was having none of it:

[...] it may be that Bean's *essential* [his emphasis] balance was not so wrong. However grim the prospects became in the Anzac area [...] the bulk of the force gritted its teeth and stuck it out. The AIF as a whole would have been defamed had the demoralised been exaggerated — just as it may be libelled by any elevation of the larrikin element to a majority of the force.[18]

Like Bean, Barrett asserts that there is an essential Anzac experience and identity, and defines away contradictory behaviour as atypical.

Another, sophisticated reworking of Bean's thesis is John Robertson's *Anzac and Empire*. In a concluding chapter which asks, 'Was It Worth While?', Robertson comes straight to the point of recent historical debate about the Anzac legend:

The notion that the 'Anzac legend' was 'created' by C. E. W. Bean or was a figment of his imagination seems to be coming fashionable among a younger generation of historians [...] Eliminate Bean's writing from the story, and the same picture emerges of bravery, recklessness, a cynical or disrespectful attitude towards authority outside battle, stern discipline under fire, and so on.

The creators of the Anzac legend were, of course, the men themselves.

Robertson claims that Australians 'simply drew inspiration from what had happened in the battlefield' for their Anzac legend. He argues that Australian soldiers were effective and distinctive because of their mateship and their close relationships with officers, and because of the national identity that arose as 'for the first time Australians made a spectacular and praiseworthy contribution [...] to the course of world history'. For good measure, he wonders 'what qualifies people who have never experienced the rigours of campaigning or the terrifying savagery of battle to belittle the valour of those who have'.[19]

Recent critics of the Anzac legend have not sought to belittle the Australian soldiers. Rather, we have argued that, by explaining the Australian experience of war in terms of national character and achievement, Bean and his successors have narrowed the range of our understanding of Anzac, and have excluded or marginalised individual experiences that do not fit the homogeneous national legend. Furthermore, by claiming that the legend was 'of course' created by 'the men themselves', and that subsequent generations of Australians have drawn direct inspiration from Anzac achievements on the battlefield, these historians neglect the ways in which the soldiers' story was regulated and shaped in particular ways by Anzac legend-makers, and in the context of Australian social and political culture.[20]

I am certainly not arguing that historians should ignore the experience and testimony of Australian soldiers. There is plentiful evidence in such testimony to make for histories representing the range and complexity of Australian experiences of war, which recognise patterns of commonality as well as difference and contradiction. I am arguing that the use of soldiers' testimony, whether taken from contemporary diaries and letters, or from memoirs and oral accounts, needs to be sensitive to the ways in which such testimony is articulated in relation to public narratives and personal identities. The different forms of participant testimony are an attractive and valuable source for historians. Anzac histories, however, have often used such personal testimony interchangeably and uncritically, and while the testimony is assumed to be a direct expression or essence of the Anzacs' experience, it is easily used in ways that suit the ideological and narrative imperatives of the historian. An alternative use of personal testimony in a history of the Anzac landing, for example, might explore the ways in which participants constructed and articulated the meaning of the event, and how their articulation changed over time and in relation to the public, national narratives of the landing. 'What actually happened' at Anzac Cove on 25 April 1915 can only be understood in relation to the articulation of the event, and personal testimony is thus part of the history of the event and not just a historical source.

One of the dilemmas of a historical approach which challenges the homogeneous, nationalist story of Australians at war is that it may threaten the personal composure that veterans have found through the legend. Histories that recall difference, difficulty and exclusion may open up traumatic and painful memories. On the other hand, new historical narratives can enable individuals to recover and explore aspects of the personal past that have been silenced or repressed, and can facilitate reparation and reconciliation with that past. Perhaps more importantly, new histories can help give voice to the experiences of individuals and groups who have been excluded or marginalised by prior historical narratives. For many years the Anzac legend neglected and thus silenced the wartime lives of most Australian women and of men who did not go to war, and it is only in recent decades that histories have begun to articulate those forgotten lives.

Issues about the relationship between personal and public histories are now of little more than academic interest to the very few surviving Australian veterans of the Great War. But these issues are relevant to veterans of other wars, and to the creation of histories and popular memories of those wars. The memory of Australian participation in the Vietnam War is a case in point, and suggests contemporary applications for the ideas and approaches outlined in this book.

For many years Vietnam veterans felt rejected or disregarded by Australian society, and their internalised trauma was a source of terrible psychological and social wounds. In the late 1980s there was a transformation in the regard with which Australians held Vietnam veterans. 'Welcome home' marches by Vietnam veterans have generated enormous media and public attention, and new books and films have portrayed the Australian soldiers' Vietnam experience in sympathetic terms. After two decades of public neglect and exclusion, this renaissance of sympathetic public recognition for Australia's Vietnam veterans — recorded in Bernard Clancy's moving account of his second 'home-coming' on Anzac Day 1987 — has been tremendously affirming for the veterans, who have returned in great numbers to Anzac Day and other forms of public remembrance. In turn, the new public forms and meanings of remembrance have helped veterans to articulate their Vietnam experiences in positive terms emphasising comradeship, masculine self-worth and national identity. In many ways this process is similar to that experienced by Australian Great War veterans in the inter-war years, as they found recognition and affirmation in the Anzac legend of the ex-servicemen's culture and public remembrance, and thus composed positive Anzac memories and identities.

Yet recent popular accounts of the Australians in Vietnam are also con-structing a new history which excludes or redefines contentious aspects of Australian participation in that war. Books, films and ceremonies make the diggers' experience the centre-piece of the Vietnam narrative, and portray

Figure 15   The Herald-Sun's *(2 January 1993) response to the author's writings which suggested the Australian 'straggling' at the Anzac landing had been censored from the British official history after pressure from Australia.* (Mark Knight)

Figure 16   *This image in the* Sydney Morning Herald *(26 April 1993) conveys concern about radical historians' questioning of the Anzac legend.* (Michael Mucci)

that experience as "The Legend of Anzac Upheld'. These new accounts emphasise a particular version of the digger experience, that of decent mates who were effective jungle fighters in a bloody but necessary war. Little is heard about the negative aspects of Australian relations with the Vietnamese, and not much information is presented from the perspective of opponents of the Australian Task Force in Vietnam and back home. The symbolic 'coming home' of Vietnam War veterans in the 1990s may make the veterans' wartime past easier to live with, but it is also reshaping the popular memory of Australian soldiers in Vietnam, and silencing significant aspects of individual and collective experience. This new popular memory of Australians in their most significant recent war matches the political agenda of conservative historians and activists, who favour a celebratory national history which excludes dissenting voices and stories of conflicting interests within the nation.[21]

There are alternative accounts of Australians in Vietnam that do recognise the complex and divisive issues that the war posed for servicemen and for Australian society. These are not easy histories to live with. Vietnam veteran Terry Burstall describes his own pain at the loss of the hope that the war had served some purpose: 'I had clung to the false hope for over twenty years that "they" had not died for nothing. Now I could squarely face the fact that they had'.[22] Through autobiographical war writings and visits to Vietnam, Burstall makes an impressive effort to come to terms with his personal Vietnam history. Burstall's project suggests that it is both necessary and possible to create a critical popular memory of Australian participation in the Vietnam War. This account would recognise the experiences of Australian ex-servicemen alongside those of other participants. It might help veterans to develop ways of remembering the war that are challenging and empowering in their lives, and it would enable all Australians to explore the lines of fracture and conflict in our past.

Part IV

Anzac memories revisited

CHAPTER 10

SEARCHING FOR HECTOR THOMSON

The stigma of mental illness

Sometimes you can't write the history that needs to be told. In 1986, in an early draft of the autobiographical introduction to the first edition of *Anzac Memories*, I wrote that my grandfather Hector Thomson contracted malarial encephalitis while serving with the Light Horse in Palestine and that as a result he was 'in and out of mental hospital' after the war. I only knew about Hector's mental illness fourth-hand, from my mother. My father, David Thomson, had never talked about it with me or my brothers, indeed he had only found out himself from an older relative after his father died. He was appalled by my reference in the draft to the mental hospital and demanded that I remove it.

The stigma of mental illness ran deep. My father insisted that he would never have been accepted into officer training college in 1942 had they known about his father Hector's illness. In the 1980s, mental illness in the family was still shameful. My father was also furious about 'the radical ideology' of my naive early efforts to debunk the Anzac legend in a chapter for a *People's History of Australia* published in 1988 and in the very different (and unpublished) first version of this book that focused on 'forgotten' radical diggers like Fred Farrall.[1] He felt I had betrayed his thirty years in the Australian army and the values that sustained him, and hoped that 'none of my old soldiers read it'. Perhaps worst of all, my writing ripped off the scab that had formed across his terrible childhood and unleashed angry, painful memories. I changed the line 'in and out of mental hospital' to the more socially acceptable half-truth that Hector was 'in an out of Caulfield Repatriation Hospital' after the war, which you will still find in the introduction to the first edition.[2]

Oral historians often make difficult choices between a responsibility to history and a responsibility to narrators who have shared a life story, as well as to their wider family. In 1986 I prioritised my father's feelings and hoped to repair our fractured relationship. Yet the whole point of the story about Hector Thomson had been to show that within families, as within the nation, some histories can be told while others are hidden or forgotten. In removing the reference to mental illness I was contributing to a selective version of Anzac history. Though no one else could have spotted the omission, I felt that I was compromising the aim of my book.

A quarter of a century later, it's easier to write about soldiers and mental illness. In recent years the stigma surrounding mental illness has begun to lift across Australian society, and historians and veterans themselves now write more readily about 'shattered Anzacs' who return from war both physically and mentally damaged.[3] My father and my family now talk more openly about Hector's troubles.

We can also access sources about war veterans that weren't available 25 years ago, including the case files for individual ex-servicemen kept by the Repatriation Department (now the Department of Veterans' Affairs), which are now available on application to the National Archives of Australia. The M (medical) and H (hospital) files for Victorian First War veterans alone comprise almost three kilometres of archive shelf space. These were once working files, initiated by the Department when a veteran claimed support for a war-related medical condition. Each veteran's file grew over the years, as the Department dealt with claims, accumulated medical reports and other evidence, and corresponded with claimants (see the block of images in this chapter for examples of the documents in Hector Thomson's file). These case files are an extraordinarily rich record of twentieth-century Australian social and medical history. Letters from returned men — and from their wives and parents — detail medical complaints that demanded treatment and a pension from the 'Repat'. Doctors, expert witnesses and Repat officials argue each case, sometimes with sympathy, sometimes with callous suspicion of malingering.[4]

Files M58164 and H58164 detail Hector Thomson's medical history, from enlistment in 1914 through to his death in 1958. Hector's repatriation story illuminates the battles of the peace that were fought in Australian homes and hospitals after the war. It shows how damaged and desperate veterans sought support and recompense from the government, and how their desperation became more acute during the Depression of the 1930s. The story also reveals contemporary medical understandings and prejudices about mind and body, how doctors and other officials struggled to balance limited resources against increasing needs, and how medical diagnoses and bureaucratic attitudes towards veterans changed over time. In these records we can read Hector's own account of the impact of war on his life, albeit an account framed by Repat guidelines and his need for a war pension. Most surprising to me, and most poignant, my grandmother, Hector's wife Nell, emerges as the tragic heroine of the tale. One man's war story becomes a family history that stretches across the decades and reverberates through the generations.

For many Australians, family history is one of the ways in which we know about war and its consequences. In part, the resurgence of interest in Australians at war in recent decades has been fuelled by family historians researching the service records and life stories of parents, grandparents and other relatives. Sometimes unexpected discoveries challenge family and national mythologies; sometimes family war history is framed by the lens of an Anzac legend of dutiful sacrifice and stoic comradeship.[5] By making awkward, troubling family histories about war and its aftermath we can disrupt simplistic national narratives and transform family remembrance. We can better understand what happened to men like Hector Thomson during and after the war, and how their return impacted upon Australian families and society.

Enlistment: 'one of the strongest young men in this district'

Prior to enlistment Hector Thomson 'was held to be one of the strongest young men in this district — as was his father before him'. Ex-servicemen seeking pensions needed to show that their medical condition was war-caused, and Hector's Repatriation files are thus crowded with accounts — this one by the local bank manager — of his 'splendid health' before the war. Hector had grown up on one of several Thomson family farms around Clydebank near Sale in the south-east Victorian region of Gippsland. Six foot tall and solidly built, he was an attractive young man. When Hector went to war in 1914 an anonymous poem had seven Gippsland girls lamenting his departure, with its opening lines:

> The world is looking cold and grey
> Just 'cos Hector's sailed away
> All the girls they sob and sigh
> And hang their hankeys out to dry.[6]

Twenty-three year old Hector enlisted from Queensland, where he had been working as a station hand. Serving in Palestine as a driver with the 2nd Light Horse Field Ambulance Unit, he was awarded the Military Medal in August 1916 for helping to rescue wounded men while under 'heavy rifle and shell fire'. Two months later he was mentioned in despatches for bravery. Hector's service record also notes several charges for the less commendable behaviour that was not uncommon in Australia's volunteer army, including disobedience, 'familiarity with natives' (an ambiguous charge which may refer to prostitutes), ill-treatment of a mule, being 'improperly dressed' and bringing intoxicating liquor into a hospital. At face value, Hector Thomson's war record of bravery and larrikinism exemplifies the two sides of the Anzac legend.[7]

In March 1917 Hector was hospitalised for a month with his first attack of malaria. Over the next year he spent another 87 days in hospitals recovering from the 'severe ague and fever and vomiting' brought on by four more bouts of malaria, and from a respiratory infection ('coryza') and severe diarrhoea which left him 'very weak'. Doctors suggested that a cooler climate might improve his health and he applied to be transferred to his brother's unit in France (his mother wrote asking the military authorities to grant Hector leave in England), but was refused because his unit was short of men. He returned to the ranks in Palestine until November 1918, when he was granted '1914 leave' to Australia with other men who had enlisted in the first year of the war.[8]

Return: a 'most serious and pitiful case'

On arrival back in Australia on Christmas Eve 1918, the army medical officer recorded that Hector had had a malaria attack in October but 'No attacks since. Never otherwise in Hospital. Feels quite well. All organs normal' — and signed him off in 'A' grade health. Perhaps Hector was, like many others, keen to get home and did not bother to report the full extent of his illness. But Hector was not well. Discharged from the army in February 1919 he returned to his parents' home in Gippsland where he suffered further malaria attacks and was 'unable to do constant work & just pottered about my father's farm'. By April he was renting a local grazing property and applying for a soldier settler's block but the attacks continued, usually accompanied by violent headaches, and at one point he was found lying unconscious beside the plough.[9]

In November the Thomson family doctor, Dr Campbell of Sale, reported that Hector had lost a stone in weight (over six kilograms) over the course of six malarial attacks in the previous six months. Hector now applied to the Repat for support. The Department agreed that his 'general weakness' was due to an infection suffered while on service and was 'not due to his default', and granted a 50 per cent pension back-dated to the point of discharge. In the summer of 1919–1920 Dr Campbell treated Hector's malarial attacks with injections of Sodium Cacodylate which, Campbell reported ten years later, 'cleared them up, except for occasional slight reoccurrences' which did not affect his work. Over the next three years, as the attacks became less frequent and less severe, the Repat progressively reduced Hector's war pension, to 25 per cent, then 16.6 per cent, then 10 per cent, until it was suspended in August 1922 when Hector failed to attend his annual Repat medical examination.[10]

Hector had decided that he was now well enough and the small pension was not worth the effort of time-consuming and intrusive medical examinations. There were other more significant changes in his life. Upon the death of an unmarried aunt he had inherited a 405 acre mixed-farming property near to the other Thomson farms around Clydebank. In February 1922 he took up residence with his new bride Nell in a four-room cottage they renovated and called 'Bungeleen', the name of an Aboriginal leader who had lived in the area. Nell was the daughter of an Anglican clergyman who had served in Sale when she and Hector were both in their teens. The loss of Hector's pension just a few months after their marriage would become a source of great regret for Nell. In 1929 she recalled that a medical officer in the early 1920s had commented to Hector 'Well bad fever has played up with you', yet 'the next report we received was that the small pension my husband had been receiving had been stopped. I wanted my husband to appeal then but he would not do so & the necessity was not great then as it is now.' In 1922 Hector probably did not want his bride to think that he was damaged by the war.[11]

Within a few years Hector's illness recurred, but in an unexpected and debilitating form. The evidence collected for a second pension claim in 1929 tracks his decline. Hector recalled that he had 'the first definite attack of lapse of memory' in 1926 when his wife was given power of attorney to manage his affairs, and 'it was only then that I realised I had been ill for a long time past because I found I had neglected many things I should have attended to many months before'. Dr Campbell reported several episodes of memory loss and violent vomiting attacks which culminated in total collapse in November 1927:

> Preparatory to harvest, he brought in [to Sale] some horses to be shod. He was in his working clothes. He did not arrive home that night, and next day realised he was in Melbourne Botanical Gardens. He immediately caught a train back. His memory of going to Melbourne or what he did there was a blank, except that he thought he stayed at a Coffee Palace. His condition after this was one of complete nervous exhaustion. He would sleep for the greater part of the 24 hours but when awake would talk quite lucidly and cheerfully. He took no interest whatever in the harvest or his affairs.

As Hector's bank manager Mr Witts reported to the Repat, this was 'a serious and most pitiful case'. Thomson had been a 'very sick man' since 1927 and 'will never work again in all probability. The slightest exertion prostrates him.'[12]

Dr Campbell sent Hector to Dr Sidney Sewell, a prominent Melbourne neurologist who was renowned for treating veterans with shell shock. Sewell arranged for brain X-rays and conducted other tests (including the Wasserman anti-body test which came up negative for syphilis) but could find no alteration from the normal in the cerebro-spinal fluid and reported that 'the history of the onset of his condition was indefinite & the diagnosis lay between an exhaustion Psychosis & a Post Encephalitis Lethargica'. Encephalitis lethargica had been first described in 1917, and between 1915 and 1926 an epidemic spread throughout the world, with symptoms including high fever and headaches, lethargy and sleepiness, with extreme cases suffering from a coma-like state. The cause was not certain, though recent research suggests that it may have been a consequence of the Spanish influenza epidemic, with an immune reaction to infection causing neurological damage.[13]

Hector later claimed that Sewell had told him he had 'a definite inflammation of the brain'. Sewell never reported any such inflammation to the Repat, but in 1929 Dr Campbell wrote to the Repat that he agreed with Sewell's diagnosis of encephalitis and suggested that it might have been caused by an infection resulting from malaria. As the Thomson family doctor, Campbell probably wanted to ensure that Hector received a pension for this new condition. He would have known that the Repat was more likely to pension a man with observable, war-caused physical damage, and he did not mention the alternative diagnosis of a psychotic condition due to 'exhaustion'.

Hector trusted Campbell, who was also an ex-serviceman (we will see this trust was ill-deserved), and from this point almost certainly believed that his ill-health was due to some form of brain inflammation (encephalitis) caused by wartime malaria. This was how Nell came to understand her husband's illness, and 'malarial encephalitis' became the explanation whispered in family oral tradition and passed on to my generation.[14]

Hector improved after a long holiday, but when he returned to the farm his condition recurred and he suffered several more breakdowns after undertaking the strenuous activity that was inevitable on a small family farm. By now, Nell was running the farm and the finances, and with two small boys underfoot and a sick husband to care for, she was desperate. In May 1929 she submitted the new pension claim for Hector, and then took the lead role in proceedings. Several letters from Nell to the Repat detailed Hector's condition and its effect on the family. She wrote in May that 'Owing to his constant ill health we have to employ constant labour and this year I am finding it very difficult to carry on. [...] It is a great strain not knowing when this loss of memory might occur again.' In September she explained that Hector had 'collapsed just at the most busy time of ploughing and putting in the crops. I had to send him into Hospital & depend on the kindness of my neighbours & had the very great expense of having to employ three extra men as well as the help of my brother in law & my husband's cousin in the endeavour to get the crops in on time.' Whilst Hector attended medical examinations in Melbourne Nell visited the Repat to press his claim and then chased up expert witnesses. By late September her husband's illness and 'all the worry I have had in connection' caused Nell to become ill so that she 'simply had to close' her home and take the children to stay with a sister across the New South Wales border in Holbrook while Hector stayed 'with his people' in Gippsland. Nell continued to write to the Repat, while stressing that they should address all correspondence 'to me personally':

> My reason for asking for a pension for my husband is that my husband is unable to work for any length of time without a complete breakdown & I cannot afford to keep a permanent man. Owing to my husband's severe illness which occasions loss of memory his business affairs have become very tangled. For three years I have been unable to keep any domestic help & I have had to manage all the business part & the running of the farm as well as the constant nursing of my husband & the care of my two very small sons aged 3 years & 4 ½ years. I feel that if my husband could receive a pension it would enable me to carry

on. [...] If only we could have seen ahead the far-reaching effects of this dreadful malaria I would have begged my husband to appeal for a bigger pension & most certainly for a renewal of it.[15]

The clergyman's daughter had mastered many unexpected responsibilities in the first years of married life. From Repat medical and soldier settlement records we know that Nell Thomson was just one of many wives of damaged veterans who were forced to manage family life and livelihood in the inter-war years.[16]

After receiving a report from Dr Sewell in August 1929, the Repat doctors, who could find no physical causes for Hector's condition, requested that Hector return to Melbourne in November for examination by another 'Nerve Specialist', Dr Clarence Godfrey, to test whether there was 'any connection between malaria and loss of memory & neurasthenia'. Godfrey, a lecturer in psychiatry at the University of Melbourne and one of the first in Australia to recognise the significance of Freudian psychology, delivered a detailed report based upon Repat records and examination of the patient. He explained that since 1927 Thomson had 'manifested symptoms suggestive of some pathological condition of the central nervous system. Prominent among these were disorientation, amnesia, and apathy, diminished power of attention and lessening of initiation (Bradyphrenia)'. He noted that Dr Sewell had 'reduced the diagnosis of the cerebral condition to post-encephalitis, lethargic signs, or to exhaustive psychosis', and then concluded that the evidence weighed 'more in favour of the former disorder':

> [....] in examining him I found him extremely candid and apparently truthful. The nervous sequelae of malaria even the malignant type, after ameliorated — as in this case — are certainly uncommon. One is forced to consider whether the 'coryza' was a mild or abortive case of encephalitis lethargica, with apparent recovery, and it is well known that after apparently complete recovery in such cases relapses may occur after long intervals, even several years. [...] as there is no history of encephalitic infection during the period intervening between his discharge in 1919 and the exhibition of the nervous symptoms in 1926 I am disposed to grant claimant the benefit of the doubt and recommend acceptance as due to War Service with an assessment of 75% incapacity.

In short, while Godfrey doubted the connection between malaria and brain damage, he speculated that Hector's coryza (respiratory) infection in 1918 might have caused encephalitis lethargica, and that he was a deserving case. The Repatriation Commission accepted Godfrey's diagnosis, agreed that three linked conditions — 'Post Encephalitis Lethargica, Cerebral Exhaustion and Loss of Memory' — were all war-caused, and approved a 75 per cent pension, back-dated six months from January 1930. Hector was informed that future treatment would be at the Commission's expense (this was a significant saving), and that he should use the Repat's medical officer in Sale, Dr McDonald.[17]

Within a year, Hector collapsed again. In the winter of 1930 he managed 'ordinary jobs' on the farm, but by summer harvest Dr Campbell noted that 'he had got much thinner and looked duller than before. I tried to persuade him to employ a stack builder, but he did not on account of the expense, with the result that he got completely knocked up, and the failing of memory and semi-comatose condition supervened.' In February 1931 Hector went back to Dr Godfrey in Melbourne, who noted that Thomson had deteriorated since their earlier meeting and was 'unable to activate his mentality and appears as if in a dream'. He concluded that the patient was 'permanently incapacitated' and 'now quite incapable of work', and recommended observation at Caulfield Repatriation General Hospital.[18]

From Caulfield, the first doubts about the original diagnosis appear in the Repat correspondence. Fourteen years after discharge from the army, Hector's condition was now being explained according to new medical paradigms. Dr Paul Dane had been an army doctor at Gallipoli and served with the 1st Australian General Hospital in Egypt, and after the war he became interested in neurology and the treatment of shell-shocked veterans. An early convert to Freudian psychology, by 1925 he had published on 'The psycho-neuroses of soldiers and their treatment' and was recommending treatment by analysis. Perhaps not surprisingly, Dane was looking for psychological causes for Hector's symptoms:

> I can find no evidence in this man's history which would suggest to me an attack of Encephalitis; there is nothing in this [sic] symptoms to suggest a diagnosis of post encephalitic disorder. There are no signs of organic disease of C.N.S. [central nervous system] but he is quite definitely an athyroidic type [affected by a malfunctioning thyroid gland] — he is also the typical manic depressive character type and has been all his life. There are occurrences in his military history which point towards slight psychotic trends or character defects. This inherent familial type of mental make up plus malarial infection and athyroidism is sufficient in my opinion to account for his present condition.

There is nothing in Hector's military records that suggests either psych-otic trends or significant character defects (perhaps Dane was referring to Hector's wartime disobedience and 'familiarity with natives'?). But Dane was not unusual at this time in explaining mental illness in terms of a flawed character and family history. In the absence of definitive physiological or psychological evidence, Dane was speculating about Hector's condition, and his speculations say as much about Dane and the limits of medical understanding as they do about Hector.[19]

Dane prescribed thyroid treatment and daily douche baths, and at first the treatment (or perhaps the hospital rest) seemed to work. Within a month Dane reported that Thomson was 'Much better since taking thyroid', and in April Hector was discharged to the care of the Repat medical officer in Sale, five and a half pounds heavier (three and a half kilograms) but still on thyroid and other drug treatments. But by July Nell reported to the Repat that Hector had collapsed 'in the usual way' and after nursing him at home for two weeks he was now in the Gippsland Hospital. In an articulate yet despairing plea, Nell detailed her efforts over many years to manage her children, the farm, and her sick husband:

> This state of affairs has been going on for so many years now that I am in great financial difficulties and am heavily indebted [...], I have absolutely nothing to live on except my pension out of which I have to pay a man 30/- a week and his keep. It is indeed a very serious position, not only for my husband, but for myself and my two children, the youngest of whom is not yet five and the elder is 6 ½ years of age. I am struggling, and have been for several years now to carry on the property with the advice of my husband's brother who lives 20 miles from me — I had to borrow on my husband's Life Insurance this year to enable me to put in a little crop, I therefore hope that my request for a full pension will be granted. Would you let me know if it would be possible for anyone else to collect our pension. I live 10 miles from Sale, and very often it is over a month before I can get in to draw my pension.[20]

On 13 November Dr Campbell phoned Caulfield Repatriation General Hospital to say that he was sending Thomson from Gippsland by car for readmission with 'Cerebral Exhaustion & loss of memory'. Admitted just after midnight on the 14th, the next morning Hector had a 'violent maniacal attack [...] of all around' and was discharged to Royal Park Receiving House, the short term admissions section of Royal Park Hospital for the Insane. There is no further detail in the files about this attack, which seems to have been out of character. Within a few days the Repat agreed to pay for the costs of admission to the civilian mental hospital, and increased the pension for Hector's family to 100%, based upon Dr Godfrey's recommendation that the new mental condition of 'Acute Mania' was due to war service. The Medical Superintendent at Royal Park advised that the outlook for the present attack 'should be fairly good', and indeed within six weeks Hector was transferred back to Caulfield, 'as he is now quite normal mentally, but requires a couple of weeks treatment for neurasthenia' (nervous exhaustion).[21]

Royal Park recommended further treatment for neurasthenia but Caulfield's admission record for Hector on Christmas Eve 1931 missed this recommendation and instead cited the previous conditions of 'Malaria and Loss of Memory'. When no indications of either of those complaints could be found, Hector was discharged from Caulfield and sent back home to Gippsland with a 75 per cent war pension.[22]

Times were hard on the farm. My father was then aged seven, and in an interview I conducted with him in 1985, he recalled 'this was right at the end of the Depression, when things were very tough, there'd been a drought'. Nell could not afford any help in the house, which 'was fairly primitive and there was a lot of work doing it'. Assertive at every stage, she successfully appealed to the Repat for reinstatement to a 100 per cent pension, back-dated to July 1931 and subject to review every six months, though in June 1932 the pension was again reduced to 75 per cent because Hector had recommenced some farm work.[23]

Single father

Nine months later, Nell was dead and Hector was a single father of two small boys aged seven and five. Nell had needed an operation on her gall bladder and Hector insisted on using Dr Campbell because he was an ex-serviceman. My father believes that Nell died on the operating table in September 1932 because his father was determined to use 'the worst local doctor, because he had been in the War'. The bond of ex-servicemen had a devastating effect. My father remembers sitting outside in the back garden after a horse-rider brought news of their mother's death, and saying to his younger brother Colin, 'I wonder what we're going to do now.'[24]

The boys stayed with their grandparents on the neighbouring farm for several months. Then, astonishingly, Hector brought them back to live with him at Bungeleen, with the support of a paid housekeeper who lived in a small cottage next door. The contrast with their earlier life was stark. Nell was remembered by my father as a bright and witty woman, 'full of laughter', who 'was quite modern' in her ways of upbringing, 'very particular about the way we dressed', a voracious reader who read classics while she was pregnant in the hope that it would rub off and was reading Dickens' David Copperfield to her boys in the week before she died. For the first couple of years after Nell died the housekeeper was a 'marvellous small English woman', but she left — quite likely because Hector was not easy to live with — and was replaced by a succession of 'dreadful females'. By this stage they were getting 'very poor indeed, there was no money, and the house was getting shabbier and the garden was neglected'.[25]

Sometime after Nell died, Hector began to drink 'and we always dreaded his return from town from stock sales'. One night he did not return, and the next day David discovered him in hospital recovering from a car crash. 'We never had a car after that which restricted our lives even more.' When the boys reached the school leaving age of 13 Hector wanted them to work on the farm, but David was keen for further education, and a new rural school bus service was the 'miracle' that made high school in Sale possible. They now lived 'very enclosed lives':

> My father by this stage was getting very withdrawn and silent and didn't socialize at all, didn't go out and quite often we'd be asked to birthday parties at other families 15, 20 miles away and he wouldn't take us, he just couldn't go. Oh as we got older we could ride, but he wouldn't take us, just withdrew … so it was a fairly tough life. By the time I was 15 things got worse, 'cause I was 15 my brother was 13, and we were informed that no longer could we afford a housekeeper, and I said, "What are we going to do?" and he said, "You'll have to look after yourselves." And he really *had* no alternative, I couldn't understand it at the time, I can now understand he was desperately … I imagine he had a huge overdraft and no income, or very little.

The boys did the shopping during the school lunch break, cooked the meals and did all the housework except the laundry, which was sent out. They also helped their father on the farm, because he could afford no paid help. 'Things were really pretty grim [....] I remember, I suppose it must have been Christmas of 1939 or '40, we spent Christmas Day making a haystack. And we had cold mutton or something for Christmas dinner. No one else to do it. We hadn't been invited for Christmas dinner by anyone else, so the three of us had it ourselves.'[26] It's not clear why there was so little family support. David and Colin enjoyed occasional holidays with Scott family relatives, and Nell's elder sister Kathleen helped out when she could, but she lived in Melbourne. The Thomson families who farmed around Sale were less forthcoming.

At that time my father did not know Hector was a very sick man. That Hector managed at all was exceptional. Most men in his situation in the 1930s found a new wife to raise the children and keep the house, or either gave up their children to a female relative or placed them in a home (single fathers often had their children taken away on the assumption that childcare needed a woman's touch).[27] Hector saved a note that Nell had written in pencil from her hospital bed the night before she died, in which she said, 'if anything happens to me, don't let the boys be separated'. My father testifies that Hector 'stuck to that, we weren't separated, we were kept together for our childhood, difficult though it was'.[28]

The Repat files record that after his wife died Hector just about managed to keep his health together while he raised his sons. In 1933 he told the Repat doctor he still suffered occasional malaria attacks and had lost two to three months' work in the past year, that his memory was better 'but still very bad', and that he slept badly and suffered headaches. The doctor agreed that reduction in the pension from 75 per cent was 'not yet advisable'. Dr McDonald, who was the Repat medical officer in Sale and thus answered to the Repat rather than the patient, was more wary of Hector's claims. In 1934 and again in 1936 he reported that the symptoms were 'all subjective'. Another doctor noted that none of the examinations since 1932 had found any indication of either malaria or mania, though 'some slight mental impairment' was still attributable to the effects of Encephalitis Lethargica. The Repat deleted 'Post Malaria' from Hector's list of conditions accepted as due to war service, and in 1936 reduced the pension to 50 per cent, which added up to slightly less than half the basic wage.[29]

Hector appealed against the deletion of 'Post Malaria' and argued that he had had four severe attacks in 1936. Two new Repat doctors now concluded that 'It is doubtful if he ever had Malaria', pointing to the evidence of wartime medical records, though it is hard to see how they came to that conclusion as those records very clearly state that Hector was admitted with malaria on five separate occasions during the war. The appeal was disallowed and Hector was advised to report to Dr McDonald during the next 'alleged attack'. McDonald believed that Thomson 'wanted a diagnosis of malaria established on account of his pension'. When Hector complained in 1939 that he was often unable to work because of the same recurring symptoms,

McDonald dismissed the complaint and recorded once again that the patient was 'mentally not very bright'.[30]

We can't know for sure whether or not Hector was still suffering from malaria. Recurrent malaria attacks tend to burn themselves out, and Hector may not have had as many attacks as he claimed. When his blood was tested a decade later in 1948 there was no sign of the malaria parasite and the attending doctor concluded that malaria seemed 'very unlikely'. It is possible Hector focused his complaints on malaria because it did not carry the stigma of mental illness. By the mid to late 1930s Hector was probably trying to work the Repatriation system to improve his war pension, playing one doctor off against another, citing symptoms that he hoped would match his accepted conditions, and seeking to have other conditions accepted. In response, Repatriation doctors and officials were increasingly suspicious and unsympathetic. The evidence of his son's memory suggests that Hector was probably mentally unwell, though it is not clear whether this was due to a physiological condition with its origin in the war, or was simply due to some form of mental illness or depression. He certainly had good cause to be depressed.[31]

I like to think that although Hector barely managed as a parent, the fact that he kept his boys with him on the farm, in dire circumstances, was an impressive achievement. While Nell was alive, Hector came to rely on her and succumbed to ill-health. Unwell, unable to provide for his family, unable to manage the finances or even conduct his pension claim, Hector almost certainly felt a failure as a husband and father, and as a man. Nell's death must have been a terrible blow, yet it also led Hector to take back control of his life, and of his farm and family, and to work through the worst of his illness for the sake of his sons.[32]

Another war

In 1941 Hector went back to war. By then seventeen-year-old David had decided that there was 'absolutely no hope of getting anywhere' on the farm, and reached an agreement with his brother Colin that Colin alone would inherit the farm. For David, education was 'the only way', and in 1941 he used a £100 inheritance from his grandmother to pay for a year at Scotch College in Melbourne, following his father and grandfather at the school. The money ran out before the year was up but the headmaster generously let him finish his Leaving year so that he could qualify for officer training college. While David was away at boarding school, Hector sent Colin to live

Driver Hector Thomson, 2nd Light Horse Field Ambulance, Cairo, 7 May 1915. Hector sent a postcard of this studio photograph to his mother in Gippsland. (David Thomson)

Hector Thomson and his wife Nell, outside their Gippsland farm house 'Bungeleen' shortly after their marriage in 1922. (David Thomson)

Hector and Nell with their young sons David and Colin in 1927, photographed on the driveway of their farm house 'Bungeleen'. (David Thomson)

Hector with David and Colin at 'Bungeleen' in about 1930, shortly before Nell died. (David Thomson)

rms
237
40

MEDICAL CASE SHEET.*

No. in Admission and Discharge Book.	Regimental No.	Rank.	Surname.		Christian Name.
3619	482	L/Cpl	Thomson		Hector
Year	Unit.		Pte	Age.	Service.
1917	2nd L.H.F.A			26	28/12

Provisional Diagnosis.

Station and Date.   Disease   Malaria

No. 14 AUSTRALIAN GENERAL HOSPITAL A.I.F.   adm ???   WARD No. Cpl. ???

2 9 MAR 1917

Pre war occupation. Station Hand.

P.H. Shivering attack 2/3 yrs ago in Q'land: only one app??

S.H.

Present illness began at Raffa on 18th inst. with shivering attack: lasts abt 2 hrs: attacks repeated rapidly for 3 days. then every other day. Attack on night of admission here 18th for 3 days.

Ot. Healthy. well nourished.

Aliment. Tongue clean & moist. B. Regular Appetite fair.

Resp. ✓ Circ. ✓

Abd. Splade??? over spleen: not palpable ??? ??? ??? ??? CH.

Urine 1016 Cic. no alb. no sugar.

Mist Quin Sulph. 10 ??? ???

2.4.17. 90°/ slight ??? ??? during night. none today.

5.4.17. Shivering attack lastng 4 hrs: no rise of temp. Quin ??? ???

7.4.17. Another slight ??? today for 3 hrs: no rise of ???

8.4.17. Fairly well: slept ??? ??? ??? ??? Quin ??? ???

14.4.17 Discharged to convalescence

19.4.17 Discharged to Unit.

Capt ??? ???
R.A.M.C ???

The Medical Case Sheet that identified Hector Thomson's first malaria attack in 1917. Together with other surviving wartime medical records, this was included within Hector's Repatriation Medical File. (National Archives of Australia, B73, M587164)

C O P Y

R 58164
C 33789

Bungalene,
Clydebank,
SALE, GIPPSLAND.
10/7/29.

To The Deputy Commissioner,
    C/33789 P.

Dear Sir,

        With regard to my letter of the 7/5/29, in
connection with the renewal of my husband's pension, I
would be glad to hear if there is anything being done with
regard to this matter.

        My husband has been very ill and is still ill.
Major Campbell D.S.O., has advised me to try and get him in
to the Caulfield Military Hospital. Will you kindly let me
know how to proceed with regard to this (of getting into
Caulfield) at your earliest convenience. I have to employ
two farm hands to carry on with the work and am finding it
more than difficult to make ends meet, so will be most grateful if the matter of the pension can receive your earliest
consideration. Major Campbell tells me he has not received
any letter asking for details of this case.

                Yours faithfully,

                    N. THOMSON,

                    per K.A.B.

P.S. This letter has been written for Mrs. H. G. L. Thomson
     by her sister as she has broken her arm.

                    N. Armerod.

*One of many letters Nell Thomson sent to the Repat on her husband's behalf. This one
in 1929 was typed by her sister because Nell had a broken arm. Repat officials often
wrote comments to each other, and for the record, around the edges of claimant letters.
(National Archives of Australia, B73, M587164)*

Form U.

COMMONWEALTH OF AUSTRALIA.

Australian Soldiers' Repatriation Act 1920-1922.

RECORD OF EVIDENCE.

Evidence given by _Thomson Hector._

of _Bungalene. Clydebank. Sale._

in respect of the claim made by _____

for a pension to be paid to _____

I am suffering from _Loss of memory_

and I make application for acceptance of the disability as

due to War Service. In the event of acceptance

I claim pension benefits as from the date of my application on

Form 20B (Revised)

For some years after leaving school I was employed on my father's farm & for the six months immediately P.T.E. was a station hand. I was in perfect health.

I enlisted on 20·11·14 in splendid health & had no troubles until early in 1917 when I contracted malaria. For some days I attended the unit R.O. but eventually became so ill that I was evacuated to Port Said. During the remainder of my service the trouble recurred at frequent intervals & necessitated constant treatment. I also had occasional heavy colds & in 1918 suffered with a severe attack of diarrhoea which left me very weak. With Each attack of malaria I was accompanied by severe ague & fever & vomiting.

I was discharged on 22·2·19 in poor general health.

(Signed) _H. Thomson_

Date _24. 7. 29_

The foregoing evidence was read by me to the person who gave it before he (or she) signed this sheet.

* Registrar of Pensions at }
* Special Magistrate at } _G. M. Tully_

Date _24·7·29_

T.355./2.27.—C.13730.—75m.   * Strike out what is inapplicable.

A Record of Evidence form completed by Hector Thomson in 1929 as part of his Repat pension claim for loss of memory. Margin notes such as 'omit', and red pencil strokes before and after the first paragraph, indicate sections to be typed up in Repat evidence documents for subsequent claims. (National Archives of Australia, B73, M587164)

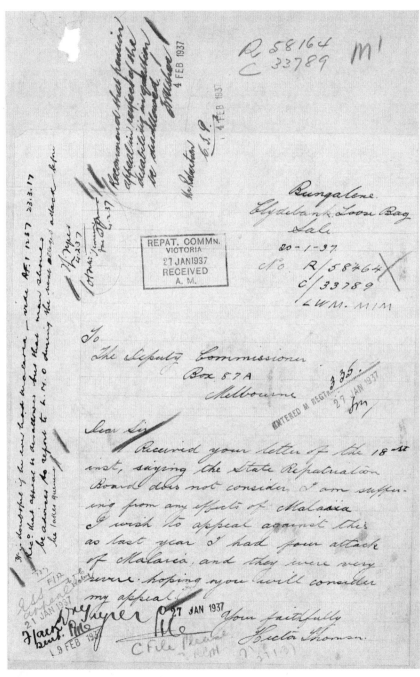

A 1937 letter to the Repat from Hector Thomson initiating an appeal against the removal of malaria as a pensionable condition. Note the marginal comments by Repat staff, one of which states that 'it is doubtful if he ever had malaria'. (National Archives of Australia, B73, M587164)

THOMSON. Hector.

Form K. contd.
Disability caused through war service - 100%.
Work on farm in a desultory way - always throws him back.
Is now quite incapable of work. Condition may improve in six
months time, but outlook is unfavourable. Recommend for admission
to Caulfield.

9.2.31 O.P. Notes.
Dr. Lemon.
Has been seen by Dr. Godfrey 6.2.31.
Is easily exhausted. Severe headaches, poor appetite.
Dr. Godfrey by telephone confirms the man's statement that
hospital treatment is necessary. To Caulfield.

9.2.31 Adm. R.G.H.C. - Report Med. Supt. 4.4.31.
Admitted - Cerebral Exhaustion.
DR. A.E. GREGG reports: C.O. loss of memory, and becoming very
sleepy and easily tired. Headache, severe, frontal region - at
times very giddy. Has noticed voice becoming weaker - at times
difficulty in expressing himself.
No tremors noticed at any time - general weakness. Unable to
carry on work. Has had similar attacks as above - first
commenced three years ago, lasting 2-3 weeks, gradually becoming
more severe and frequent. Did not have any illness preceeding.
General health good.
Wife states that patient is subject to lapse of memory - on one
occasion left Sale, and was found in Melbourne - did not remember
leaving - sets out to finalize deals (wheat, sheep etc.) and
forgets to carry out same.
O.E. Memory at present not good - appears tired and drowsy.
      Answers questions very slowly.
      Color and nutrition satisfactory - Teeth ✓
      Heart, Chest and Abdomen: N.A.D.
      C.N.S.: N.A.D.
      Urine: 1010 Ac. No alb. No sugar.
              Mic. N.A.D.
      Wassermann (Blood): Negative.
      X-Ray (Skull): N.A.D.
      Eyes (Sir James Barrett):
              18.2.31 R.V. = 6/6   L.V. = 6/6
                       Wearing +.75 R. & L.
              21.2.31 Under H. & C.
                       R.V. +1   L.V. +1   = 6/6
                           +.5 cyl = 6/6   +.5 cyl

                                    115
              25.2.31 Ord. for reading as above. N.D.W.S.
20. 3.31 Dr. Dane: "I can find no evidence in this man's
          history which would suggest to me an attack of
          Encephalitis; there is nothing in his symptoms to
          warrant a diagnosis of past encephalitic disorder.
          There are no signs of organic disease of C.N.S. but he
          is quite definitely an athyroidic type - he is also the
          typical manic depressive character type, and has been
          so all his life. There are occurrences in his military
          history which point towards slight psychotic trends or
          character defects. This inherent familial type of
          mental make up plus malarial infection and athyroidism
          is sufficient in my opinion to account for his present

*An extract of the Evidence file compiled by the Repat in response to Hector
Thomson's 1937 appeal against the removal of malaria as a pensionable condition.
Evidence files like this comprised typed extracts from the claimant's Repat medical
file. This page includes the extract from 1931 in which Dr Dane doubted that Hector
had ever had an encephalitis attack and suggested he was a 'manic depressive
character type'. (National Archives of Australia, B73, M587164)*

with relatives on a nearby farm and travelled to Melbourne to enlist in the Second AIF. In Melbourne he could get away with lying about his age (39 instead of 50), birthplace (Glasgow instead of Sale) and surname (he added a 'p' to Thomson) so that his over-age status and Repatriation record would not be discovered. On the medical history form Hector admitted an appendix operation but wrote 'no' to each of a long list of ailments which included 'fits of any kind'. The medical officer was suspicious about Hector's stated age — the photo taken at enlistment shows a ravaged face which looks older than 50 years — but let him in. Though the First War may have caused the ruination of Hector's health, he had no grudge with war service itself, which offered a welcome escape from the hardships of his farm and family life. By early 1942 he had returned to the Middle East, this time as a private with the 2nd/6th Cavalry Regiment, and at least one photo shows him looking content back among soldier mates and within the security of the army. Later that year Hector suffered serious petrol burns to his arms and legs, and it is likely that his real age was discovered. He was returned to Australia and discharged, in October 1943, ostensibly because he was required for work in the reserved occupation of farming.[33]

In 1946, with both of his sons now serving in the army of occupation in Japan, Hector sold Bungeleen without telling them (after he died they discovered he had spent the proceeds which were intended to get Colin another farm). Both David and Colin were furious about the sale, and in the following years they had only limited contact with their father, who was now living in boarding houses in Melbourne. Through a family friend he found a job at the Goldsborough Mort wool store; to the Repat he said he was a wool classer but he was probably a clerk. In 1947 he made his first visit to a Melbourne Repatriation clinic since 1931, and a doctor reported that he looked 'more than his age'. After he tested negative for malaria in 1948 he makes no further appearances in the Repat files until a stroke in 1955 sent him to the Repatriation Hospital at Heidelberg. He was now totally incapacitated and his pension was increased to 100 per cent because the stroke was regarded as a sequel to his 'accepted cerebral conditions'. When my father and Nell's sister Kathleen sought in 1956 to place Hector on a waiting list for an Anzac Hostel, a Medical Social Worker reported that 'Mr Thomson has some difficulty with his speech, but he is able to make himself understood and is a pleasant, cooperative type of patient'. Hector never left Heidelberg. By 1957 he was in a 'poor mental state' and broke his hip, and in January 1958 his Repat file ends with death by broncho-pneumonia, aged 67, two years before I was born.[34]

Family history and remembrance

In 1992 my father wrote to me that at war Hector was 'a hero and a successful soldier':

> It was his civilian life which was painful and not discussed. We make his war illness an excuse for his failings, but he may have failed in any case. Perhaps if Nell had lived he would have been different, but in some ways I feel that had she lived she would have had a very unhappy life. I know that Aunt Kar [Kathleen] believed that she should never have married Hector — apparently another man she loved was killed in WW1. Perhaps we were all victims of the war.[35]

That was an astute judgement by a son who felt failed by his father and who idolised the memory of his mother. But as a boy my father never knew the nature or extent of his father's health problems. Nobody explained that their father was ill. When he wrote that letter to me in 1992 my father was still not privy to the history which unfolds in the Repat files, and he was struggling with his memory of a broken father and traumatic childhood.

In recent years my father has declined with Alzheimer's. He can't remember yesterday and speaks very little, but he still recalls that his father was 'damaged'. Over the Christmas of 2012 I explained about the Repat files and gave him a draft of this chapter. He spent hours slowly reading each page, with an intensity of concentration that he rarely manages nowadays. His eyes narrowed and creased with pain as he recalled his childhood and said that Hector's physical and mental condition was worse than I described. For a lucid, fragile moment I think that he, too, came to a new understanding about the cause and extent of his father's illness and about his mother's tenacity. He agreed that Hector probably did the best he could in the circumstances and consented to the publication of this chapter.

When, as a young man, I researched and wrote the first edition of *Anzac Memories*, and sought out old war veterans and their stories, I may have been searching for Hector Thomson all along. The topics of historical research often have personal roots, even when we are not aware of them. It's been good to find Hector in the files and explain the story behind a painful family secret. What I hadn't been expecting was to my find my grandmother Nell. The Repat files helped me understand the impact of the war on soldiers' families and the critical postwar role of wives like Nell, both themes that were understated in my interviews with ex-servicemen and in the first edition of *Anzac Memories*.

One of the ways in which Australians connect to twentieth-century wars is through their own family history.[36] When my undergraduate students at Monash University conduct original research in Australian history, many of them turn to relatives who served in the armed forces. It's easy to investigate family war history because it's easy to access extensive war records, because war stories are often preserved in oral tradition, and because there is a vast historical literature about Australians at war. Students, like other family members, are drawn to compelling and poignant war histories; they hope to understand an elderly grandparent or a long-lost relative. They often start by framing their family history through the comforting lens of the Anzac legend, though they sometimes make jarring discoveries, such as a grandfather who contracted venereal disease at war, or a relative who exaggerated his military service or success.

How families deal with complicated and even challenging war stories says something about the processes of family remembrance, and about how the Anzac legend works, and sometimes does not work, at the intimate level of the family. For example, my father and his brother Colin knew that Hector had won a Military Medal for bravery and they wore his medals to school on Anzac Day (this was one of the few stories about Hector which they later shared with their own children). Hector did not talk to the boys about his war, and we don't know exactly when they were told that he had contracted malarial encephalitis at war, though we do know this became the accepted (and faulty) family explanation of his illness. Within the family that diagnosis probably concealed more disturbing concerns about psychosis or depression, and it carried the legitimation of a war-caused illness. We know that my father only learnt after Hector's death about the time in a mental hospital. Aunt Kathleen carried much of the family oral history and she guarded its secrets carefully. My mother did not meet her future father-in-law until just before her marriage in 1955, and did not learn about his illness till many years later (perhaps my father feared she would baulk at marriage had she known about the family history of mental illness). As a child I grew up with a mixture of heroic stories, half-truths and silences about my family war history. At that time, the bitterness and pain of my father's memory of Hector was such that he could hardly talk about him at all. The story of Hector's cousin, the soldier poet Boyd Thomson who died on the Somme and was commemorated in a memorial book of verse, was easier to tell, and to hear.

We need to take care and risks with family history. Broaching secrets and breaching confidences can hurt people we love and disturb the equilibrium of family life and relations. But secrets and lies can be more damaging than confession, and family historians who delve hard and deep can not only make better histories, they can also generate better family understanding. In an Australian context, where the Anzac legend can underpin superficial, limiting stories of Australians at war, family history has an especially important role. Taken seriously, using all the evidence that is now available and questioning family mythology, we can create family histories that illuminate complex, multi-faceted military experience, of bravery and fear, of loss and achievement. More than that, we can show that it is not just military men (and military women) who are affected by war. We can explore the family context and consequences of war, the postwar impacts of war service, and war's reverberations across the generations. In short, through family history, researched carefully and written with searing honesty and a critical eye, Australians can help create a different type of war history.

CHAPTER 11

REPAT WAR STORIES

'In my opinion, ex-soldier does not exhibit anything more than a constitutional incapacity to meet his difficulties, with the accompanying feeling of not having got a fair deal. He is, from the same cause, introspective and somewhat hypochondriacal. [...] Rejection of Neurosis recommended.' (Dr Godfrey, reporting on Fred Farrall for the Repatriation Department, 24 July 1939)[1]

'Well, you see, the pensions in the 1920s, unless you had an arm off or a leg off or a hand off or something like that, it was almost as hard to get a pension as it would be to win Tatts. There was no recognition of neurosis and other disabilities [....] they treated the diggers as they interviewed them and examined them as though they were tenth rate citizens.' (Fred Farrall interview, 14 July 1983)[2]

The Repat

When soldiers come home their war comes with them, but at home there are new battles and the war can make very different sense. The Repat case files open up ways of understanding the postwar lives and war stories of Percy Bird, Bill Langham and Fred Farrall that were not available when I wrote the first edition of *Anzac Memories*. Their medical files cover the years from enlistment until each man's death (Percy in 1990, Fred in 1991, and Bill in 1994 just after our book launch), though the files after 1983 were (in 2013) still closed to researchers because of the 30 year rule. In their pension claims and correspondence these men told their stories of war-caused injury or illness to improve their chances of a pension, and that telling was influenced by what they understood, at the time, about their condition — just as Repat medical examinations and decisions were framed by professional imperatives and attitudes. Before I bring new perspectives from the Repat files to the three men's memory biographies, it will be helpful to explain aspects of Australia's postwar repatriation system that shaped the relationship between the Repat and ex-servicemen.

Since *Anzac Memories* was published in 1994, several histories have examined the impact of the war-caused damage that blighted the postwar lives of so many veterans and their families and required support from the Repat; in the interwar years between a quarter and a third of returned soldiers were receiving war pensions for incapacity. Though the Repat's primary relationship and concern was with returned servicemen, repatriation was a family issue. Families struggled to live with damaged men and suffered through their loss of income and esteem. Wives like Nell Thomson played a crucial role in the family economy and were forced outside their expected domestic roles; sometimes they became an advocate for their men's rights. Children like David and Colin Thomson bore the lifelong effects of fathers who never recovered from the war. Though Hector Thomson was not a soldier settler, his story also points to the difficulties for families on the land dealing with drought and falling prices and with men who could not manage the arduous demands of rural labour.[3]

Historians disagree about the generosity of Australia's repatriation system. In the interwar years, before the expansion of the welfare state in the 1940s, Australians battling poverty or ill-health were often forced to look to charity for support. War pensioners received state benefits that were not available to other Australians, and the full war pension was significantly higher than the old age and invalid pensions. By 1938 over a quarter of a million Australians were being assisted through war pensions (only slightly less than the number receiving old age and invalid benefits), which comprised just under a fifth of total Commonwealth spending. The 'Repat' became the colloquial name for the 'vast medical and welfare bureaucracy' of the Commonwealth Repatriation Department, which administered not just war pensions but also war gratuities, vocational training, war service homes, veteran hospitals and hostels for long term care, and which funded soldier settlement schemes managed by the states. As the official historians of Repatriation concluded in 1994, 'Without the Repat the quantum of human wretchedness, physical pain, mental anguish and poverty in the Australian community over three quarters of a century would have been incomparably greater.'[4] The war pension that Hector received throughout the 1930s almost certainly enabled him to keep his farm and support his sons.

Yet the interwar Repat was not beneficent nor its war pensions as generous as Australian politicians liked to claim. In 1930 the official war medical historian A.G. Butler calculated that when one factored in higher Australian costs of living the purchasing power of the Australian war pension was less than that of Britain, France, New Zealand, Canada and the USA. Though a standard Australian war pension was higher than the old age and invalid pensions, it was lower than the Australian basic wage. War pensioners were thus consigned to the 'working poor'.[5]

Strictly speaking, the war pension was neither a pension nor an entitlement. All returned servicemen had received the war gratuity based on a payment per days of service. But until a war service pension was introduced in 1936 (for veterans aged over 60 and permanently unemployable), the war pension was only paid to men with a war-caused disability. It was calculated in percentage terms according to the effect of the disability on the man's earning capacity (thus Hector Thomson's pension varied over the years between 10 and 100 per cent), with additional sums paid for maintenance of a wife and dependent children. Though the pension compensated for loss of earning capacity, the Repat's aim, in common with other welfare efforts in the first half of the twentieth century, was to create self-reliant citizens who were not dependent on charity, and breadwinning husbands and fathers who could support a family. Repat officers were vigilant against returned men who were perceived to be abusing the system or becoming dependent on it, and were quick to reduce the percentage granted as a man's employability seemed to improve. Nearly two-thirds of the 72,760 pensions paid in 1924, for example, were set at less than 50 per cent.[6]

Assessment of incapacity was fraught with difficulties. For some conditions the assessment was reasonably clear-cut, with the loss of a limb or an eye awarded a standard percentage allocation, though even in these cases the pension might vary as a man's earning capacity improved. The medical definition of incapacity excluded consideration of social and economic factors such as discrimination against disabled workers. For most men — like Hector Thomson — incapacity was difficult to judge. Repat officials made a subjective determination, which was often contested by veterans, based on medical opinion and references from employers or reputable public figures.

Just as difficult to determine was whether or not an incapacity was due to (or, from 1922, aggravated by) war service. One problem was that many of the AIF's most important wartime medical records had been mistakenly destroyed in London in 1919. Though soldiers' bodies and minds had been scrutinised and recorded in great detail from enlistment through to discharge, the surviving paper-work offered a partial and often inadequate account of a soldier's wartime medical history. The medical records that had survived were treated as gospel, so that a suggestion of a pre-war condition or a hint of wartime malingering might haunt a pension claim across the decades. The regard for medical records matched the immense power of the medical authorities, and especially the Repat doctors, whose judgement about war-relatedness and incapacity was usually decisive. The power of the Repat doctors was one expression of the professionalisation of medical practice in the early decades of the twentieth century and its increasingly close relationship with the governments seeking to record, improve and control public health.[7]

In the early twentieth century medical understanding of war damage was often indeterminate and, as we saw with the diagnoses of Hector Thomson's condition, doctors could make very different sense of the same set of symptoms, depending on training and expertise, professional context, and personal attitudes. For example, in the mid-1920s cases of war neurosis were three time more likely to be accepted in Victoria than they were in New South Wales.[8] Proving war-relatedness was especially difficult for intangible damage — such as that caused by gassing or 'shell shock' — which manifested through internal or psychological symptoms that were difficult to assess, let alone understand, and which had different effects over time. Medical explanations were also influenced by prevailing ideas about the relative impact of environment (both wartime and postwar) on the one hand, and of heredity and character on the other. The Repat files are crowded with moral judgements about family traits and mental weakness that affected pension decisions, particularly in cases of mental ill-health such as Hector Thomson and, as we shall see, Fred Farrall.

In the interwar years most Repat staff, including the doctors, were themselves war veterans, but any sympathy they might have had for fellow returned men was tempered by their bureaucratic role and by their social background and values. Repat doctors were usually former officers from upper-middle-class backgrounds, and as historian Kate Blackmore argues, they mostly shared a set of conservative and 'militaristic' social values about hierarchical authority, personal responsibility and moral character. Repat medical decisions were 'seriously compromised by the values and class interests of senior doctors, by the characteristic indeterminacy of medicine as a body of knowledge and by the social constitution of disease or illness'. The responsibility for the well-being of ex-servicemen was also compromised by a public service commitment to bureaucratic efficiency and financial accountability. In 1943 a Repat doctor wrote to official historian Butler that 'of course a departmental M.O. [medical officer] cannot be a "nice kind doctor" by giving away public moneys, he has to be like all Public Servants "a careful custodian of the public exchequer" [...] Moreover the majority of departmental clients were not "heroes" but plain men and many of them were not as much "wounded" as they wished to be.' By contrast, as we saw in the case files for Hector Thomson, family doctors who were not employed by the Repat, and specialists who were paid for their opinions but who were not public servants, might be less influenced by bureaucratic and financial imperatives.[9]

The interwar Repat case files record a protracted battle between the Repat and veterans and their supporters. The stress caused by the adversarial process of pension claims took its toll, as we saw in the increasingly desperate tone of Nell Thomson's letters. Over time the terms of engagement would change, as effective lobbying by ex-service organisations in the 1930s and 1940s gradually shifted the burden of proof in pension claims so that rather than the veteran having to prove that a condition was war-caused the Repat had to prove 'beyond reasonable doubt' that it was not. With the return of World War II servicemen and women, and with a financial context and social attitudes that enabled more generous spending on veterans, the period from the 1950s to the 1970s saw the sustained enhancement of repatriation entitlements.[10]

In the 1980s and early 1990s when I wrote the memory biographies of Percy Bird, Bill Langham and Fred Farrall that appear unchanged in this book in chapters 3, 6 and 9, I was mostly reliant upon the men's memories to explain their earlier lives and to suggest how their war stories developed over time. I used their interview accounts about the significant influences and changes in their lives, and about their story-telling in the past — in the trenches or behind the lines, at unit reunions or war commemorations, as they read new war books or went to films like Gallipoli — to explore the development of their war and repatriation stories, and to gauge the impact of different factors. Memory biography was, necessarily, a speculative process.

The Repat files provide a contemporary source which reveals change in these men's lives and attitudes as it happened, albeit mediated by the aims and processes of the Repatriation Department. The following additions to my memory biographies tap this new source. The files expand our understanding of Percy, Bill and Fred's postwar lives and of the repatriation process. The Repat records test and stretch my understanding and use of veterans' memories and show how their relationship with the Repat was one context in which these men created war stories and drafted the memories which they later shared with me.

Percy Bird

Percy Bird's Repat medical file is very thin. It's important not to overlook such thin files.[11] Browsing amongst the Repat medical records the historian is readily drawn to the fat files generated about veterans with serious on-going health concerns like Fred Farrall. The thin files of men like Percy Bird represent veterans who were not so badly affected by war and its aftermath. Compared with many World War I veterans Percy had an easy and healthy return to civilian life, and had little need for state support. There was no recurrence of the tubercular neck gland that caused his discharge in 1917. He had a successful postwar career and was an active sportsman for much of his life. A Repat medical examination in 1967, when Percy was 78, concluded that he was a 'very fit active intelligent man' for his age.[12]

The most intriguing aspect of Percy's Repat file is the record of the neck condition that ended his war service. In 1983 Percy explained to me that in April 1917, while working behind the lines at Lagnicourt in northern France, he got a 'touch of gas' and his 'TB glands started to swell'. Wartime medical records confirm that Percy first noticed the swelling on the left side of his neck about May 1917, and that 'tubercular glands' were removed by operation on 24 August. Percy had a condition known as 'scrofula', a bacterial infection of the lymph nodes in the neck which in adults is often caused by the tuberculosis bacteria and is usually observed in patients with a weakened immune system. TB infection was rife in the crowded and insanitary conditions of the trenches, and was a major cause of casualties. Both the doctor who operated on Percy in 1917 and the Medical Board that decided he was 'permanently unfit for General Service', agreed that Percy had been predisposed to TB by the 'climate in which he was serving', in other words by conditions on the Western Front. They noted that although the operation was successful, Percy looked 'white & is not as well as he should be'. This was not unexpected. Scrofula caused by TB is often accompanied by symptoms such as fever and weight loss. Those symptoms were not in themselves sufficient reason for medical discharge from the army, and it is most likely (though not recorded in the surviving records) that Percy was returned home and discharged because of the fear that he might be infectious.[13]

There is nothing in the wartime medical records to suggest that Percy's 'tubercular neck glands' might have been caused by gassing. Indeed, the army doctors stressed that his condition was 'not directly due to active service' but was contracted whilst Percy was on service and could be attributed 'more or less to climate'. Like most soldiers serving in the appalling conditions of the

Western Front, Percy was probably immuno-suppressed and thus vulnerable to TB infection, but gassing would not have caused a bacterial infection and the relationship between the gas attack and the subsequent swelling of Percy's lymph nodes was probably no more than coincidental. But during and immediately after the war there was a widespread popular belief that gas poisoning could trigger a latent TB infection. Percy probably assumed that the tubercular swelling that followed so soon after his gassing was caused by the gas, and it may be that medical staff or even fellow soldiers suggested that connection, though it was never recorded in the files.[14]

It is not surprising that after the war Percy explained his premature return and discharge as due to the effects of a gas attack. The doctors on the homecoming ship and during a subsequent week of observation at Caulfield Repatriation Hospital established that he was fully healed and that the TB had not infected his lungs and he was not infectious. But Percy would have known of the social stigma of the dreaded 'white plague', and as a young man returning to work and marriage he may not have wanted to talk about his TB diagnosis. Gassing, on the other hand, was a respectable war wound. During the war Percy had felt guilty about leaving his 5th Battalion mates to take on clerical duties on the ship in 1916 and behind the lines in 1917, and though he was 'glad to get away' when he was repatriated to Australia he was almost certainly uneasy about leaving his battalion. By explaining his medical return as the result of a gas attack Percy had a story that made a legitimate and reassuring sense of his premature discharge. It was a story that stuck.[15]

Percy made very few claims on the Repat. In February 1918, when he fronted a Medical Board in Melbourne, Percy explained that he was about to return to his pre-war clerical job with the Railways and that after the long sea voyage he felt 'all right'. The Board concluded that he was not eligible for a war pension because he was 'not now incapacitated as result of war-like operations'.[16] Almost 50 years later, 13 years into retirement, Percy reappears in the Repat file in 1967. He had been experiencing dizziness and stomach pains, and an old school friend who was a Repat Board member 'repeatedly requested me to be examined by your department but this is the first time I have made application'. In a hand-written statement Percy described his gassing at Lagnicourt, the swelling it caused in his neck, and the operation that he believed saved his life, using almost exactly the same words as our interview in 1983. Of his recent stomach trouble he wrote, 'I cannot say whether or not this is due to the war but the 1916/1917 winter in France was the worst winter for 30 years and the conditions the soldiers had to put up with must have some affect [sic] on them now.' Like many aging veterans Percy was wondering whether the ailments of later life might have had an origin in the war. The Repat accepted that the 'Healed Cervical Adenitis' (from his neck operation) was war-caused but decided there was no incapacity to warrant a pension. Percy's other claims for anaemia and dyspepsia were rejected because there was no documented connection to war service. Percy probably agreed with that decision. He was not in poor health and had only been persuaded to apply by a mate who thought Percy deserved recompense for war service. A few years later he wrote that 'I have never had any serious illness at any time in my life so I wasn't surprised when I was notified after a thorough examination that I was not eligible for a pension'.[17]

Percy Bird was not a man who sought undue benefit from the Repat system. He had no great need until he moved into an RSL nursing home in the mid-1980s. Unlike many other veterans, he had come through the war without lasting physical or mental damage, and had enjoyed a healthy and successful postwar life. In 1980 Percy's son-in-law asked the Repat to provide Percy with a medical alarm system, now that he was widowed and living alone and might suffer a fall. Percy was a sprightly 91 year old — just as I remember him from our first meeting in 1983. The Repat inspector who visited Percy in Williamstown wrote a thoughtful and moving report about Percy and the quality of support that he enjoyed from family members and neighbours. Percy told him that he did not need the alarm.

Bill Langham

Bill Langham's Repat file confirms that the detail of his remembering in our interviews was often impressively accurate. Many of the incidents which Bill related in 1983 and 1987 are recorded in the files, though as we will see, the Repat did not always represent Bill's emotions and understandings in his terms. Bill's stories about taking absence without leave in the months after the war ended are corroborated by punishments recorded in his AIF service record. The details of his head wound are confirmed by medical reports, as are the decisions and consequences of his Melbourne discharge in 1919, his failed attempts to win a pension in the late 1920s and early 1930s, and his eventual success in 1976.

Though there are errors and contradictions in Bill's Repat file, the only significant discrepancy between the official records and what I took from our interviews leaps out of a 1933 pension application, which records Bill as a 'Widower — one child'. I met Bill's wife at our second interview in 1987 and made the mistake of assuming she was the same woman he married in the 1920s. Bill did not talk much about his family in the interviews, and said nothing about the death of his first wife or about being a single father. I've since discovered that Bill's first wife Vena died in the late 1920s after the birth of their second child (who also died) when their surviving son John was two. Bill's widowed mother moved into the family home in Yarraville to keep house and help raise the child, until Bill remarried in 1937. Perhaps for Bill the death of his first wife was a difficult topic that he preferred not to mention; perhaps he felt it was not relevant in interviews about the war and its impact on his life. I was focused on the war and its effects and did not clarify Bill's family history, which was an important part of the story. The lapse is a sharp reminder for interviewers to record vital personal details, to follow up on the clues and to explore significant silences.[18]

A careful analysis of the Repat files also confirms that Bill Langham was — as he claimed in our interview — badly treated by the Repat in the interwar years. When Bill and his horse were shot down on 1 October 1918, the shrapnel wound on the right side of Bill's scalp damaged the muscles that controlled the movement of his right eye. Though his eye was not affected, the doctors at Colchester Eye Hospital in England reported that Bill was suffering 'diplopia' (double vision) because he could not look to the left with the right eye. In 1918 the damage to the eye muscle would have been untreatable, though by 11 October the medical notes reported the diplopia was 'improving', and by the 28th the doctors concluded that Bill was fit to leave because there was 'No fracture. No disability. Feels well.'[19]

Back in Melbourne, in March 1919 a Medical Board recorded that Langham said he felt 'quite well'. On examination they concluded there was 'Nil abnormal' and discharged Bill in 'A' grade health, a decision that would prove significant for Bill's pension claims. The doctors' transcription of 'feels quite well' misrepresented what Bill said and thought at the time, in an incident he recalled with vivid detail in our interview. Taken straight from the ship to Melbourne's Sturt Street barracks, he had seen his mother and little brother waiting for him outside the gates. Desperate to join them, after a two hour wait for his medical inspection he told the doctors that he 'didn't give a continental' what they graded him as long as he could get out to his family. Told that he had 14 days to appeal the decision, Bill missed the deadline while he was enjoying his homecoming, and then assumed that he had missed his chance and let it slip. After several years at war the eye problem was a minor inconvenience and he was just keen to get on with his life.[20]

By 1927 Bill decided that his eye was causing enough trouble to warrant a pension. He wrote to the Repat explaining why he had not appealed the Medical Board decision in 1919, and that he had been 'suffering from the effects of the wounds in my head ever since I returned home'. He told Dr Craig that the wound caused headaches and 'frontal attacks of giddiness' which made him stagger and fall over at work. Craig noted that Bill's right eye could not rotate properly to the left. A second doctor, O'Brien, confused the issue by ignoring the double vision and giddiness (a likely consequence of diplopia) and instead diagnosing defective eyesight. O'Brien concluded that this disability was 'aggravated by war service' and required glasses. Bill did have impaired eyesight due to astigmatism (a defect in the shape of the lens) but this would not have been caused by his war wound. Ignoring Craig's report, and dismissing O'Brien's erroneous finding that astigmatism was aggravated by the war, the Repat decided that Bill's head wound was war-caused but rejected the claim for defective vision (and associated complaints of 'pain above right eye', nasal catarrh and right frontal sinusitis) as not due to war service. It didn't help Bill's cause that he failed to attend two medical examinations and was not able to press the case for recognition of the effects of diplopia. As a soldier and a veteran Bill was a serial absence-without-leave offender (in this case perhaps he did not fully comprehend the seriousness of the examinations or the consequences of his absence). But even without Bill present the Repat should have seen the records from both 1918 and 1927 that linked war-caused diplopia to dizziness and headaches.[21]

In 1933, now a widower in Yarraville supporting a young son, Bill tried again. Probably influenced by Dr O'Brien's suggestion that his poor eyesight was aggravated by war service, this time Bill made the mistake of claiming for defective vision caused by astigmatism rather than for the double vision. The Repat rightly ruled that astigmatism was unrelated to Bill's diplopia and was not war-caused, and that in any case the incapacity was negligible because astigmatism could be corrected with glasses. Bill was refused a pension on the grounds that there were 'no after effects' of the head wound and no pensionable incapacities. In our interview Bill angrily recalled that he was required to return glasses that he had previously received from the Repat. Though there is no record of that request in the file, Bill's conviction that the Repat made mistakes about the position and nature of his war wound is corroborated by the written records, including his aggrieved correspondence of the time. Without realising that he had claimed for the wrong complaint, and without any advice to that effect from the Repat, Bill gave up on a pension for the next 40 years. The whole sorry episode generated a bitter memory of the postwar Repat and confirmed Bill's belief that 'When we want you to go away and fight we'll give you the world, but when you come back we'll take it off you again'.[22]

In 1975 the President of Bill's local RSL club in Yarraville decided that his case 'surely' warranted a medical review and wrote to the Repat that 'Like most soldiers at the end of their war service' Bill was 'in a hurry to finalise his Military Service and did not receive the repatriation service he was entitled to. He tried and had two or three rebuffs and did not carry on. I would like to add he received a head wound and the scar still shows, his sight has been impaired and a health deterioration [sic].' The RSL helped Bill complete a new claim in which he explained that as a result of the head wound he had 'suffered with headaches for many years', had been 'embarrassed by being cross-eyed', and yet had received no compensation from the Department. At first the Repat rejected the claim, arguing that the headaches were probably caused by an age-related degeneration of the spine and not by his old wound. But on appeal a new doctor examined Bill's Repat records more carefully and realised that Bill had been treated for diplopia in 1918, that the eye muscle damage which caused double vision was sustained on active service, and that Bill could not see properly to his left side. The incapacity was judged at 10 per cent and Bill finally got his pension, though as he ruefully remarked to me, it came 56 years too late.[23]

By comparison with some veterans, the medical consequences of Bill's war wound were inconvenient and distressing but did not have a serious impact on his health (he was, the Repat noted in 1980, a 'remarkably fit octogenarian').[24] Yet Bill's battle for a pension, revealed by Repat records used in tandem with our interviews, illustrates several features of the relationship between veterans and the Repat. It demonstrates the mistake that many veterans made at discharge in not fully reporting medical conditions for fear of delaying their release from service. The records show how the interwar Repat was suspicious about 'false' claims and might not give a veteran the benefit of the doubt when medical evidence was incomplete or inconclusive, and that its staff were quick to deduce that a condition was not war-caused and slow to help a veteran develop a more convincing claim. The episode highlights the power of the Repat bureaucracy and the use of medical jargon that bewildered and disempowered veterans. Making a pension claim was confusing and stressful for veterans, and Bill was right to be upset about his treatment and to bear a lifelong grudge against the Repat.

Fred Farrall

Of all the veterans featured in *Anzac Memories*, Fred Farrall has had the greatest impact on readers and reviewers. People seem to be drawn to the shell-shocked soldier who became an anti-war activist, to a story that cuts across Anzac expectations. My writings about Fred have been republished in anthologies and reprinted in student readers. Most recently, in 2010 the *Socialist Alternative* magazine and website featured Fred as a centrepiece of an article attacking the 'celebration of war' on Anzac Day. Fred's story, argued Kyla Cassells, proved 'the absolute lie' of the defence that 'Anzac Day doesn't promote militarism, but recognises those who fought and died in past wars'.[25] Fred had wanted me to promote an anti-war moral through his interview. As a novice historian and occasional peace activist in the 1980s, was I too quick to accept Fred's story and mythologise his radical life? New sources enable me to confront that question and add fresh layers to my account of Fred's life and memory.

Fred Farrall's Repat medical files are 10.75 centimetres thick. Five fat files cover eight health conditions that were accepted as war-caused, the first in 1920 and the last in 1981. None of the conditions were medically straightforward and all were the subject of endless claims, appeals and arguments. The files present rich evidence of the on-going physical and mental impact of the war, and of the postwar battles between the Repat and a damaged veteran. They illustrate contradictions and changes in Repat medical understandings, and how the Repat characterised 'vexatious' applicants like Fred and used moral judgements to inform pension decisions.

Apart from the Repat files, Fred's own papers are now available at the University of Melbourne Archives. There are 69 archive boxes of files related to Fred's role in the Federated Clerks' Union and the Communist Party, as a Prahran City Councillor and Mayor, and as an enthusiastic member of the Henry Lawson Society, St Kilda Football Club and other organisations. Scattered throughout the collection are several unlisted boxes with files about the war, including press cuttings, notes about books on the war and material from our interview. There are also several files of Fred's correspondence with the Department of Veterans' Affairs from the late 1970s, which include copies of some of the Department's documents about Fred in the 1980s that are still closed in the National Archives.

Stashed within Fred's papers is a sheaf of five letters Fred wrote to his mother and siblings from English military hospitals after he was wounded at Polygon Wood in 1917. Fred's letters, which he did not show me in the 1980s, are suggestive about his wartime attitudes and about his postwar story-telling. In the letters, Fred explained to his family that this correspondence, unlike letters from active service, was not censored, though it is clear that he did not want to upset his mother or represent his own actions in too negative a light. To his mother he wrote that before the attack 'every man was anxious to get a crack at the Boche' [Germans] and explained that 'I got my crack' [wound] during heavy shelling as they charged the German line. To his sister Laura he wrote that he would never 'forget the 26th of September 1917 in a hurry […] We enlisted to fight didn't we & now we are getting it & Ypres is the worst ever the Anzacs were in. I don't care though Laura. I won't lose any sleep about going back to the firing line.' Bravado does not entirely conceal anxiety about the immediate and long-term future. 'There is a very big chance of me getting trench feet again this winter. But I'll have to chance it, & what a useless lot of cripples we'll be in a few years with rheumatics, etc. We'll be always patching up & messing about with doctors and hospitals, that's just what it will be Laura.'

Fred wrote the most detailed account of the battle to his older brother Sam, who was back on the farm at Ganmain. This eight-page letter from 1917 is very close in factual detail to the interview account from 1983: waiting in shell holes for the order to attack while suffering a terrible bombardment, with wounded and dead men 'everywhere'; losing his rifle and then advancing behind a British barrage; copping a 'Blighty' wound in the leg ('I was as happy as could be Sam, when I got knocked'); hobbling back five miles to safety 'under shell fire all the time' past the wounded and dead ('that's what hurts most Sam, to see our dead & wounded lying everywhere & can't help them'). Fred was 'done in' when he finally reached the Field Ambulance in Ypres; his 'leg was cold' and he 'couldn't move a yard'. There is, however, one significant difference between the battle stories told in the letter and in the interview. In the interview Fred described sheltering in a concrete pillbox where German prisoners were helping Australian and British wounded while a couple of miles further up 'they were carrying on with the job of killing one another as fast as they could'. It was this scene, Fred recalled in the interview, which planted the seeds of his questioning about the war and 'how silly can we be'. The pillbox shelter is not mentioned in the letter to Sam, though Fred does tell his brother it was 'a pleasure to see the German dead & to see the prisoners coming in & how cunning they are too the rotters. They fight till the last & then throw up their hands & expect mercy.' The prisoners, wrote Fred, made themselves 'very useful as stretcher bearers & helping our wounded out. They know that our chaps won't hurt them if they are doing that.'[26]

The letters from 1917 suggest that Fred had little if any pacifist or internationalist feelings at that time, and confirm that Fred's experience with German prisoners took on a different meaning later in his life. Perhaps the seeds were 'planted' in 1917, as Fred suggests, but the pacifist story took root and flourished with Fred's political conversion in the late 1920s. This incident, with its political lesson, would become one of Fred's defining life stories. At his 90th birthday celebration in 1987, Fred told the pillbox story to an audience which included the Consul from the German Democratic Republic, and concluded, once again, that this was the moment when his loyalty for warfare was shaken because he realised 'how stupid it all was'.[27]

Fred Farrall's Repat files, including the set within the Farrall Papers, show that he was not a passive victim of the war or the Repat, but that he became, in time, a determined and articulate activist for his own cause and for other veterans. Fred Farrall learnt how to work the Repat system to best advantage. He probably believed that his growing list of ailments were all war-caused, and he certainly believed that he deserved compensation for the effects of the war, even where cause and effect were difficult to determine. Over the years he listened carefully but selectively to doctors who examined and treated him, and where a diagnosis made useful sense he incorporated it into his war and Repat story. As I explained in earlier chapters, Fred's story of the war and its aftermath was shaped by his distinctive experiences of service and injury, by his bitterness about postwar mistreatment, by the socialist politics he developed in the 1930s, and by new ways of understanding war and its soldier victims that became available by the 1970s. The Repat files reveal details that Fred did not relate in our interviews but they also show how parts of Fred's interview story had already been articulated through his adversarial relationship with the Repat.

Like most veterans applying to the Repat, Fred insisted that he was 'always in good health' prior to enlistment, and there is no reason to doubt that claim. The record of Fred's enlistment age (18 years and three months), height (five foot eight inches, about 1.73 metres) and weight (126 pounds, about 57 kilograms) support his memory that in December 1915 he was a callow and skinny recruit by comparison with his perception of the tough bushmen on the 'Kangaroo' recruiting march. Wartime medical records corroborate his memory of the dates and details of each of the injuries and ailments he suffered on the Western Front, and show that during three years of military service he spent about six months in three separate periods on active service, with the balance of time in military hospitals and then recuperating in army bases in England.[28]

Fred had joined the 55th Battalion in France in late September 1916. In November he had four days treatment out of the line for rheumatism, and then on 17 December he was reported with trench foot and invalided to hospital in England. Trench foot was caused by the damp, cold and insanitary conditions of the front. Constricting footwear and wet socks affected the blood supply to the feet and caused swelling, blisters and open sores which could lead to infection and gangrene if not treated. Most of the hospital records that would have detailed Fred's condition and treatment were lost, probably in 1919. Fred later claimed that the injury was exacerbated by an incompetent officer who ignored the first symptoms and sent him to an outpost in no man's land for another four days until his feet were so bad that three mates had to carry him behind the lines for treatment. A surviving hospital report from January 1917 recorded the 'anaesthesia' of Fred's toes which were 'very cold' and 'discoloured'. Another report from July 1917 reported that when Fred left the line 'the skin was broken' on his feet, and though the sores had now healed, his feet were 'still stiff & rather painful' and required further treatment by massage. Six months later he wrote to his sister Laura that the cold December weather was 'playing havoc with my old feet', which suffered him 'a few hours torture every morning'. Fred would later argue to the Repat that a stay of more than six months in English hospitals for trench foot was evidence of the seriousness of a condition that he blamed for much of his postwar ill-health.[29]

In August 1917 Fred rejoined his Battalion in Flanders and within a month suffered the wound to his left knee and went back to hospital in England. The flesh wound caused no bone injury and was 'practically healed' by the end of October when Fred left hospital, yet for reasons that are not recorded Fred was listed as unfit for active service and remained in English army camps until he was finally declared fit to return to the Battalion in France in October 1918. In our interview Fred recalled that it was during this second period of hospitalisation in England that he showed the first signs of a nervous condition. There is no record of 'nerves' in any of the surviving wartime medical documents or in his wartime letters, but the length of time that Fred was kept in English hospitals and camps suggests there may have been tacit official recognition that he was suffering from more than the physical wound.[30]

When Fred arrived back in Sydney in July 1919 he reported that his feet were still painful on walking and was admitted to a military hospital in Randwick for a further six months of treatment, for which no records survive. Upon discharge from both hospital and army in January 1920 he was still complaining of pain in his feet as well as back pain and was granted a war pension at 25 per cent. Told that his foot and knee conditions would not interfere with 'any of the callings for which he desires training', Fred was classified fit for moderate work and commenced Repat vocational training in motor body trimming. In October he had a minor throat operation (possibly for infected tonsils which were not accepted as war-caused) and a Repat doctor concluded there was 'no objective sign of trouble' with his feet and there 'should not be any disability now'. In November Fred's war pension was cancelled.

Over the next few years Fred continued to complain about his feet and knee to his local 'lodge' doctor, Dr Graham (like many Australians, Fred paid for medical care through membership of a Friendly Society or 'lodge'). In 1926 Graham reported to the Repat that in 1921 Fred's condition 'did not appear to be serious yet he was emphatic regarding the discomfort of his feet and knee joint'. Fred later recalled that in 1921 Dr Graham encouraged him to appeal to the Repat for his ailments, but the Repat advised that a claim was unlikely to be accepted. In 1953 Fred wrote that after that rebuff in the early 1920s he 'was annoyed and did not bother further' with the Repat, and in 1983 he told me the same story.[31]

Most intriguing, and most frustrating, is Fred's statement to the Repat in 1939 that in 1920 and 1921 he was treated for 'nerves' by Dr Arthur of Macquarie Street, Sydney. There was nothing in the 1920s Repat records to support Fred's statement, and a Repat investigation in 1939 revealed that Arthur's practice had closed after his death in 1932 and his case files could not be located. Fred did not recall this treatment in his interview with me, but it seems unlikely to have been an invention. Richard Arthur was a prominent doctor and public figure who treated nervous disorders including shell shock at his Macquarie Street practice in the 1920s. Arthur's missing files might have proved that Fred was suffering from some form of nervous condition when he returned from the war, and would have prevented the Repat from concluding, as they would in 1939, that any such condition was caused by Fred's inability to deal with postwar circumstances.[32]

Fred's employers did provide the Repat with evidence that his health was not good in the early 1920s. In 1939 his 'old Foreman' at the Meadowbank Manufacturing Company told the Repat that when Fred was a motor trimmer from 1922 to 1924 he did 'not appear to enjoy best of health but always worked well' and was a 'very good man'. Mr Da Silva, the owner of an upholstery company which employed Fred from 1924, reported in 1926 that Fred's health had been 'very indifferent', and that he suffered from 'dysentery', copious nose bleeds, and pain in his legs and feet. Fred had lost a few days off work with his gastric complaints, and because upholsterers had to do 'almost all their work in a standing position' there were, explained Da Silva, other days when he should have been 'at home instead of at work'. In spite of these ailments, after 1921 Fred did, as he later recalled, stay away from the Repat until his health collapsed in 1926.[33]

By the mid-1920s Fred was living in a war service home in Rockdale, was married (in 1923) to Sylvia with a daughter, Valerie, born in February 1925. He was also getting involved in union and Labor Party politics. In our interview Fred recalled that in 1926 he suffered a 'general breakdown' which in retrospect he explained as a result of a war-caused nervous condition such that 'You don't appear to have any energy or any, you know, desire to stir yourself'. The escalating paperwork in Fred's Repat file confirms his deteriorating health, though the records — including Fred's own written testimony — focus on Fred's physical rather than mental health. In October 1926 Fred began to frequent the Outpatient Clinic at Randwick Repatriation Hospital and submitted a formal appeal against the cancellation of his pension in 1920. He argued that his feet and legs had 'never been quite right since I contracted trench feet on the Somme', and that during the past 18 months he had 'suffered more pain and inconvenience than previously'. He could no longer play sport and his legs ached after much walking or standing, and it was increasingly difficult to carry on at a job which required him to stand. He also claimed that in the previous four years he had suffered stomach problems which were worst at the start of summer and caused him to lose days off work. Upon request from the Repat his employers now provided evidence about Fred's ill-health, and the lodge doctor Graham confirmed that Fred had lost weight over the past two years, that he suffered from diarrhoea, and that the pain in the feet had become 'more pronounced'.[34]

The Repat doctors who examined Fred noted the wartime documentation about trench foot, rheumatism and naso-pharyngeal infection (a nose and throat condition with cold-like symptoms), accepted that all three conditions were war-caused, and agreed to reinstate Fred's pension, back-dated to April 1926. But the decision that incapacity due to those conditions was only 15 per cent suggests that although the Repat felt obliged to restore Fred's war pension the doctors were not convinced there was very much wrong with him.[35]

Fred was furious about the decision, and wrote from his parents' property in Ganmain that the pension of six shillings and three pence per week was 'quite inadequate'. It was 'absolutely impossible' to continue at work because of his legs and feet and he had now come to the country to 'live with my people, & seek better health' (it is not clear if he was accompanied by his wife and child, whom he almost never mentioned in claims or correspondence). He complained that he had wasted many days and much income attending Repatriation examinations, but that 'except for the paltry pension' he had received 'no treatment & no satisfaction' from the Department, whose doctors 'evidently do not value time'. He concluded, 'I do not wish to draw a pension, would rather be fit & well. But when my health is not what it should be, & I am refused treatment for a war disability, I consider I am at least entitled to a decent pension.'[36]

Over the next decade, at regular examinations and in response to repeated appeals, the Repat acknowledged that Fred had 'a long list of complaints' but they could find no 'objective' evidence of significant war-caused incapacity. Foot X-rays showed that Fred had 'marked' flat feet but 'no other bony changes', and that the flat feet 'satisfactorily' explained pain in the feet on standing which was thus not due to war service. There was no comment on whether Fred's feet had nerve damage or poor blood circulation, which were more likely consequences of trench foot than bone damage and might still be causing pain. In fact, a confidential report written by a private doctor in 1927 in support of an application by Fred for life insurance, which was obtained by a Repat investigator in 1939, had noted that the pain in Fred's feet probably was due to the disturbance of circulation caused by the wartime damage. The Repat doctors did not make that connection.[37]

In the late 1920s and early 1930s there was widespread Australian publicity about a phenomenon that became known as the 'burnt out digger'. Increasing numbers of veterans were reporting to the Repat and to the press with multiple health complaints which they claimed had their origins in the war and which were now recurring as their bodies aged and they suffered the economic hardship of the interwar years. Many veterans were certainly suffering the physical and mental effects of war. Some may have exaggerated their ailments or invented the war connection to get a better pension in hard times. Repat doctors were suspicious of fraud and malingering, and with the Department seeking to reduce public spending its doctors were unlikely to accept doubtful claims. Between 1930 and 1934 the case load of Repat enquiries and claims increased but the number of accepted claims fell dramatically.[38]

Throughout the 1920s and 1930s — as Fred made frequent complaints and appeals — it is quite clear that the Repat doctors thought he was exaggerating or inventing his symptoms to get a better deal. One remarked in 1927,

'I doubt if he will confess improvement until pension decided'. The doctors' medical conclusions were profoundly influenced by their judgement of Fred's physical and mental character. He was 'a thin, rather slow type of man'; he was a 'small poorly nourished and anaemic man with nasal voice — slow and dull witted — has a long list of complaints etc'; he was a 'spare nervous man' with 'wide set ears'; he looked 'gloomy' and did not 'seem to have enough sense to bend his knees when asked'. These characterisations — with their hint of eugenics (the social movement that sought to correct undesirable human hereditary traits) — implied that Fred was unwell because he was a feeble specimen in both mind and body. Nor were the doctors impressed by what they perceived to be Fred's verbose and argumentative manner. Fred was 'loquacious' and 'querulous'; he had 'a knack of hindering exam, and questioning'; he gave a 'most unhelpful history'. As a unionist and Labor Party member from 1926, a Labor-nominated Justice of the Peace in 1929, a committed Communist from 1930 and Communist Party candidate for the New South Wales elections in 1932, Fred had learnt how to make a case and deal with authorities, and the doctors did not like his challenge to their authority.[39]

Repat doctors were quick to dismiss Fred as a hypochondriac and attribute his complaints to moral and bodily weakness, but they did not countenance that his ill-health might have psychological causes with their origin in the war. Fred himself did not make that connection or even mention his 'nerves' in his correspondence with the Repat before 1938. If it is true that Fred had seen Dr Arthur for nerves in 1920 it is likely that he suspected it was war-caused, but perhaps he believed he had a better chance of success with the Repat if he emphasised physical symptoms that could be directly linked to conditions listed in his wartime medical records. Yet within the Repat records of the 1920s and 30s there is ample evidence of Fred's mental ill-health. Fred was described as 'lethargic' and 'neurasthenic' (a listlessness and fatigue due to nerves); he looked 'gloomy'; he was a 'nervous man'; he had a 'slight impediment in speech'; he 'wrinkled' his forehead and displayed 'almost tic-like movement'. Farrall was, concluded one Repat doctor in 1927, 'neuropathic' (a person of abnormal nervous sensibility or affected by nervous disease). Yet none of the Repat doctors considered whether or not these symptoms might have been war-caused. Indeed, one asked in 1927 if there was 'any record of his mental condition before' but concluded that Fred's 'mental dullness might be constitutional or due to ill-health' caused by his 'record of work since the war'. If the nervous condition was not war-caused then the Repat was not responsible for its treatment, and none was offered.[40]

Fred badly needed a war pension in the 1930s. Early in 1930 the Don Brothers furniture factory closed down and Fred lost his job. State Labour Exchange and Relief Bureau records, obtained by the Repat in 1939, show that after a short-lived and unsuccessful attempt to start his own upholstery business, Fred was unemployed until the end of the decade and survived off the 15 per cent pension, food relief and occasional relief work (he was classified as physically unfit for relief work in 1934 and again in 1938). Shortly after joining the Communist Party in 1930, Fred separated from his wife and child. There was certainly a political disagreement between Fred and his wife's more conservative family, but perhaps Fred's mental and physical state also contributed to the break up. As explained in the earlier chapter, Fred now teamed up with fellow-Communist Dot Parker, who would be his life partner until her death in 1979. Repat records suggest that Fred paid some or all of his pension to Sylvia and Valerie until at least the 1940s. Fred must have been struggling financially and needed a better pension from the Repat. In 1935 he enlisted the help of the RSSILA and in 1937 he asked for support from government Minister Billy Hughes: 'We all know that you have always been a very good friend to the Diggers when they have a genuine grievance, and I consider that I am being unfairly treated in the matter of pension. I am unemployed and through my war disabilities, I am unable to do manual work.' There is no record of a response from Hughes, but the Repat steadfastly refused to raise Fred's pension above 15 per cent.[41]

At the end of 1938 Fred was still unemployed and drawing food relief, and in December he moved to Melbourne with Dot to look for work. He had lost his war discharge certificate and now wrote to army records requesting a copy because his 'prospects of suitable employment' required that he 'possess this document'. The radical Melbourne lawyer and politician Bill Slater, another veteran of the Western Front, was trying to get Fred a job as a lift driver in a hotel, and they needed evidence of his service to make the case for soldier preference. Though Fred recalled to me that by this time he was politically alienated from the army, the RSSILA and the patriotic celebration of Anzac, the letter to army records, along with his use of the League and indeed of Billy Hughes in his pension claims, shows that he had no qualms about asserting his ex-service status when it suited him.[42] To Fred's dismay the hotel opted to employ young women, who were cheaper. As a new arrival in Melbourne he was ineligible for dole relief and relied on Dot's wages and casual work helping relatives who ran a café, until he finally got a job as a cleaner in a bank, followed by wartime clerical work with the public service.

Repat records confirm that at the time of his move to Melbourne Fred was suffering a breakdown in his health comparable to that in 1926. In 1938 he submitted two separate appeals against his pension level on the grounds of 'increasing disability due to trench feet, bad nerves and general health' evidenced by rheumatism in his shoulder, nasal catarrh, knee pain, giddiness, nausea and headaches. This was the first time he had mentioned nerves in an appeal, and on examination he explained that he was 'upset by excitement and noise', had 'frequent dreams' and was easily fatigued. Both the 1938 appeals were rejected because, according to the Repat doctors, there was 'no evident disability' in his feet. Though the doctors had noted Fred's 'nervous manner', the Repat decision did not mention the complaint about nerves.[43]

In May 1939 Fred tried again, in his first correspondence with the Repatriation Department in Victoria. This time, rather than appealing the level of pension paid for his three accepted conditions, he made a new claim for acceptance of a 'nervous condition as due to war service'. He explained that he experienced 'irritability' for things which did not 'worry the average man', 'sleeplessness and shakiness with any excitement', and that he had had 'these nervous symptoms, more or less, since war service' (Fred did not mention the stammer he also suffered at this time). He told the Repat's Dr Klug that he had received no medical treatment for nerves (it is not clear why he did not mention Dr Arthur from 1920–21) but that a doctor at the Randwick Repat Out Patients' Clinic had told him in 1927 that he was 'neurasthenic'. This was true, and is confirmed by a Repat medical record from 1927 in which Dr Minty described Fred as 'lethargic and neurasthenic'. But Minty had not accepted that Fred's symptoms were war-caused, and his confidential written report implied that Fred's character was the problem. In the 1939 examination Dr Klug added that Fred's problems might be due to his current personal circumstances, though he noted that Fred claimed not to have financial worries despite his unemployment and said he was not worried 'to any great extent' by the separation from his wife. Klug found Fred to be a 'man of introspective appearance' and 'small physique', and recommended further examination by the psychologist Dr Godfrey to ascertain '?any nervous condition'.[44]

Repat officials took this claim seriously, and it was at this point that references and records were sought from the doctors who had treated Fred outside the Repat system since 1920, from his employers and the State Labour Exchange, and from companies to which Fred had applied for life insurance. The insurance records — a medical examination in 1927, and a form he completed in 1938 — would be a problem for Fred. On both occasions he had understated his ill-health to ensure acceptance for insurance, never expecting that the Repat would find and use the records. The 1927 examination by a Dr Arthur (it's not clear if this was the same Macquarie Street specialist) had concluded that there was no evidence of brain or nervous affections, that Fred's constitution was 'sound', and that he should be accepted as a '1st class' life at ordinary rates. On the 1938 form Fred lied that he had received 'no medical advice or attention' during the previous seven years, and claimed that he did not suffer from any of a long list of disabilities.[45]

Dr Godfrey — the psychologist who had examined Hector Thomson in 1929 — was presented with a 36-page typed evidence file comprising these reports and relevant extracts from Fred's medical records from 1915 to 1938. From the file he noted Fred's six interrupted months of field service, the various ailments listed in the wartime records and the many postwar pension claims. He highlighted Dr Minty's conclusion that Fred was 'querulous' and 'neurasthenic', that the 1927 insurance examination showed no evidence of brain or nervous affection, and that in his letter to Billy Hughes Fred had not mentioned a nervous disorder. This selective reading of the paper trail gave Fred little chance. On examination, Godfrey found Fred to be a 'rather depressed, neurotic looking man' but was not convinced that Fred was suffering from headaches, poor concentration, giddiness or fainting as claimed. Godfrey concluded that this ex-soldier did not 'exhibit anything more than a constitutional incapacity to meet his difficulties, with the accompanying feeling of not having got a fair deal'. He was, 'from the same cause, introspective and somewhat hypochondriacal'. He recommended rejection of neurosis, and even suggested that Fred's pensionable incapacity be reduced to 10 per cent. A second doctor concurred that the 'very definite evidence' of the files did not support Fred's claim, hinted that Fred's 'rather brief' service in the trenches would not have caused significant damage, and concluded that 'the neurosis present is obviously due to post-war causes only'. The Repatriation Board accepted these recommendations and wrote to Fred that although the doctors agreed he was suffering from neurosis it was not attributable to war service and was thus ineligible for benefits.[46]

Recent histories of Australian postwar responses to 'shell shock' help to explain Fred's diagnosis by the Repat doctors. The term was used early in the war when it was first thought that shelling had a physical effect on the brain which caused symptoms such as shaking and stammering, inability to communicate or move limbs or, at worst, a catatonic state. That explanation was mostly displaced when it became clear that not all sufferers had been exposed to shelling. The phrase 'shell shock' stuck in popular usage, though in the 1920s 'war neurosis' became the more usual official term. During the war, soldiers claiming shell shock were often regarded as cowards and malingerers, but the numbers of men affected and the significant proportion of officers forced the authorities to take the problem seriously, both during and after the war. The Australian official medical historian of the war estimated that 80 per cent of the medical aftermath of the war was caused by veterans' mental troubles. Between 1924 and 1940 the number of successful pension claims for neurosis increased by 27 per cent by comparison with a five per cent rise in all accepted pensions. [47]

In interwar Australia there was widespread popular recognition of the mental damage caused by the war, and both the RSSILA and the 'Diggers' paper' Smith's Weekly supported ex-service pension claims for neurosis. Yet medical opinion was divided. Some doctors, probably the minority between the wars, believed that mental damage was caused by the extreme conditions of war. Among these doctors there were further disagreements about whether the damage resulted from the physical consequences of mental exhaustion, or was a consequence of repression caused by the mental conflict between the pressure to fight and the desire to flee (the new Freudian psychoanalysis was beginning its influence and the treatment of shell shock victims would have a major impact on the development of psychology and psychoanalysis in Australia as in other combatant nations). Yet the majority of medical practitioners between the wars still believed that postwar neurosis was caused by a 'neurasthenic personality' that was predisposed to breakdown in wartime conditions or, more likely, under the stresses of postwar life. Diagnosis of war neurosis was plagued by disagreement about causes, the lack of effective diagnostic tests and by the vague and shifting categories of mental health (for example, some doctors used 'neurasthenia' to refer to a legitimate psychological condition while others used it as catchphrase for character failings). Where there was no wartime medical record of mental damage or plausible physiological explanation of neurosis, the Repat doctors were reluctant to accept that 'nerves' were war-caused. Thus while Godfrey could accept that Hector Thomson's mental condition was likely due to wartime illness and thus pensionable, the documentary evidence about Fred Farrall's condition convinced Godfrey that Fred's character predisposed him to a neurosis that had been brought on by his inability to manage the travails of postwar life rather than by the effects of war.

But Repat policies and medical opinion were gradually changing, and by the end of the 1940s Fred was vindicated. After the 1939 decision, Fred was sick of his humiliation by the Repat and decided to seek treatment outside the system (he later complained that looking across the table at a pension hearing was just as bad as looking across no man's land on the Western Front). In our interview Fred recalled that in 1940 the Melbourne psychologist Reg Ellery taught him breathing techniques which reduced the tension in his mind and body and cured the stammer he had suffered since the war. Ellery, a prominent progressive psychologist with a shared interest in the Soviet Union, had met Fred through radical networks and provided the treatment for free. A 1944 note in Fred's Repat file confirms that Ellery was still treating Fred outside the system 'for nerves'. It is not clear whether Ellery told Fred his symptoms were linked to the war.[48]

In 1949 the Repat finally made that link after Fred claimed a war pension for a duodenal ulcer. At a Repat medical examination Fred told Dr Freedman that he still suffered 'occasional nightmares' about the war. Freedman recommended reopening the case for neurosis because it was, he suggested, a 'potent factor' in the formation of duodenal ulcers. Freedman saw 'no reason to doubt' Fred's statement about treatment by Dr Arthur in 1920 and 1921, which Godfrey had ignored in 1939. Moreover, because of the 'broader view' of the recently amended Repatriation Act, which now gave a veteran the benefit of any reasonable doubt, he argued that Farrall had had 'considerable nervous stress on Service' and that 'surely' trench foot and a leg wound would have aggravated any 'tendency to Neurosis'. There is no record in the file of disagreement with this new diagnosis, and the Repat now accepted both the ulcer and neurosis as war-caused and increased Fred's pension entitlement to 40 per cent.[49]

Though he overcame his stammer in the 1940s, Repat records show Fred continued to suffer from attacks of 'nerves' throughout his life. With the condition now accepted by the Repat, he was admitted to the Repat Hospital in Heidelberg for three weeks after a breakdown in 1950, and then again for six weeks in 1961. After the 1961 episode he took early retirement on medical grounds from his public service job, and in 1962 he was granted the general service pension for unemployable war veterans. In 1983, just a few months before our first interview, Fred reported to the Repat that his nerves were in 'a shocking state'. He was startled by the phone, upset by memory lapses and worried about appointments, had blacked out in the street and did not feel confident outside his home. When I wrote this chapter in 2013, Fred's post-1983 Repat records in the National Archives were still closed, but Fred's own set of Repat files confirm that he was hospitalised for five weeks after another breakdown, with similar symptoms, in 1987. We conducted our final interview a month after he came out of hospital. Fred told me that his health had been bad lately though did not specify the cause. During the interview when I expressed concern that he might be tired and need a break, he responded that 'all I can do now is talk' and pressed on with the interview. Fred was determined to tell his war story and showed no obvious signs of distress while doing so.[50]

Though Fred and the Repat agreed after 1949 on the diagnosis of war-caused neurosis, Fred did not stop campaigning for a better pension that would cover all the physical and mental ailments he believed were caused by war. In 1953 he won acceptance for anal irritation that he said had been set off by gastric problems in Egypt in 1916. During the 1950s he applied unsuccessfully three times for an increase in his pension, citing all the usual complaints, until finally, after his hospitalisation in 1961, it was increased to 50 per cent, with another increase to 70 per cent in 1971 (30 per cent for the neurosis, 20 for the ulcer and 20 for rheumatism). In 1977 a hernia was accepted as war-caused, though arthritis and cervical spondylosis (a degenerative condition of spine) were rejected as due to old age. In 1981, citing 'a general worsening of all conditions', including a throbbing in his knee, the gunshot wound was added to the list of accepted disabilities and the pension increased to 80 per cent. Eighty per cent was not good enough for Fred, who now appealed to the Repatriation Review Tribunal. With the help of a sympathetic Repat doctor who argued that 'the strain of trench warfare' had left its mark on Fred's psyche, and using his own careful critique of the Repat medical records, Fred won over the Tribunal who concluded that the previous assessment for an 80 per cent pension was 'illogical', 'contradictory' and 'frankly and bluntly ludicrous'. The Tribunal adjudged Fred 'totally and permanently incapacitated' (TPI) and lifted his disability pension to 100 per cent. Fred lost significant battles along the way, but in 1983 he finally won a comprehensive victory in his long war of attrition against the Repat.[51]

Fred used his skills as a political activist to assert his own rights, but in his later years he increasingly linked his own concerns to those of other war veterans. In 1955 he enlisted federal Labor politician Frank Crean in a campaign to win an extra pair of surgical shoes for veterans like him with damaged feet. In 1961, after his stay at the Heidelberg Repat Hospital, he wrote a letter to the *Age* newspaper (copied to the Repat) praising the care staff and singling out one of the doctors (the 'department would be richer with more medical officers of this calibre') and, in a typically egalitarian gesture, the pantry-maid Beryl ('the most overworked person, I think, in the hospital, surely this can be remedied'). The Repat had 'a policy of doing things on the cheap' and the Minister should ensure that facilities were improved and a better bus service instituted for the hard-working staff. In 1968 (the year he was elected as a Pensioner Candidate for Prahran Council) he wrote directly to the Minister with a list of complaints. The means test on the general service pension was 'an insult to the third class of citizens (Pensioners)'. The Minister should reinstate the chocolate drink Akta-Vite on the list of subsidised treatments (in 1965 Milo had been approved as an appropriate treatment for Fred's ulcer and for nutrition problems caused by his false teeth) as it 'ill becomes the department, or the government, to still further reduce the old soldiers' standard of living'. A Department decision not to subsidise sandals had left Fred 'with a feeling that trench feet is a minor disability' and should, he argued, be overturned. Fifteen years later, in 1983, Fred finally got his Repat sandals.[52]

Though Fred won most of his later exchanges with the Repat, Repat doctors continued to have mixed feelings about Fred and his conditions. Some of them felt that their hands were tied by changes in the onus of proof that required them to prove that a disability was not war-caused. Responding to Fred's complaint about anal irritation in 1953, one doctor remarked that he did 'not think it could be disproved (there is of course no documentary proof)'. Others were suspicious — just as they had been in the 1930s — of Fred's 'multiple grouses' and suggested that he was exaggerating his symptoms and working the system because he was 'pension motivated'. In the 1950s and 1960s Fred was variously described as 'very hypochondriacle' and 'pernickety in speaking of symptoms'. In 1961 his Local Medical Officer reported to the Repat that Fred was 'playing ducks & drakes with his dept & is a loquacious old gentleman that takes a day & a half to listen to'.[53] By the late 1960s Repat doctors would have been well aware of public controversy about veterans' rorting the system and winning pensions for conditions not caused by war. In 1963 a group of Repat doctors in South Australia resigned after their complaints (which highlighted 'pampering' in Repat psychiatric wards) were ignored, and in 1969 one of them published a fictionalised account of the Repat. Be In It, Mate was a scathing attack on corruption, inefficiency and false claims. The press picked up on the criticisms which sparked a series of reviews of the Repat system and a gradual shift in emphasis from compensation to rehabilitation.[54]

It is difficult to reconcile the Repat doctors' changing and contradictory diagnoses with Fred's version of his conditions and their cause, and impossible to judge the extent to which Fred was exaggerating his symptoms and working the system. Clearly he suffered significant physical injury and illness during the war, though the extent to which his trench foot or wounded knee caused on-going problems is unclear. Though there was plenty in the war that Fred — and in due course some of the Repat doctors — could blame for his nervous condition, there was also much in his civilian life that might have caused anxiety. He struggled to find work in the early 1920s and was unemployed for most of the 1930s. He separated from his wife and child during the Depression. As an active member of the Communist Party he was under constant suspicion and surveillance (government intelligence agencies started a file on Fred in 1933 that likely continued throughout his activist life). He was a central figure in the bruising battle in the 1950s and 1960s between Victorian left-wing unionists and the right-wing Industrial Groupers, and was famously taken to court for burning what he believed to be tainted union ballot papers (it is not surprising that during the court case in 1950 Fred told a Repat doctor that he had 'had a good deal of worry lately' which aggravated his neurosis). Elected to Prahran Council on a Pensioners' ticket, in 1973 he became the Communist Mayor of a Council which included the silvertail suburb of Toorak. He had stuck with the Communist Party through the fallout after the uprisings in Hungary in 1956 and Czechoslovakia in 1968 and held onto his Marxist-Leninist faith through the collapse of the Soviet Union and its European satellite states (at his 90th birthday celebration in 1987 the German Democratic Republic Consul presented Fred with a large bouquet of flowers). Given such a troublesome and troubling life, there's every chance that Fred's peace was as damaging to his mental health as his war.[55]

Yet Fred came to understand and explain all his physical and mental health problems as caused by the war. That was a story with practical pension benefits and an instructive political moral. Socialist politics and histories of shell-shocked Australian soldiers helped Fred to see himself as a damaged victim of war. But it was through his exchanges with the Repat that Fred developed a medical language of explanation and put the jigsaw pieces of his war and postwar life together in a way that made satisfying and useful sense. Perhaps Dr Arthur in 1920 had planted the seed of a link between trench war and nerves, which was then nourished by postwar representations of shell-shocked veterans. In 1940 the politically-sympathetic Dr Ellery probably suggested a connection between the war and Fred's anxiety. In 1949 Dr Freedman made it official and pensionable when he recorded that Fred's war neurosis had contributed to an ulcer, and by 1950 Fred had taken on that official explanation and was reporting to other doctors that his neurosis set up the 'distressing irritation in the stomach and bowel'. In his 1977 claim for a war-caused hernia, Fred wrote a three-page life history of his war and its aftermath, and concluded that the hernia 'could be related to the duodenal ulcer, which in turn was related to the neurosis condition which I would like to elaborate on when interviewed'. When I interviewed Fred in 1983 he told the same story: that his war-caused nerves developed into an inferiority complex such that after the war he could not speak without stammering because 'inside me everything had got into a knot, and that went on for years and years and years'.[56]

By then, Fred was very good at elaborating on his conditions and their causes. The Repat doctors had complained for many years that Fred took for ever to tell a life history of complaints. The man who once stammered every time he tried to speak in public was, by the 1970s, a confident public speaker and story-teller (Reg Ellery had done his job well; indeed, Fred's slow and measured speech in our interview was probably due to Ellery's teaching). In 1971 Fred published an article about 'Trade Unionism in the First AIF' that included many of the well-worn anecdotes that he repeated in his interviews with me in 1983 and 1987. Also in 1971, a Repat Local Medical Officer reported that Fred went 'on and on about his complaints' but did 'not appear unduly distressed or affected by them'. Indeed, the doctor noted that Fred 'enjoys the narration' and that 'his florid narration of his illness' suggested 'some features of hysteria'. The following year a Repat psychiatrist concluded that although Fred seemed fit for his age he had a 'personality disorder' and his 'perseveration of thought and speech' (persistent repetition) indicated 'early senile cerebral impairment'.

I doubt the diagnosis of senile cerebral impairment. Twelve years later, when we met for our interview in 1983, Fred was living by himself very successfully and his mind was still razor sharp. And though Fred suffered anxiety attacks throughout his life, I am not convinced that Fred's determined and precise narration was a sign of psychiatric disorder. Fred's extraordinarily detailed and deliberate story-telling about the war and its consequences — evident in the interviews where my questions were usually interruptions — might equally be understood as a way of making positive and useful sense of life's difficulties in both war and peace. In old age, after a lifetime of trying to understand his war and its impact, Fred's life history had incorporated and combined socialist politics, social history and medical explanation into a compelling explanatory narrative with a strong political purpose.[57]

Before I saw the Repat files, I concluded in the first edition that for Fred in his later years this 'composure' of his war story also provided psychological reassurance, a past that he could live with. I now know from the Repat records that despite that reassuring story, Fred never fully overcame his nervous condition. In his review of the first edition of *Anzac Memories*, historian Michael Roper rightly pointed out that there is a risk of overstating 'the healing power of narration'. Jerome Bruner argues that narratives can console, but not necessarily by solving problems. They offer 'not the comfort of a happy ending but the comprehension of a plight that, by being made interpretable, becomes bearable'.[58] Composure may never be fully achieved and though an explanatory life narrative can be useful and comforting it may not, by itself, cure life's ills.

As I was reading Fred Farrall's Repat file, and saw how he was working the system and shaping his medical history to best advantage, I also wondered if perhaps Fred had worked me in the same way in our interviews. Was I a naïve and ingenuous young oral historian, too ready to accept Fred's words at face value? On reflection, I think not, well not much. I'm convinced Fred believed the story he told me, and indeed much of it is borne out by the extensive documentary record of his life, in the Repat records but also personal papers, trade union archives and even a government intelligence file. In our interview, Fred recalled many significant events from his war and its aftermath in remarkable and impressive detail: the Kangaroo recruiting march in 1915; his wounding at Polygon Wood in 1917; unemployment and illness in the 1920s and 1930s, and so on. Yet the Repat files and other contemporary records such as his wartime letters, when used alongside the interviews, show how some of the meanings of Fred's war story were refashioned over time, affected by changes in his own life and attitudes, and by a lifetime of living with and sometimes against the shifting public narratives of Anzac, of being excluded and silenced by those narratives but also sometimes finding recognition and affirmation. Our interviews were just the latest incarnation of a lifetime of remembering the war and its meanings.

Written records and oral history

Historians who use oral histories alongside other sources are often alert to the strengths and weaknesses of each.[59] The Repat records have the great value of showing the chronological development of a veteran's postwar health and providing very precise detail from particular points in time that is often lost to memory. But just as the contemporary written record might point to a lapse or embellishment in memory, the interview can point to errors in the contemporary record. Examples include Bill Langham's oral testimony exposing the egregious inaccuracy of his medical assessment as 'A1' despite the damaged eye muscle which would be a life-long problem; or the Repat doctor taking Fred Farrall's life insurance application as decisive evidence about good health. A compelling example of memory testing old documents comes from 1982, when Fred received from the Repatriation Department a 48-page set of Repat medical reports dating back to the 1920s, which were to help him prepare a pension appeal. Fred stapled yellow paper notes (using the back of pages of *The Socialist* magazine) onto key extracts and then wrote a 24-page critique of the 'insulting manner' of Repat officials (a reference to an exchange in 1937) and of the errors and assumptions in the reports (in response to Dr Gilani, who reported in 1971 that Fred was 'garrulous and aggressive', Fred noted that the doctor had 'no idea whatsoever of what effect trench warfare might have on a man').[60] In our interview the following year Fred did not go into as much forensic detail, but he made the same criticisms of the interwar Repat and its faulty diagnoses.

The Repat paper trail is also flawed by significant gaps. For example, crucial wartime medical records that might have evidenced Fred Farrall's wartime nervous condition were destroyed after the war, and Dr Arthur's case notes from 1920 could not be found. Further, the doctors' conclusions were shaped by, and are revealing about, the policies and processes of the Repatriation system, the medical understandings of the time and the social attitudes and prejudices of the profession (of course ex-service claimants were also influenced by institutional practices and expectations and they made claims that maximised their chances in the system). The records also highlight differences between medical practitioners, and reveal changes in policy and understanding over time. We can see this most clearly in responses to mental health. In short, the Repat records, like any official records, need to be read with a wary eye.

For all their faults, the Repat records (and Fred Farrall's private papers) confirm one of the central arguments in *Anzac Memories,* that how individuals represent and remember their past life develops over time: as an effort in the first instance to make a story and to make sense of significant war or postwar experience; drawing upon (and sometimes silenced by) available and changing cultural meanings and expressive forms; affected by the intimate relationships within which stories are shared and affirmed (or not); attempting to comprehend the jagged edges of experience and compose a bearable past; responding to new life circumstances that suggest different ways to think about that past. As Italian oral historian Alessandro Portelli famously explained, remembering may be less about events than their meaning. Memory is not a passive depository of facts but is rather 'an active process of creation of meanings'.[61] That process operates in the letter written a month or so after a battle;

in the medical history of symptoms and causes delivered to a Repat doctor in the 1930s; and in an oral history interview recorded in the 1980s. The interview is rarely the first or the last word in remembering.

The Repat records illuminate one set of circumstances and relationships in which meaningful war stories were fashioned over time. We can see, for example, how Percy Bird created a story about being wounded by a gas attack which to his mind was a legitimate war-caused explanation for discharge and homecoming; or how Bill Langham's disillusionment with the Repat was generated through a series of negligent mis-diagnoses; or how Fred Farrall's conviction that trench warfare damaged his 'nerves' evolved through his battles with the Repat across many decades. These examples confirm the importance of what the Popular Memory Group defined in the 1980s as the 'particular publics' of remembering, and which more recent theorists have labelled 'communicative memory'.[62] Life stories are articulated in story-telling relationships, such as the family gathering, the veterans' reunion or a medical examination, and of course the oral history interview.[63] In these relationships people share, rehearse and hone their stories, often responding to and sometimes resisting others' versions and questions. One person's story may gain expressive power and coherence in response to interest and affirmation; or another's might be cast into silence because their story does not fit.

Historians using recorded memories work with the paradox of oral history. On the one hand, remembering involves an active creation of meanings in a social context. On the other hand, memory research suggests that long-term memory is remarkably robust.[64] While the short-term memories of the mundane minutiae of everyday life are transient and mostly lost within a few hours or days, we create long-term memories about events which are particularly significant: because they have an emotional charge, are novel, dramatic or consequential, or because they are signposted. It is not surprising that Bill Langham had such detailed memories of the moment his horse was shot from under him, or of the day of his return to Melbourne. Story-telling, too, is central to the creation of long-term memories. The creation and repetition of the story about an event converts that event into a meaningful experience and consolidates it in memory. The story is never fixed — every time we return and remember the event for a different audience it might change in subtle or even significant ways and take on new meanings — but much of the fundamental detail will be retained. The challenge for the oral historian is to live with this paradox and make the most of the memories with which we have the privilege to work. Our opportunity in oral history is to study both the past and the uses and meanings of the past in the present.

POSTSCRIPT

ANZAC POSTMEMORY

What happens to historical remembrance when the last surviving witnesses pass away? There are almost no living Australians who can remember the First World War, and in a few years there will be none who can remember the Second. In *Anzac Memories* I have explored how Australian First World War veterans remembered their war, and how their remembering was shaped by wartime and postwar experience and by their use of Anzac narratives which changed across the years. I've also shown how the memories of war veterans about their wartime and postwar lives can be used to complicate the national story of the Anzac legend.

The passing of these witnesses to history poses a new set of questions. How has Australian war remembrance changed as we leave the era of personal memory? Will it flourish 'more luxuriantly' as it is freed 'from the limitation of historical fact and the human frailties of surviving representatives'.[1] As individual memories fade away will it be cultural representations of the war — in film and fiction, for example — that endure and 'eventually overtake the private and familial myths'.[2] How and why does a twenty-first century Anzac legend work — or not work — for new generations of Australians? How does the relationship between public historical representation and individual historical understanding work when the personal resources for making sense are second- or third-hand family stories combined with popular history and official commemoration? How can historians best contribute to Australian war remembrance? With the centenary of the Anzac landing looming in 2015, historians and other commentators have been taking these questions very seriously. My reflections in this postscript are a contribution to the debate.

The Anzac revival of the 1980s and early 1990s which I outlined in chapter 8 has gathered pace in subsequent years. Indeed, historian Mark McKenna argues that the years of the Howard government from 1996 to 2007 saw a 'revolution' in Australian war commemoration as Anzac became a key element in Australian political nationalism.[3] I'm not convinced by the revolution thesis — Anzac has been conscripted for political use and abuse ever since 1915 — but over the past two decades there have been significant changes in institutional support for Australian war commemoration, and in the forms and meanings of Anzac remembrance.

As the vast cohort of veterans from the two World Wars has diminished the RSL has become less influential in commemoration. Filling that gap, more surprisingly, the Department of Veterans' Affairs (DVA) has expanded beyond its repatriation role to become a key player in official commemoration. The Department now provides schools with extensive educational material about Australians at war. DVA sponsorship, alongside support from other levels of government as well as local communities, has contributed to a proliferation of war memorials and renewed efforts to renovate existing memorials. As Ken Inglis noted in the 2008 edition of *Sacred Places*, his classic history of Australian war memorials, in the ten years since the first edition there had been 'more making and remaking of war memorials [...] than at any time since the decade after 1918'.[4]

Following government support for the visit by Gallipoli veterans to Anzac Cove in 1990 to commemorate the 75th anniversary of the landing, a sequence of officially sponsored events has galvanised popular interest in all things Anzac. In 1993 the interment of the Unknown Australian Soldier — anonymous remains dug up from a First World War cemetery in France and entombed in the Hall of Memory at the Australian War Memorial — captured the national imagination. During the event people queuing to pay their respects initiated a more personal ritual, as they wedged the metal stems of red poppies on the Rolls of Honour next to the names of family members. In 1995, Con Sciacca, Italian migrant and Minister for Veterans' Affairs in the Keating Labor government, instigated 'Australia Remembers', a year-long popular festival of publicity and events to commemorate the 50th anniversary of the end of World War II and to acknowledge the service and sacrifice of men and women who contributed to the war effort.[5] When the last Australian World War I veterans died in the early 2000s they were honoured with state funerals. In Hobart in 2002, Alec Campbell was eulogised by Prime Minister Howard as representing a story of 'great valour under fire, unity of purpose and a willingness to fight against the odds that has helped define what it means to be Australian'.[6]

All the while, newspapers reported that attendance at Anzac Day events, and especially the Dawn Service, continued to grow, with the World War generations of servicemen and women increasingly replaced by younger people with no direct personal experience of war. The Anzac Day protests of the 1980s are now a distant memory, and the trend towards inclusiveness has been sustained, perhaps most obviously in media attention on Australian servicemen from non-Anglo backgrounds and on migrants marching with contingents representing allied forces (including the Turks, allies in the Korean War and now welcomed as friends rather than First World War enemies). In her Prime Ministerial speech at Anzac Cove in 2012, Welsh migrant Julia Gillard remarked that Anzac belonged 'to every Australian', 'not just those who trace their origins to the early settlers but those like me who are migrants and who freely embrace the whole of the Australian story as their own', and to 'Indigenous Australians, whose own wartime valour was a profound expression of the love they felt for the ancient land'.[7] There is now widespread recognition of the significant role of indigenous servicemen and women, though the Australian War Memorial has resisted calls for a gallery to represent black Australians who fought against European settlement. Some local communities have moved in that direction. In the Queensland mining town of Mount Isa, for example, commemoration of the Kalkadoon people's 'heroic, desperate and failed' resistance in 1884 has challenged settler historical narratives and reshaped the local Anzac story.[8]

Since I returned to live in Melbourne in 2007 I've taken groups of university students to a Dawn Service that is very different to the one I recall from my first attendance with student friends in 1982, or indeed the 1987 Service I described in chapter 8. Then it was a small affair with several hundred participants in a simple ceremony focused on the reading of the Ode and the bugler's Last Post and Reveille. After the ceremony we quietly filed through the Shrine of Remembrance and then enjoyed a cup of tea provided by volunteers and chatted with veterans. Now tens of thousands pack the Shrine forecourt and reach back to St Kilda Road, a master of ceremonies explains the origins and meanings of the day through loud speakers, there are speeches by servicemen and women and student winners of Anzac essays, and we flinch as the crack of a rifle salute interrupts a choir singing 'Abide with Me'. Though participants make their own meanings — in the two minute silence my thoughts drift to my family at war — we are also told what to think, and more so than in the 1980s. By contrast, where my parents live in Bateman's Bay on the New South Wales south coast, the Dawn Service has a more local and informal feel, there is almost no military presence, and the clergyman who speaks keeps it short and simple. There is no universal Anzac Day template.

Critics argue that changes in Anzac Day are symptomatic of a wider politicisation of Anzac commemoration and a shift in emphasis from mourning and remembrance to national pride and even celebration.[9] When Carolyn Holbrook interviewed Malcolm Fraser in 2012 about his attitude to war commemoration when he was Prime Minister from 1975 to 1983, Fraser remarked that if he had 'gone to Anzac Cove for Anzac Day, people would have said "what the hell is Fraser doing?"'.[10] By 1990, Prime Minister Bob Hawke was telecasting live to the nation from the Dawn Service at Gallipoli and praising the soldiers whose exploits defined 'the very character of the nation'. Two years previously, the bicentennial celebration of European settlement had been plagued by protests about the white invasion of 1788, and Hawke and his speech-writers found in Anzac a more positive and inclusive national story.[11] Hawke's successor Paul Keating also made good use of war remembrance for political purposes, though his republican leanings favoured World War II when, he argued, Australia was fighting for its survival and asserting its independence from Britain. As Prime Minister from 1996 to 2007, John Howard refocused attention on World War I, in which his father had fought, and his Anzac tradition of mateship and egalitarianism redefined those attributes in conservative terms that emphasised national unity over social dissent. Anzac became a central plank of Howard's critique of 'black armband history' and his appeal for a more celebratory Australian history.[12] Howard's Anzac tradition also had a military aim, as he invoked the 'great tradition of honourable service by the Australian military forces' in support of military intervention in Iraq and Afghanistan. In this new 'war on terror', Howard claimed in 2002, Australia was fighting 'for the same values the Anzacs fought for in 1915: courage, valour, mateship, decency [and] a willingness as a nation to do the right thing, whatever the cost'.

It's clear the level of public investment in and political usage of Anzac remembrance has escalated in recent decades. It's equally clear that significant numbers of Australians — as evidenced by Anzac Day crowds, backpackers at Gallipoli and the commercial success of military history books — are drawn to war commemoration and fascinated by Australian war stories. Yet interest in and understanding of Australians at war is not simply created by politicians and journalists. A study of how Australians relate to the past suggests that people are most distrustful of versions of the past propagated by politicians and in the media, and more likely to trust the histories produced within their own families and by museums.[13] As historian Jay Winter notes, 'politicians have not succeeded in telling people how to remember'.[14] Yet how people remember Australians at war — or rather, how they understand and relate to a past that is beyond lived experience and personal memory — is framed by cultural narratives. And some agencies, including branches of the state, have more power than others to generate those narratives.

The 'popular memory' theory which informed the first edition of *Anzac Memories* is still, to my mind, a useful approach to understanding Anzac remembrance, though it now needs a postmemory twist.[15] The theory focuses on remembrance at two levels, public (or collective) and private (or individual), and the interconnections between these levels. At the public level there is on-going contestation about how the past is represented. Some social agencies are more powerful than others and more effective in promoting their version of the past. Branches of the state may be especially powerful in this regard, and state-sanctioned commemoration — such as Anzac Day — may be especially significant in framing the forms and meanings of remembrance. Yet official versions of the past are only successful in as much as they are meaningful and resonant for a significant proportion of the population. In this book I've argued that a major reason for the success of the Anzac legend — from wartime inception through to Anzac revival — was that it worked for the men whose story it told. For the most part, and for most veterans, it resonated with at least some aspects of their war experience and provided a positive way to make sense of their war and postwar lives, albeit through stories that muted some discordant memories. (Other historians have shown how Anzac remembrance also needed to work for the bereaved family members who were the other important stakeholders in war commemoration, though their commemorative aims often cut across those of the more powerful ex-service lobby.[16]) Had that not been the case, had most soldiers resented their representation in the *Anzac Book*, or in Bean's official history, or on Anzac Day, then we might have had a very different Anzac legend.

The relationship between public remembrance and private memory has changed with the passing of the generation who witnessed the war, who could draw upon their own experience to support, complicate or contest Anzac narratives. The notion of 'postmemory', coined by North American writer Marianne Hirsch, is useful here.[17] Hirsch uses the term to refer to how the Holocaust is transmitted by survivor witnesses to their children. Postmemory includes not just spoken stories, but also potent silences and the ways in which the Holocaust experience impacted upon cultural practices in everyday life, such as the use of food or clothing. For the second generation, postmemories can loom so large in family life and have such emotional affect that they seem, at times, to be their own personal memories. (Hirsch is clear that postmemory is not the same as memory of one's own experience, though it bleeds into the lives and identities of subsequent generations.) Family postmemory, like individual memory, draws upon the collective narratives of the wider culture to create meaning about the past, but with different consequences.[18] On the one hand, without direct knowledge of the remembered past, family postmemory may be more vulnerable to cultural mythologising. On the other hand, the next generation may have less need for a reassuring collective narrative and be more able to bring a critical perspective to the past.

Family history connections loom large in remembrance of the twentieth-century world wars, which had such dramatic effects on so many families. The centenary of the outbreak of the 1914–18 war will be profoundly significant in Australia and many other countries because it is so deeply embedded in family history, though for Australia we might sound two cautionary notes.[19] Over 300,000 Australians served overseas with the First AIF, but a narrow majority of men of eligible age did not volunteer; some of their descendants may have an equivocal response to commemoration.[20] We also need to better understand how post-World War II migrants and their descendants relate to Anzac. Participation of migrant contingents at Anzac Day marches suggests an attempt by some at least to embrace Australian ritual and identity and perhaps adapt it in their own ways. On the other hand, perhaps the reinvigoration of Anzac Day and nostalgia for the two world wars (which, according to John Howard, 'helped define what it means to be Australian') shores up an idealisation of the old white Australia in the face of increasing ethnic diversity.[21]

Australians who do have a family connection to the First World War often make sense of the war through what they know of their forbears' war stories. As I suggested in chapter 10, sometimes that family war history proudly celebrates the Anzac qualities and achievements of its central characters; while sometimes the awkward details of the story complicate any simplistic national narrative. In a recent study of family members' publication of letters and diaries from the First War, Bart Ziino shows how 'the nexus between family remembering and the public myth of Anzac remains mutually constitutive: Anzac frames and affirms family histories, while at the same time it is proving adaptable to the expanding variety of experiences that emerge in family histories'.[22]

Though some family historians bring careful research and a critical eye to their war history, to what extent can we say the same of young Australians who flock to Dawn Services and backpack to Gallipoli? Bruce Scates argues that young Australian Gallipoli pilgrims often use personal connections to make sense of Anzac. His research shows there is often a heartfelt attempt to connect to dead ancestors, and to the war dead more generally, and to emphasis the folly and futility of war rather than patriotic flag-waving.[23] Critics respond that the hunger for ritual and meaning among young Australians is 'manufactured by the prevailing political and commercial imperatives in contemporary Australia', and that lessons learnt through the 'growing commemoration of the Anzac Legend in the classroom' can tend to be 'automatic rather than analysed'.[24]

Part of the problem with this debate is a confusion about contemporary Australian war narratives. Though young Australians may well take pride in their military forbears, they are more likely to pity the terrible experience of war than celebrate a warrior hero. The decline of its warrior elements is a striking change in the Anzac legend. As I argued in chapter 8, this change was already apparent in the 1980s and had been prefigured by influential social histories from the 1970s, such as Patsy Adam-Smith's *The Anzacs* and Bill Gammage's *The Broken Years*. They sought to unhitch the story of the First AIF from a militaristic nationalism and instead depict war as anything but glorious and soldiers as ordinary Australians who suffered the horrors of war yet demonstrated remarkable (though still distinctly Australian) qualities of endurance. Historian Christina Twomey has shown how this was an Australian variant of an international transformation of understandings about war and war service, and about suffering, trauma and remembrance more generally. Drawing upon Didier Fassin and Richard Rechtman's study *The Empire of Trauma*, she argues that it is only since the 1980s 'that it has been widely accepted that "a person exposed to violence may become traumatized and so be recognised as a victim"'. In the 1970s, Holocaust remembrance and cinematic representations of damaged American Vietnam War veterans contributed to this change; the American psychiatric profession's creation of the category of 'post-traumatic stress disorder' in 1980 was a 'signal moment'. Twomey notes that just as Australian feminists in the 1980s were protesting about male wartime violence against women, veteran groups and the media — incensed by the protests — articulated an alternative narrative of ex-servicemen as themselves victims of war. Trauma thus became a 'point of entry for empathetic identification' with old Anzacs and helped bring an audience back to Australian war commemoration.[25]

This analysis helps explain Fred Farrall's reconciliation with Anzac in the 1980s, as he enjoyed newfound public recognition and sympathy for an old digger. It also explains why Australians can readily combine criticism of war and pride in Australian soldiers. Yet just as the radical edge of Fred Farrall's war story was tempered by his embrace of a more affirming Anzac legend, the construction of a 'universal victimhood of the Broken Years'[26] can mute the diversity of Australian war experience. We see this in family war histories that show the stoic courage of broken veterans but will not, quite understandably, represent grandad as a wartime killer. We see it in blockbuster war histories by journalists which are critical of military folly and sympathetic to the long-suffering serviceman, yet portray Australian soldiers, and explain Australian success, in terms of national character.[27]

One of the reasons for the success of the Anzac legend is its plasticity; the story and its meanings stretch and shift with the times and in different contexts and this malleability helps ensure popular support. The versatility of the legend is not always welcome, for another concern about the Anzac resurgence of recent years, articulated by historian Marilyn Lake and her co-authors in their book *What's Wrong with Anzac?*, is that the dominant presence of Anzac has caused a 'militarisation of Australian history'. Other important topics in Australian history, such as the achievement of Federation, the struggles for women's and workers' rights and the social democratic advances in the years immediately before World War I, quite simply get less air play than war history. That's a fair point. It was not too difficult to find a publisher for a new edition of this book about three old Anzacs, but I expect my history of four migrant women — with its focus on domestic life in post-World War II Australia — won't get a second life.[28] Not all topics are equal in the history marketplace.

Historians do need to keep researching, producing and promoting histories about less favoured topics. But that doesn't mean we should leave the study of Australians at war to the journalists and politicians. As Lake argues, we need history to 'run counter to myth-making'.[29] Nor should we underestimate the extent, if not the impact, of rigorous critical histories about Australians at war. The booming Anzac marketplace creates opportunities as well as challenges. For example, many recent military histories have confirmed that Australian military success (or failure) has little to with national character and natural talent, and much to do with training, leadership, logistics and support.[30] Political histories such as Neville Meaney's study of Australian diplomacy during and after World War I have debunked 'one of the most widespread misconceptions in Australian military history', that Australians have often been 'fighting other people's wars'.[31] Other studies have reported the 'Bad Characters' as well as the good among Australian servicemen and women ('many were Anzac heroes.

Some were criminals. Some were both'), and represented indiscipline, fear and brutality alongside courage, endurance and comradeship.[32] This is not disrespectful, concludes Craig Stockings, editor of two recent books which challenge the 'zombie myths of Australian military history' (historians can chop off their head but they keep on coming). It simply recognises the diversity of military experience and that Australian combatants are less distinctive than we might like to think.[33]

Just as important in recent decades has been a flourishing of histories about war's aftermath. These have illuminated the postwar experiences of war veterans, with new sources, such as the Repatriation files, facilitating new historical understanding.[34] Importantly, these histories have also focused attention on the immediate and long-term impact of war upon women and families.[35] They show that war history need not be military history, and indeed that histories that start with war can and must explore multiple issues in the wider society. The trick for historians is thus to take advantage of the militarisation of Australian history, of publishers or documentary-makers looking for war stories, but to take that history in other directions and thus shine light beyond the war and battles and onto less favoured historical topics.

The study of remembrance has been central to this history of war's aftermath, in Australia and abroad. From the intimate remembrance of the bereaved through to the 'sacred places' of war memorials, from school history lessons to popular film, many important recent histories have explored how individuals have made sense of loss and grief, how social agencies have promoted commemorative forms and meanings and how the state has engaged in Anzac. These histories illuminate the complex interplay between individual memory (and postmemory) and public representations of the past.[36] This work points to a dual role for historians of such a potent and contested subject as Australians and war: first, to scrutinise and explain the past; second, to investigate how and to what effect that past is remembered and represented.

More than that, because scholarly writings and lectures have only limited and gradual impact on popular understandings of the past, scholarly researchers need to engage as public historians in the creation and contestation of Australian war histories. The long centenary of 1914–18 is creating ample opportunities. Historians are joining centenary committees, working with museums to develop new exhibitions, making radio and television documentaries, and debating what's wrong and right with Anzac.[37] In this context, 'historian' includes not just academic and professional historians, but also the amateur and family historians who combine thorough research with careful interpretation.

Witness accounts — war diaries and letters, memoirs and oral history — are important sources for war history, though of course there is much that soldier witnesses like the men in this book cannot know, for which we need other sources and other types of history. We are living in 'the era of the witness', in which our society valorises first person testimony — the soldier's story, the survivor's evidence — as the most direct and authoritative account of past events.[38] Historians who use witness accounts need to take care, in both senses of the word 'care': by respecting the narrator yet also bringing a careful critical reading to the account. Perhaps care in that latter sense is especially necessary when the story passes out of living memory. After the last witnesses have passed on it may be easier to distort their stories, neglect the awkward edges and enlist them for other agendas. When Prime Minister Howard eulogised the 'last Anzac' Alex Campbell at his funeral in 2002, he did not say that Campbell was an active trade unionist who came to oppose war.[39]

The distinction between 'common memory' and 'deep memory' suggested by Saul Friedlander and other historians of Holocaust remembrance may be useful here.[40] For all its frailty and forgetting, the 'deep memory' of survivor witnesses offers a rich and heterogeneous account of historical experience which can complicate and disrupt the conventional account of the 'common memory'. But as witnesses pass away and the traces of deep memory fade, the common memory, Friedlander argues, 'tends to restore and establish coherence, closure and possibly a redemptive stance'. That risk is perhaps greatest when the common memory is a national story, such as the Anzac legend, backed by the power of the state and promoting a selective version of national history.

With oral history, we can at least preserve deep memories. Thirty years ago, when I recorded my interviews with working-class World War I veterans, I was struck by how their stories cut across the conventional expectation of the Anzac legend yet had been affected by a lifetime of living with, and sometimes against, the legend. I used those memories in two, interconnected ways: to illuminate the men's experience of war and its aftermath, and to understand their remembering in the shadow of the legend. In this era of Anzac postmemory, I like to think that the memories of men like Percy Bird, Bill Langham and Fred Farrall might continue to be used, with due care, to counterbalance and disrupt the national mythologising of Anzac's 'common memory'.

Appendices

APPENDIX 1

ORAL HISTORY AND POPULAR MEMORY

This appendix explores some of the issues about oral history posed by my Anzac project. I detail my two different approaches to oral history and outline the underlying debates on oral history theory and method. I also explore the impact of the oral history relationship upon remembering, the ethical and political dilemmas posed by a popular memory approach, and the writing of memory biographies.*

The Anzac oral history project

I decided to locate my oral history project in the western suburbs of Melbourne (known as 'the west') because I had worked on other history projects in the area, and because the region suited my focus on working-class diggers. Initially separated from the rest of the city by geographical features, in the late nineteenth century the western suburbs had developed as an industrial and residential area of working-class communities. Most of the men I interviewed had lived in the west for much or all of their lives. They enjoyed its distinctive and often proud working-class identity, and had been active participants in the region's cultural, sporting or political organisations, and in the local RSL sub-branches. Yet although the west had been an important part of these men's lives, the main features of their Anzac experiences, both during and after the war, were not significantly different from those of working-class veterans in other parts of Victoria or in other States.

Class was a more significant defining feature in their lives and identities. With a couple of exceptions, the men I interviewed all defined themselves as working class. Few of them had had more than primary school education,

---

*   New edition note: These methodological reflections present the intellectual and
     personal context of the 1980s and early 1990s when I created and interpreted the
     interviews. I have not changed the first edition text.

and most of them worked throughout their lives as factory, office or farm workers. Significantly, none of them became officers in the AIF; their stories were those of the other ranks. I had suspected that this would be the case when I decided to locate my project in the western suburbs. It was unlikely that many AIF officers would have moved to the region after the war, just as it was relatively unusual for working-class men from the west to become officers.

To find the surviving western suburbs' Great War veterans, I first contacted the RSL sub-branches in Footscray, Williamstown and Yarraville, and was given a list of First World War members who were still in good health. Although my first contacts were all RSL members, and I may well have missed men who were not in the RSL, and who might have had a very different relation to their experience as soldiers and ex-servicemen, some of the men I interviewed passed me on to friends who were not RSL members. I also interviewed a number of veterans who were not from the western region, including the grandfathers of two of my friends, a member of the Gallipoli Legion and, through a contact in Melbourne's Labor History Society, three diggers who had become active in the socialist and peace movements.

Between May and September of 1983 I conducted interviews with eighteen Great War veterans, fourteen of them in the western suburbs, and recorded twenty-seven ninety-minute tapes (I conducted three other interviews in 1982 and 1985). Most of the interviews were initiated by an introductory letter followed by a phone call to arrange a meeting. At the first meeting with each man I described my project in more detail. I explained that I had a rough outline of questions and issues for discussion — about their life before, during and after the war — but that it was only a guide and prompt sheet, and that they should feel free to tell the stories that were important to them. I also explained how the interview would be used, in the first instance for my university thesis and perhaps a book or a public talk, and subsequently as a resource for other researchers at the library of the Australian War Memorial.

After these introductions, which also gave me an opportunity to set up my tape recording equipment, I began with an open-ended question like, 'Where did you start off in life?' Depending on whether the man found it easy to talk or required or expected prompts, I then tried to allow him to follow his own, usually chronological flow, but also brought the narrative back to my questions if a man strayed for long beyond the scope of my interests. As we approached the ninety-minute mark I decided whether a second tape was justified, and whether we should continue the interview now or in a subsequent session. Sometimes I was more exhausted than the interviewee, who seemed to be revitalised by remembering; in other interviews the man was quite drained by the experience. If I decided against a second tape the remaining recording time was sometimes awkward and hurried as I tried to focus on my main interests. If we agreed to make another tape then it was easier to allow the narration to occur in its own time and with its own emphases.

When taping finished I recapped about the use of the tape, and helped the man to fill in a 'conditions for use' form. One man stipulated that he did not want any names, including his own, revealed in publication, but all the others were happy for the tape to be used by me and by bona fide researchers as we wished. We then usually relaxed over tea and biscuits, occasionally accompanied by family members. At this point the man sometimes told stories that had not been committed to tape, which I tried to jot down. I then took my leave, promising to be in touch with a copy of the tape. As soon as I got home I wrote up an interview summary about the occasion and about the life story I had recorded. Then, within a few weeks, I copied the tape and sent the copy with a letter of thanks to the interviewee, promising to return with a copy of the transcript when it was ready.

Soon after completing these Anzac interviews I left for a year of study in England. Back in Melbourne in 1985 I revisited each of the men I had interviewed to give them copies of their transcripts, which had been pro- duced in my absence by a typist with considerable skill in converting the nuances of spoken language into the written word. Some of the men had died, but others were delighted to see me and to receive the transcript, and told me how much their families had enjoyed the tapes. During the year I wrote about the lives of the radical diggers I had interviewed (Fred Farrall, Ern Morton, Sid Norris and Stan D'Altera) in the manuscript 'The Forgotten Anzacs'.[1]

Contesting 'the voice of the past'

At the end of 1985 I returned to England, where I commenced a doctorate on Anzac memories. In order to understand the relationship between Anzac memories and the legend, I began to explore some of the debates about oral history theory and method, and developed the following critique of mainstream oral history. The 1970s' oral history revival in Britain and Australia was profoundly influenced by the criticisms of traditional documentary historians. The main thrust of the criticisms was that memory was unreliable as a historical source because it was distorted by the deterioration of age, by personal bias and nostalgia, and by the influence of other, subsequent versions of the past. Underlying these criticisms was concern about the democratisation of the historians' craft being facilitated by oral history groups, and disparagement of oral history's apparent 'discrimination' in favour of women, workers and migrant groups. Goaded by the taunts of documentary historians, the early handbooks of oral history developed a canon to assess the reliability of oral memory (while shrewdly reminding the traditionalists that documentary sources were no less selective and biased). From social psychology and anthropology they showed how to determine the bias and fabulation of memory, the significance of retrospection and the effects of the interviewer upon remembering. From sociology they adopted methods of representative sampling, and from documentary history they brought rules for checking the reliability and internal consistency of their source. The new canon provided useful signposts for reading memories, and for combining them with other historical sources to find out what happened in the past.[2]

However, the tendency to defend and use oral history as simply another historical source to discover 'how it really was' led to the neglect of other aspects of oral testimony. In their efforts to correct bias and fabulation some practitioners lost sight of the reasons why individuals construct their memories in particular ways, and did not see how the process of remembering could be a key to understanding the ways in which certain individual and collective versions of the past are active in the present. By seeking to discover one single, fixed and recoverable history, some oral historians tended to neglect the many layers of individual memory and the plurality of versions of the past provided by different speakers. They did not see that the 'distortions' of memory could be a resource as much as a problem.

These more radical criticisms of oral history practice were taken up and developed in the early 1980s by the Popular Memory Group at the Centre for Contemporary Cultural Studies in Birmingham. The group drew upon debates in film and television studies about screen representations of the past, and upon more general cultural studies of the significance of the past in contemporary culture. They were also influenced by the small but growing number of international oral historians — such as Ronald Fraser and Luisa Passerini — who were beginning to probe the subjective processes of memory, but whose work was still largely neglected in Britain.[3]

In *Making Histories,* published in 1982, the group outlined its initial, relatively crude alternative for oral history, which required investigation of the construction of public histories and of the interaction between public and private senses of the past. Members of the group then experimented with their theories in a number of case studies of the British memory of the Second World War. The choice of that war enabled the group to combine popular memory work with a second, related interest in popular nationalism, and thus to investigate the ways in which British national identity draws upon particular versions of the national past. During the Falklands War members of the group were astonished by the apparent degree of popular support for the Task Force, and concluded that this popularity was in part due to the ways in which the Falklands War revived selective memories of the Second World War that were deeply fulfilling for many British people. The Popular Memory Group was focusing on the forms in which people articulate their memories (adapting theories about narrative from literary criticism), and upon the relationship between memory and personal identity, when it broke up in 1985.[4]

Some oral history practitioners have been wary of the Popular Memory Group's approach, which seems to represent a view of memory being created from the 'top down'. More justifiably, they point to the group's apparent neglect of different kinds of memory (for example, memories of aspects of life that have not been so overlaid or reworked by powerful public accounts), and of the processes of ageing and remembering in later life. Nevertheless by the late 1980s, British and Australian oral historians were increasingly influenced by the ideas of Popular Memory Group members and of international oral historians who were exploring issues about memory and subjectivity. My own work, which began as a critique of mainstream oral history theory and practice, is now part of a growing movement towards more sophisticated approaches. As Paul Thompson commented in the editorial of *Oral History* in Autumn 1989:

> Our early somewhat naive methodological debates and enthusiasm for testimonies of 'how it really was' have matured into a shared understanding of the basic technical and human issues of our craft, and equally important, a much more subtle appreciation of how every life story inextricably intertwines both objective and subjective evidence — of different, but equal value.[5]

From the writings of international oral historians and the Popular Memory Group, I developed the theory of memory composure which has informed my study of Anzac memories. This new theoretical framework prompted me to reconsider my initial Anzac oral history project, and to reflect upon features of the project that had influenced the remembering of the men I interviewed, including the interview relationship itself.

The oral history relationship

On reflection, it was clear that the stories I was told were the product of a particular age cohort of Great War veterans, who were well into their eighties when I first met them in 1983. In chapters 7 and 9 I explored how remembering was influenced by the social and psychological experiences of old age, and by the resurgence of interest in the Anzacs during the 1980s. Age was also significant because the men who were still alive to be interviewed by me were, on the whole, very young when they enlisted in the AIF. The experiences and memories of young recruits were in some respects quite different from those of older soldiers. They were less experienced in the ways of the world, but may have been physically fitter. They were less likely to be leaving a wife and children behind in Australia, or to have training, employment and a family to return to after the war. Because war was also their youth, remembering often emphasised this coincidence, either bitterly as the loss of innocence or nostalgically as a period of excitement and adventure. The fact of survival into old age is itself significant. Although physical survival in wartime was as often as not a matter of chance, the diggers I interviewed were men who eventually coped with the traumas of the war and postwar years. Some of their mates committed suicide or drank themselves to death, and the interview project obviously did not include such men who could not live with the scars of the war and its memory.

The relationship that was established between myself and each of the men I interviewed also influenced the remembering. Each interview constituted its own particular public, affected by the ideas that each of us held about the other and about how we should behave and represent ourselves. The introductory letter which I wrote to all of the men whose names I had received from the RSL gave them their first sense of me and what I wanted, and was the first way in which I contributed to their remembering. I introduced myself as 'a tutor in history at the University of Melbourne' (I was teaching part-time), and did not mention that I was a postgraduate student. I wanted to represent myself as a bona fide researcher and historian — not just a student with a passing interest — and may well have made an impression as an authority who knew about the past, but also as someone who would listen to their stories and use them as authoritative history. I explained that I was interested in their wartime and postwar experience, but emphasised the focus upon their 'experience of readjustment to civilian life after the war'. Clearly I wanted to know about the war and its impact upon their lives, and not about other experiences and memories — even though they may have had a significant impact on personal identity.

In the letter I also explained that it was important to conduct these interviews to ensure that 'the stories of Australian servicemen will not be lost', and so that 'future generations of Australians will remember their experiences'. I thus appealed and contributed to a possible self-image of the older citizen passing on his histories to the children of the nation. I also mentioned that I would provide each man with a copy of both the tape and transcript of the interview, 'so that you will be able to share your experiences with your family'. In practice that proved to be very well received, and may have encouraged participation, but the sense of a family audience may also have shaped and limited the nature of remembering, and even stopped some men from being involved.

A couple of men rang on the day they received the letter to ask me to come and talk with them, and their enthusiasm suggested a strong personal interest in the recording of their war memories. When I rang the other men, most of them said that they would be glad to see me, but several did not want to talk because they were unwell or felt that their memories were failing, and a few declined because they just didn't want to recall the war. These few may have been wary of the history tutor, or of recalling the war or the past in general because it was still painful. In terms of my subsequent interest in the relationship between identities and war memories, these men are significant by their absence.

Other features of the interview relationship became significant after we met and began to talk. Although my tape recording equipment was compact and quiet, and I gradually became more adept at unobtrusive use, the presence of recording equipment undoubtedly affected the remembering. Some men became so engrossed in their narrative that they were relatively unaffected, but others were quite conscious that they were speaking 'for the record', and adopted a more formal and 'historical' tone when the machine was switched on. Occasionally they asked me to stop taping and spoke off the record about a delicate or embarrassing subject. On-the-record remembering was, by contrast, shaped in accordance with perceptions of what was appropriate for a wider public audience of family and nation.

Sometimes the family or neighbourhood friends had a more direct affect on the remembering. When I arrived for an interview I was often welcomed by wives, adult children or neighbours who were interested in my visit. In a number of cases these people stayed in the room when the interview commenced, and their presence tended to inhibit the questions and responses (for example about sexuality or violence). I found that I developed more intimate relationships with single men, partly because of the emotional needs created by loneliness.

Although I sometimes developed a particularly close relationship with men who were living alone, in every interview the oral history relationship between the elderly man and myself had an effect on the stories I was told. The remembering of most of the men was influenced by two main perceptions of me, as a young man and as an historian. I was twenty-three in 1983, and looked young and healthy. My youth had not been apparent from my letter, but it became an important part of most of the relationships as soon as we met. At one level it is possible that because I was about the same age as they had been when they were soldiers, my youthful presence touched off memories of that youth, and perhaps facilitated an unconscious transference to me of feelings about themselves as young men.

This transference worked in both directions. In the course of the interviews I began to develop an emotional involvement or 'investment' in the men I was interviewing. I was particularly affected by one of my first interviews, with James McNair in Brunswick. I had been delighted by the pleasure and performance of his remembering, and by his interest in me and my project. I was impressed by the detail of his memory and the forgotten stories it revealed, but I also enjoyed being welcomed into his home, his life and his memories. I found that I liked the company of old men, and that I was particularly interested in working-class lives and lifestyles with which I had had little contact in my own upbringing. Perhaps my emotional investment in these old, working-class and 'forgotten' men was an indirect way of rediscovering my missing grandfather Hector (although Hector had certainly never described himself as working class). It may also have been unconsciously related to the significant nurturing role that one elderly, semi-retired army batman played in my very early years. Apart from historical and political motivations, my own emotional investment provided psychological fuel for the oral history project.[6]

My age had another more obvious and even explicit effect. Some of the men remarked that young people today were not interested in the lives or memories of old people. Bill Bridgeman commented that, 'Old men don't mix with young men. You understand that, you're a young man. You don't mix with old codgers'. The interview and my listening therefore gave most of the men a great deal of pleasure, and sometimes encouraged the development of an intense relationship in which I fulfilled an important need for lonely, frustrated and yet enthusiastic old men. That relationship may have occurred whatever my age; most of these men just relished having someone to talk with. But my obvious youth also contributed to the men's adoption of the role of elders relating their experience to a young person and the younger generation in general. From my point of view this relationship was useful because it encouraged the men to open up to me, although it could also be limiting. Stories that might have been told to older men or other veterans, about sexuality or brutality for example, were perhaps deemed inappropriate for my callow ears. More frequently, however, I felt that the men were relating experiences which they had been reluctant or unable to talk about in the past, and that my encouragement and apparent understanding helped make this happen.[7]

The perception of my role as historian, for whom their stories were of general historical interest, also facilitated this openness. My interest and my questions suggested that aspects of their life which may have been difficult to talk about were of historical significance, and in certain cases helped to affirm the value of such memories. For example, wartime fear or guilt and postwar despair were subjects that some of them had rarely talked about before. Several commented that I was the first person they had told in any detail about their war. The interview had helped them to overcome that silence and was an important event for the articulation and affirmation of their war memories.

Sometimes the interview was used as an opportunity to 'set the record straight' about personal or collective digger experiences. As an historian with a prospect of publication, I provided an opportunity for these men to feel that their stories could be heard as history, and to tell their stories in relation to this imagined public audience. Conversely, there may well have been some subjects (perhaps aspects of their personal lives) that the interviewee deemed to be historically insignificant, and kept to himself. Clearly the nature of the recognition available from my oral history interviews had an important effect on the type of remembering that was possible.

The interviews also involved a cross–class relationship; I was a middle-class man interviewing mainly working-class men. I've already explained why this was important for me — both in terms of the needs of my project, and because of my personal interest in Australian working-class lives — but I suspect that this aspect of our relationship was less significant in reverse. I self-consciously dressed for the interviews in a way that I thought would be easily acceptable to the men (my clothes were neat and casual and I had my hair cut because I wanted to represent myself as a particular type of youth). I usually kept my own background to myself, and if I was asked about it talked mainly about my grandfathers and their wars. Inevitably my accent gave me away, and my class background was probably assumed because of my position as a university historian. That assumption may have provoked an ambivalent relationship to public authority, respectful but also wary, but I think that in most cases my relationship with them as a young man and as a historian was more influential, and generally encouraged intimacy and trust.

The framework and content of my questioning also suggested some ways of remembering and closed off others. For example, my interest in aspects of the Anzac legend sometimes led an interview away from topics and events that were of more direct relevance to the interviewee. Yet there were many times when the men rode roughshod over my questions and asserted their own interests and emphases. This helped me to see the many varieties of digger experience, and undermined any preconceptions about the thematic neatness of the legend or of oppositional accounts of Australians at war.

More importantly, my interview focus on three distinct periods of the subject's life — pre-war (briefly), wartime and postwar (specifically the period of reintegration into civilian life) — asserted the centrality of the war in a man's life and played down other significant chapters of life history. The focus was partly due to the practicalities of the interview project; I didn't have the money or time to do detailed life history interviews of the years after the 1930s' Depression, and decided that that period was the least essential for my study. In the interviews in which I perceived a shift during middle age — such as a major change in employment or re-enlistment in 1939 — I did try to discuss new developments; but most of the interviewees, sensing my focus upon the 1914–18 war, talked about aspects of their lives relating to that war, and did not open up about their later lives. In retrospect, it would have been better to have investigated the middle age and later life of my interviewees more closely in order to analyse the events that significantly affected their identities and their remembering of the war.

Nevertheless the interviews suggested that for many of these men there were no great changes in the pattern of their lives once they had settled into a job, a family, a house and a community. Not all of the men settled in these ways, but the pattern does seem to have been common among working-class men of that generation. More importantly, most of the men perceived their middle age as a time of continuity, perhaps involving a gradual material improvement, but with few momentous events that were fixed in memory.

Their memories of that period were often generalised and vague. In contrast, the First World War was a disruptive and momentous experience, both exhilarating and traumatic, for all of them. In most cases it also coincided with a personal transformation from youth to manhood. Not surprisingly, this period was highlighted in the men's memories. In some cases this process was reinforced in retirement when men lost the affirmation of work-place identities, and in old age when they sought to recover a more vibrant identity from their youth.

A popular memory interview approach

The remembering that takes place in an oral history interview will inevitably be influenced by the interview context and relationship, by the situation and identity of the narrator, and by public representations of the past that is being recalled. The fact of such influences does not invalidate oral history; rather it suggests the need for an interview approach that is sensitive to the processes of remembering. In my second set of Anzac interviews I tried out a 'popular memory' interview approach.

In 1987, while on a two month research trip to Australia, I wrote a letter to each of the men I had interviewed in 1983, saying that I would like to 'fill in some of the gaps' of the previous interview. Most of the men had died in the intervening years, but five of them — Percy Bird, Ern Morton, Bill Williams, Fred Farrall and Bill Langham — responded to a follow-up call and seemed delighted at the prospect of a reunion and a second interview.

Careful rereading of my initial life-story interviews had revealed suggestive material about how each man had constructed and related a particular sense of his life and his identity. They showed that a life-story interview could be read in that way, and not just for information about the soldier's experience. In these new interviews I focused on how each man composed and told his memories by exploring four key interactions: between interviewer and interviewee, public legend and individual memories, past and present, and memory and identity. The personal information that I had already gained in the first interviews made it possible for me to tailor my questions specifically for each man in terms of his particular memories and identities. If I had not done the original interviews I would have needed to integrate the life-story approach with the popular memory approach.

The relationship that I had developed previously with each man also facilitated the new interview approach. All five were men who welcomed my interest and enjoyed participating in the project. Their trust and enthusiasm made it easier for me to ask difficult, searching questions, which often cut across the ways in which they told their lives. On the other hand, through talking about their lives in the first interview (as well as other occasions), and from their use of the tapes and transcripts, some of the men had 'fixed' certain stories or themes in their memories, which they subsequently repeated to me and other interviewers. The tapes and transcripts had become an active constituent of individual and collective remembering, elevating the memories of my interviewees within their families and social circles, and prioritising certain aspects and versions of the past.

Sometimes I could explore the nature of these 'fixed' stories; and in every case the process was revealing about the nature of remembering. In a similar way, some of the men resisted my thematic questioning, preferring to retell their stories in their own form and sequence. This was understandable — my new approach was potentially undermining for men who had composed a memory that they did not want to question — and that response was therefore equally instructive. In contrast, others welcomed the new questions and the opportunity for a more thematic discussion.

Whatever the response, with each man I tried to make the interview, and the interview relationship, a more open process. I tried to discuss how my questions affected remembering, and what was difficult to say to me (and to my tape recorder). To encourage dialogue instead of monologue I talked about my own interests and role. In some ways this change in my role (limited by the fact that I never gave up the powerful position of interviewer) affected the remembering. Sometimes it encouraged a man to open up to me and reconsider aspects of his life, though others resisted that opportunity. The explicit introduction of my attitudes into the interviews may have made it easier for men to tell stories for my approval — to project what they thought I wanted to hear — although I usually felt that it facilitated discussion and provoked dissent as much as agreement.

The second key interaction that I wanted to explore in the new interviews was between public and private memories. To investigate that relationship I made the public legend a starting point for questions: what was your response to various war books and films, past and present, and to Anzac Day and war memorials? How well do they represent your own experiences; how do they make you feel? We also focused on specific features of the legend: was there a distinctive Anzac character; how true was it for your own nature and experience? Were you so very different from the soldiers of other armies? How did you respond to military authority, and did you feel that the Anzacs deserved their reputation (whatever that was)?

I asked each man to define certain key terms in his own words — 'digger', 'mateship', 'the spirit of Anzac' — and discovered that some of the men who were uncritical of the legend had contrary and even contradictory under-standings of its key words. Others stuck determinedly to a conventional portrayal of the war, even though aspects of their own experience seemed to challenge it. The negotiations between public and private sense worked differently for each man, often including spaces in which a man could make dissident sense, although all accounts were framed by the themes of the dominant legend. It also became clear that the memories of some aspects of their lives, such as the return from the war, were less reworked by layers of public meanings. As a follow up to this section of the interview we discussed recent (as well as past) battles over the legend, such as the attempts by feminists and Aboriginal activists to make their pasts live on Anzac Day.

Another section of the discussion focused on experience and personal identity: how did you feel about yourself and your actions at key moments (enlistment, battle, return)? What were your anxieties and uncertainties? How did you make sense of your experiences and how did other people define you? How were you included or excluded, what was acceptable and unacceptable behaviour (what was not 'manly'), and how and why were some men ostracised? Of course these memories, and the relative composure of memory, have shifted over time (the past–present interaction), so we discussed how postwar events — such as home-coming, the Depression and the Second World War, domestic change and old age, and the revival of Anzac remembrance in the 1980s — affected identities and remembering. The new interview approach showed me that what it is possible to remember and articulate changes over time, and that this can be attributed to shifts in personal identity and public attitudes.

The new interviews also focused upon the ways in which memories are affected by 'strategies of containment', the methods we use to deal with frustration, failure, loss or pain. This required a sensitive balance between potentially painful probing and reading between the lines of memory. What is possible or impossible to remember, or even to say aloud? What are the hidden meanings of silences and sudden subject changes? What is being contained by a 'fixed' story? In what ways are deeply repressed experiences or feelings discharged in less conscious forms of expression, in past and present dreams, errors and Freudian slips, body language and even the humour used to overcome or conceal embarrassment and pain. Discussion of the symbolic content and feelings expressed by war-related dreams suggested new understandings of the personal impact of the war, and of what could not be expressed publicly. My interview notes about facial expressions, body movements, and the mode of talking often revealed emotive meanings of memories that were not always apparent in interview transcripts.

The popular memory approach raised ethical dilemmas for me as an oral historian. Interviewing which sometimes approached a therapeutic relationship could be rewarding for the interviewer but damaging for the interviewee. It required great care and sensitivity, and a cardinal rule that the well-being of the interviewee always came before the interests of my research. At times I had to stop a line of questioning in an interview, or was asked to stop, because it was too painful. Unlike the therapist, as an oral historian I would not be around to reconstruct the pieces of memories that were no longer safe.[8]

Oral history work that uses a popular memory approach poses a second ethical dilemma with a political dimension. It is relatively easy to cooperate on the production of a history that gives public affirmation to people whose lives and memories have been made marginal, and that challenges their oppression. This has been the usual aim of community-based oral history projects in Britain and Australia, including projects in which I have been a participant. It was also the aim of the 'Forgotten Anzacs' project which I conducted in 1985 with the four radical diggers I had interviewed.

In *Anzac Memories,* however, I have used oral testimony to explore and question a legend that provided a safe refuge for many of the men I interviewed, and as such they might not have agreed with all my conclusions. I tried to share those conclusions in 1983 by speaking about the project at a Footscray Historical Society event which was attended by several of my interviewees. I also showed some of the men I interviewed — especially the men of whom I wrote memory biographies — excerpts of my writing based on the interviews, and asked them for responses and suggested amendments. However most of the old diggers had died before that was possible, or were not well enough to maintain an interest in the project, and after I went to live in England I lost contact with all but a few.

If I was to initiate a similar project today I would pursue the community history approach, which involves at least some of the narrators in both the interview and publishing stages of an oral history project. Such collective work would not necessarily resolve the tension between an approach that seeks to explore remembering, and the fact that participants may not feel able or willing to interrogate their own lives and memories in this way. Indeed, a collective project might make that tension explicit and thus generate difficulty and pain. Yet, as is often the case in participatory history projects, the collective exploration of life histories might also help people to recognise and value experiences that have been silenced, and to come to terms with difficult and painful aspects of their past lives.

I hope this book will add to the growing awareness among oral historians of such ethical and political dilemmas. My own Anzac interviews were empowering for some veterans and may well have been challenging and difficult for others. The history I have produced — which seeks to do justice to the diggers' experiences and memories, while also exploring the legend of Anzac and the impact it has had on their lives — may well have the same mixed effects.

Writing memory biographies

The 'memory biographies' that I wrote about Percy Bird, Bill Langham and Fred Farrall explore the particular ways in which these veterans composed their memories of the war in relation to the legend, and in relation to their own shifting experiences and identities. The writing of memory biographies posed issues that are revealing about the processes of remembering and the uses of memory. Although it is relatively easy to glean information from an interview about a man's past locations, activities and social networks (though even these details may be hidden from an interviewer), using interviews to speculate about past identities is much more problematic because stories about those prior identities are affected by subsequent events and viewpoints. Unlike positivist oral historians, for whom this retrospectivity is a problem to be isolated and excluded, for me it was an important aspect of my study, and suggested two different ways to write memory biographies.

One chronological approach is to trace the construction of memory over time, as new layers of meaning are added and old identities are re-worked or shed. The value of this approach is that it reveals how new experiences and understandings, and shifting public contexts, create changes in our remembering. The problem with this approach is that the evidence for these changes is contained in stories that are related in the present, and that are inevitably overlaid with retrospective meanings. An alternative approach is to focus on the memory of a particular experience, and then to peel away the layers of meaning that have been constructed around that experience over time and in different social contexts. In effect this means to start with today's memory and work back through earlier articulations of the same experience. Sometimes this approach can be facilitated by answers to direct questions about changes in identity and memory, but often it requires a careful reading of the sedimentary layers of memory. This approach can be richly rewarding for an understanding of the ways in which memories have been composed. It can, however, reduce understanding of the individual's life as a whole, and of memory and identity changing over time in the context of the life course.

I sought to integrate these approaches in the memory biographies of Percy Bird, Bill Langham and Fred Farrall. For the most part I used the chronological approach, noting where the stories I used as evidence for past meanings were redolent with retrospectivity. I broke the chronological flow at certain key points to explore the layers of meaning in the memory of a particularly significant experience. The balance of the two approaches also differed between the case studies because of differences in the ways in which each man remembered. For the most part, the Percy Bird study uses the more simple, chronological approach because Percy's remembering clearly shifted over time. In contrast, Bill Langham's remembering maintained and expressed the layers of meaning he constructed over time, and I tried to show this in the writing.

APPENDIX 2

BRIEF DETAILS OF INTERVIEWEES

Percy Bird (born 1889) grew up in Williamstown and was a clerk with the Victorian railways when he enlisted in 1915. He served on the Western Front with the 5th Battalion until he was wounded and repatriated to Australia in 1917. Upon returning to Williamstown he rejoined the railways and was a senior officer in the Auditing Department at retirement.

Harold Blake (c. 1898) was working in a chemist's shop in Ballarat when he joined the navy in 1914. After service on *HMAS Australia* during the war he was based at Victoria Barracks in Melbourne until the end of his seven year's naval service, and then worked in a chocolate factory.

Charles Bowden (1888) grew up in the Gippsland bush and then joined the Victorian railways. After enlistment in 1916 he served on the Western Front with the Australian Broad Gauge Railway Operating Division of the Royal Engineers. Back in Australia he worked with the railways and the State Public Service.

Bill Bridgeman (1893) went to sea in 1912 and served on *HMAS Sydney* for the duration of the war. Between the wars he was employed by the Harbour Trust in Williamstown, and in 1939 he had a compulsory call-up for the navy.

E. L. Cuddeford (1897) grew up on his parents' sheep station near Albury and was apprenticed to a Sydney engineering company when he enlisted in the early years of the war. He served as a despatch runner with the 9th Battalion on the Western Front, and returned to skilled factory and engineering work after the war.

Stan D'Altera (1897) was an apprentice fitter and turner in a Yarraville factory when he enlisted in 1915. He served with the 23rd Battalion on Gallipoli, and with his brother in the 7th Battalion on the Western Front, until an injury required him to return to Australia in 1917. In the inter-war years he mixed casual labouring work with occasional journalism, and became active in unemployed politics and the Yarraville Citizen's Club.

Leslie David (1896) was employed as a clerical worker with the Victorian railways when he enlisted in 1917. He spent the war working as a clerk in an Australian military camp in England, and returned home to a successful career with the railways.

Fred Farrall (1897) grew up on his parents' small-holding in the Riverina and joined a 'Kangaroo' enlistment march in 1915. He served with the 55th Battalion on the Western Front, where he suffered a number of wounds and illnesses. In and out of factory work in the inter-war years, he became active in the Labor movement and a leading figure in Sydney and Melbourne radical politics.

Jack Flannery (c. 1898) was working as a farm labourer in Tasmania when he joined the 12th Battalion for service at Gallipoli and the Western Front. After the war he moved from farm work in Tasmania to quarry jobs in Melbourne.

Percy Fogarty (1897) grew up in a single parent family in rural Victoria and Yarraville, and enlisted in 1915. On the Western Front he served in the 22nd Battalion and the 5th Pioneer Brigade. After many different postwar jobs he eventually settled at labouring work in a flour mill in Kensington, where he remained until retirement.

Jack Glew (c. 1894) was a farm worker in the Western District of Victoria when he joined the Light Horse Brigade. After service on Gallipoli he transferred to the infantry and served on the Western Front. In the inter-war years he was often unemployed between labouring jobs.

Doug Guthrie (1901) worked with his father on the land and in road and rail-making contract work in Tasmania before he joined the navy just after the Armistice in 1918. Between the wars he worked with the Harbour Trust in Melbourne and the Metropolitan Gas Company, and in 1939 he was called up for further service in the navy.

Bill Langham (1897) ran away from his home in rural Victoria to work as a stable-hand at Caulfield Racecourse. During the war he served on the Western Front as a horse driver with an artillery unit of the 8th Brigade. Wounded just before the Armistice he returned to intermittent work in Melbourne until he settled at a job with the Melbourne City Council.

Albert Linton (1899) grew up in the Tasmanian bush and came to Melbourne for factory work and football. After enlistment in August 1914 he was discharged for being under-age and then re-enlisted in Tasmania. He served with the 31st Battalion on the Western Front until he was wounded at Ypres in 1917. After government retraining in Melbourne he returned to factory work.

A. J. McGillivray (1898) enlisted from rural Victoria in 1916 and served with the 29th Battalion on the Western Front until he was wounded in July 1918. After the war he took up a successful soldier settlement in Gippsland, but re-enlisted in 1939 and became a Japanese prisoner of war after the fall of Singapore.

Ted McKenzie (1890) was an apprentice carriage builder when he enlisted in the early months of the war to serve with the 24th Battalion at Gallipoli and on the Western Front. He completed his apprenticeship after the war and worked for forty-three years at H. V. McKay's Sunshine Harvester Works.

James McNair (c. 1891) was working with the Melbourne Post Office when he joined reinforcements for the 14th Battalion in 1916. Following service on the Western Front he returned to work with the Postmaster General's Department until retirement in 1956.

Ern Morton (c. 1896) grew up at Dookie Agricultural College and was a farm worker when he joined the 6th Light Horse Brigade at the outbreak of war. After service on Gallipoli he transferred to a Machine Gun Company of the 2nd Battalion and fought on the Western Front until he was wounded in 1918. After the war he trained to become a town clerk in a number of Victorian towns, and for a time was the Labor representative for the seat of Ripon in the Victorian Parliament.

Sid Norris (c. 1894) joined the 19th Battalion when he could no longer find work as a farm labourer in southern New South Wales. He served on the Western Front until he was wounded near the end of the war. Unable to find work upon his return to Sydney he went cane cutting in north Queensland, where he became active in the Australian Worker's Union and Communist Party politics.

Alf Stabb (1895) was employed by the Victorian railways when he enlisted in February 1916. After service on the Western Front he returned to Melbourne with a wartime bride who had been the sister of an English soldier friend, and took up his pre-war employment.

Bill Williams (pseudonym) worked with his father's Victorian land agency before joining a unit of reinforcements for the 23rd Battalion and serving at Gallipoli and on the Western Front. After recovering from wounds he established himself as a self-employed businessman.

NOTES

Where an entry in these notes consists solely of a name and page number (e.g. Stabb, pp. 6–7.), it refers to a transcript of an interview. A number following the name in such entries (e.g. Langham 2, p. 6.) indicates a particular interview in a series. See the bibliography for details.

Introduction to the first edition

1 Boyd C. Thomson, *Boyhood's Fancies*, Brown, Prior & Co., Melbourne, 1917, p. 8.
2 Alistair Thomson, 'Known Unto God', *Ormond Chronicle*, 1981, pp. 31–4.
3 Alistair Thomson, 'Gallipoli — A Past That We Can Live By?', *Melbourne Historical Journal*, 14, 1982, pp. 56–72; *Age*, 20 and 24 June 1966.
4 Alistair Thomson, 'O Valiant Hearts — How the Mighty Have Fallen', *Farrago*, 61, 22 April 1983, pp. 10–11.
5 Alistair Thomson, '"They were finished with you and you were finished": Memories of Western Suburbs Veterans of the Great War of 1914–1918', *City of Footscray Historical Society Newsletter*, 1984, pp. 389–90.
6 References to debates about oral history and memory, and to the writings of the Popular Memory Group, are cited in Appendix 1.
7 Graham Dawson & Bob West, '"Our Finest Hour"? The Popular Memory of World War Two and the Struggles over National Identity', in *National Fictions: World War Two in British Film and Television*, Geoff Hurd (ed.), BFI Publishing, London, 1984, pp. 10–11. For this theory of composure, see chapter 1 of Graham Dawson, *Soldier Heroes: Britishness, Colonial Adventure and the Imagining of Masculinities*, Routledge, London, 1994.
8 Dawson & West, '"Our Finest Hour"?', pp. 10–11.
9 ibid., p. 10.
10 ibid., p. 28; Richard Wollheim, *Freud*, Fontana, Glasgow, 1971, pp. 65–107.
11 The stories of Percy's written testimony were repeated in the interview and adopted the same basic form and punchlines, although my questions sometimes provoked more background detail.
12 Bill Langham, Interview 2, 30 March 1987, p. 33.
13 Langham 2, p. 2; Langham, Interview 1, 29 May 1983, p. 7; Langham 2, p. 4. For an analysis of the 'populist' ideology of 'the people' against 'the power bloc', see Ernesto Laclau, *Politics and Ideology in Marxist Theory*, New Left Books, London, 1977.
14 Langham 1, pp. 1, 5 and 7; Langham 2, p. 25; Langham 1, p. 4.
15 Langham 2, p. 35; see ibid., Précis notes, p. 1.
16 Langham 2, pp. 12, 24, 35.
17 Fred Farrall, Interview 1, 7 July 1983, Précis notes, p. 1.
18 Farrall 1, Précis notes.
19 See, for example, the story about Bill Fraser's warnings about the war in Farrall 1, p. 9; Fred Farrall, 'Trade Unionism in the First AIF 1914–1918', *Recorder*, 54, October 1971, p. 34.

20 Farrall 1, pp. 8–9. In fact Andrew Fisher made his famous 'last man and last shilling' speech as an election pledge in July 1914 just before he became Prime Minister of a new Labor government.

1 The diggers' war

1   C. E. W. Bean, *The Official History of Australia in the War of 1914–1918, Vol. vi, The AIF in France: May — The Armistice* 1918, Angus & Robertson, Sydney, 1942, p. 1098.

2   C. E. W. Bean, *The Official History of Australia in the War of 1914–18, Vol. i, The Story of Anzac: The First Phase*, Angus & Robertson, Sydney, 1921, p. 15; *Canberra Times*, 7 April 1990.

3   Richard White, 'Motives for Joining Up: Self-sacrifice, Self-interest and Social Class, 1914–1918', *Journal of the Australian War Memorial*, 9, October 1986, pp. 3–16; Denis Winter, *Death's Men: Soldiers of the Great War*, Allen Lane, London, 1978, pp. 1 and 32–6.

4   Stan D'Altera, Interview, 8 May 1983, pp. 4–8.

5   Alf Stabb, Interview, 7 September 1983, pp. 6–7.

6   James McNair, Interview, 27 May 1982, pp. 3–5.

7   Jack Flannery, Interview, 1 May 1983, p. 12.

8   E. L. Cuddeford, Interview, 6 September 1983, p. 5.

9   Ern Morton, Interview 1, 21 June 1985, p. 3.

10 Sid Norris, Interview, 13 September 1983, pp. 2–3.

11 Jack T. Glew, Interview, 6 September 1983, pp. 2–3.

12 Charles Bowden, Interview, 2 July 1983, pp. 14–17 and 40.

13 Leslie David, Interview, 27 May 1983, pp. 14–23.

14 Albert C. Linton, Interview, 12 June 1983, pp. 7–8.

15 Ted McKenzie, Interview, 29 April 1983, p. 10.

16 D'Altera, pp. 7–8.

17 Suzanne Brugger, *Australians and Egypt, 1914–1919*, Melbourne University Press, Carlton, 1980.

18 R. Lewis, 'The Spirit of Anzac — Myth or Reality', *Journal of History*, xi, 4, 1980, p. 4; Kevin Fewster, 'The Wazza Riots', *Journal of the Australian War Memorial*, 4, April 1984, pp. 47–53; Morton 1, p. 5; and see Norris, pp. 16–17; Stabb, p. 20; Flannery, p. 29; and Bowdon, p. 24.

19 Jeffrey Williams, Discipline on Active Service: The 1st Brigade, First AIF, 1914–1919, B. Litt. thesis, Australian National University, 1982, p. 72; and C. E. W. Bean, *The Official History of Australia in the War of 1914–1918, Vol. iii, The AIF in France 1916*, Angus & Robertson, Sydney, 1929, pp. 56–62. See also C. E. W. Bean, *Two Men I Knew*, Angus & Robertson, Sydney 1957, p, 165; Michael McKernan, *The Australian People and the Great War*, Collins, Sydney, 1984, pp. 116–49; Bean, *Vol. vi*, pp. 15 and 1085; C. E. W. Bean, *Anzac to Amiens*, Australian War Memorial, Canberra, 1946, p. 191; Peter Charlton, *Pozières: Australians on the Somme* 1916, Methuan Haynes, North Ryde, 1980, p. 145; Bean, *Vol. iii*, p. 699.

20 D'Altera, pp. 11–12.

21 Stabb, pp. 18 and 26.

22 Bill Williams, Interview 2, 1 April 1987, pp. 10–11.

23 Linton, pp. 8–9; Morton 1, p. 12; Stabb, p. 12.

24 Winter, *Death's Men*, pp. 46–9.

25 Currey O'Neill (ed.), *Bill Harney's War*, Currey O'Neill Press, Melbourne, 1983, p. 20; Glew, p. 18.

26 Ern Morton, Interview 2, 26 March 1987, pp. 1–2; Flannery, pp. 22–3.

27 Bill Bridgeman, Interview, 1 May 1983, p. 26.

28 McNair, pp. 9–11.

29 Bowden, pp. 27–9.

30 David, pp. 21–4 and 37.

31 For discussion of the military effectiveness of the AIF, see Jane Ross, *The Myth of the Digger*, Hale & Iremonger, Sydney, 1985, pp. 28–30; Winter, *Death's Men*, pp. 46–9; Jeffrey Grey, *A Military History of Australia*, Cambridge University Press, Cambridge, 1990, p. 111; cf. Bean *Vol. vi*, p. 1078.

32 Cuddeford, p. 11.

33 Flannery, p. 24.

34 Morton 1, pp. 7–8.

35 Linton, pp. 10–11.

36 McNair, pp. 7–8, 13 and 19.

37 See, among others, Paul Fussell, *The Great War and Modern Memory*, Oxford University Press, London, 1975; John Keegan, *The Face of Battle*, Jonathan Cape, London, 1976; Denis Winter, *Death's Men*; Eric Leed, *No Man's Land: Combat and Identity in World War 1*, Cambridge University Press, Cambridge, 1979. For feminist analyses, see: Elaine Showalter, *The Female Malady: Women, Madness and English Culture 1830–1980*, Virago, London, 1987, pp. 167–194; M. R. Higonnet et al. (eds), *Behind the Lines: Gender and Two World Wars*, Yale University Press, New Haven, 1987; Michael Roper and John Tosh (eds), *Manful Assertions: Masculinities in Britain since 1800*, Routledge, London, 1991.

38 R. G. Lindstrom, Stress and Identity: Australian Soldiers During the First World War, MA thesis, University of Melbourne, 1985, p. 96. For this type of Anzac experience see, for example: O'Neill, *Bill Harney's War*, p. 26; A. Baxter, *We Will Not Cease*, The Caxton Press, Christchurch, 1968, pp. 226–7; Morton 1, p. 12; McNair, p. 16; Farrall, Interview 2, 2 April 1987, p. 14. See also the findings of the Australian official medical historian A. G. Butler, *The Australian Army Medical Services in the War of 1914–1918, Vol. iii, Special Problems and Services*, Australian War Memorial, Canberra, 1943, pp. 77–91.

39 For references about Australian self-inflicted wounds and desertion see Thomson, The Great War and Australian Memory: A Study of Myth, Remembering and Oral History, D. Phil. Thesis, University of Sussex, 1990, p. 77. See also Butler, *The Australian Army Medical Services*, pp. 79–80.

40 Stabb, pp. 12–16.

41 Morton 1, pp. 10 and 18–23.

42 Cuddeford, p. 1.

43 Farrall 1, pp. 25–6. On the Australian mutinies of 1918 see Bean, *Vol. vi*, pp. 933–40.

44 Stabb, p. 11; Norris, pp. 11–13; McKenzie, p. 13.

45 Glew, p. 12; A. J. McGillivray, Interview, 12 September 1983, pp. 6 and 7. For a superb account of British responses under the strain of trench warfare, see Winter, *Death's Men*, pp. 137–40.

46 Leed, *No Man's Land*, p. 26; Winter, *Death's Men*, p. 137.

47 Langham 1, pp. 13, 18.

48 Linton, p. 13.

49 D'Altera, p. 16. See also Bill Gammage, *The Broken Years: Australian Soldiers in the Great War*, Penguin, Ringwood, 1975, pp. 84–8; Lindstrom, Stress and Identity, p. 230.

50 Bean, *The Official History of Australia in the War of 1914–1918, Vol. iv, The AIF in France 1917*. Angus & Robertson, Sydney, 1937, p. 144; A. G. Butler, *The Digger: A Study in Democracy*, Angus & Robertson, Sydney, 1945, pp. 17–20. See also Farrall 2, pp. 11–12; Flannery, p. 23; and Graham Seal, 'Two Traditions: The Folklore of the Digger and the Invention of Anzac', *Australian Folklore*, 5, 1990, pp. 37–60.

51 Williams 2, pp. 10–11. See Farrall 2, p. 42; Percy Bird, Interview 1, 8 September 1983, pp. 10 and 18. See also Robin Gerster, *Big-noting: The Heroic Theme in Australian War Writing*, Melbourne University Press, Melbourne, 1987, pp. 148–58.

52 For analysis of Australian troop-ship and trench papers, see Graham Seal, '"Written in the Trenches": Trench Newspapers of the First World War', *Journal of the Australian War Memorial*, 16, April 1990, pp. 30–8; and David Kent, 'Troopship Literature: "A Life on the Ocean Wave", 1914–19', *Journal of the Australian War Memorial*, 10, April 1987, pp. 3–10.

2 Charles Bean and the Anzacs

1 Patsy Adam-Smith, *The Anzacs*, Nelson, Melbourne, 1978, pp. 357–8.

2 For details of Bean's early life, see Dudley McCarthy, *Gallipoli to the Somme: The Story of C. E. W. Bean*, John Ferguson, Sydney, 1983, pp. 1–50.

3 Bean, *Anzac to Amiens*, p. 9.

4 C. E. W. Bean, *With the Flagship in the South*, William Brooks, Sydney, 1909, pp. 129–30.

5 Bean, *With the Flagship in the South; Flagships Three*, Alston Rivers, London, 1913; *On the Wool Track*, Alston Rivers, London, 1910; and *The Dreadnought of the Darling*, Alston Rivers, London, 1911.

6 *Sydney Morning Herald*, 13 July 1907.

7 See Bean, *On The Wool Track;* Graeme Davison, 'Sydney and the Bush: An Urban Context for the Australian Legend', *Historical Studies*, 18, 71, October 1978, pp. 191–209; Richard White, *Inventing Australia*, Allen & Unwin, Sydney, 1981, pp. 63–84 and p. 126.

8 Bean, *Anzac to Amiens*, pp. 6–7; and see Bean, *Vol. i*, p. 550; Bean, *On The Wool Track*, p. 152. For debates about the radical bush legend and the conservative pioneer legend, see Russell Ward, *The Australian Legend*, Oxford University Press, Melbourne, 1958; J. B. Hirst, 'The Pioneer Legend', *Historical Studies*, 18, 71, October 1978, pp. 316–37.

9 C. E. W. Bean, *The Dreadnought of the Darling*, Angus & Robertson, Sydney, 1956, pp. 218–19.

10 C. E. W. Bean, *What to Know in Egypt … A Guide for Australian Soldiers*, Société Orientale de Publicité, Cairo, 1915; Bean, *Two Men I Knew*, p. 43. See Brugger, *Australians and Egypt, 1914–1919*.

11 *Argus*, 20 January, 1915, p. 9.

12 AWM Private Records: K. S. Mackay, 25 February 1915; F. M. Rowe, 28 February 1915.

13 Quoted in McCarthy, *Gallipoli to the Somme*, p. 332.

14   Kevin Fewster (ed.), *Gallipoli Correspondent: The Frontline Diary of C. E. W. Bean*, Allen & Unwin, Sydney, 1983, p. 39.

15   *Argus*, 8 May 1915.

16   See Kevin Fewster, 'Ellis Ashmead-Bartlett and the Making of the Anzac Legend', *Journal of Australian Studies*, 10, June 1982, pp. 17–30.

17   C. E. W. Bean, *Australians in Action: The Story of Gallipoli*, W. A. Gullick, Government Printer, Sydney, 1915, p. 28; C. E. W. Bean, Diary 6, 29 April 1915, p. 11, Bean Papers, AWM, 3DRL 606/6; Fewster, *Gallipoli Correspondent*, p. 103; Bean, *Vol. i*, pp. 547–52. Unless otherwise cited, references to Bean's diaries are to the copies held in the Bean Papers, AWM, 3DRL 606.

18   Quoted in McCarthy, *Gallipoli to the Somme*, p. 328.

19   Fewster, *Gallipoli Correspondent*, pp. 153–4.

20   See McCarthy, *Gallipoli to the Somme*, p. 276; C. E. W. Bean, *Letters from France*, Cassell & Co., London, 1917, pp. 11–12; Bean's diary of 13 June 1916, quoted in Tim Morris, The Writings of C. E. W. Bean in France during 1916, BA thesis, University of Melbourne, 1985, p. 20.

21   See Ross McMullin, *Will Dyson: Cartoonist, Etcher and Australia's Finest War Artist*, Angus & Robertson, Sydney, 1984, pp. 147–51 and 175–6.

22   Fewster, *Gallipoli Correspondent*, p. 135.

23   ibid.; and Bean, *Vol. i*, p. x. See C. E. W. Bean, 'The Writing of the Australian Official History of the Great War — Sources, Methods and Some Conclusions', *Royal Australian Historical Society, Journal and Proceedings*, xxiv, 2, 1938, p. 92. The term 'cutting edge' was used by Bean in 'Our War History', *Bulletin*, 27 May 1942, p. 2.

24   Quoted in McCarthy, *Gallipoli to the Somme*, p. 177.

25   Bean, Diary 4, p. 8 and Diary 5, p. 23.

26   Denis Winter, '*The Anzac Book*: A Reappraisal', *Journal of the Australian War Memorial*, 16, April 1990, p. 58. See also Fewster, *Gallipoli Correspondent*, p. 65; David Kent, '*The Anzac Book*: A Reply to Denis Winter', *Journal of the Australian War Memorial*, 17, October 1990, pp. 54–5.

27   Winter, '*The Anzac Book: A* Reappraisal', p. 58.

28   See Fewster, *Gallipoli Correspondent*, pp. 163–4; and McCarthy, *Gallipoli to the Somme*, pp. 233 and 104.

29   Fewster, *Gallipoli Correspondent*, p. 16.

30   Bean, Diary 17, 26 September 1915, pp. 24–34; Bean, *Letters from France*, p. 189; and see Kevin Fewster, Expression and Suppression: Aspects of Military Censorship in Australia during the Great War, PhD thesis, University of New South Wales, 1980, pp. 100–6.

31   Bean, Diary 17, 26 September 1915, pp. 24–34. Denis Winter claims (in '*The Anzac Book: A* Reappraisal', p. 61) that this vital passage was added to the diary when Bean reworked it in 1916 or 1924, and was thus influenced by experiences in France. There is no textual evidence to support this claim.

32   C. E. W. Bean, *The Official History of Australia in the War of 1914–1918, Vol. ii, The Story of Anzac: From May 4, 1915 to the Evacuation*, Angus & Robertson, Sydney, 1924, p. 427; and see Peter Stanley, 'Gallipoli and Pozières: A Legend and a Memorial: Seventieth Anniversary of the Gallipoli Landing', *Australian Foreign Affairs Record*, 56, 4, April, 1985, p. 287; Fussell, *The Great War and Modern Memory*, pp. 21–3.

33 McCarthy, *Gallipoli to the Somme*, p. 247; Bean, *Letters From France*, p. 115.

34 Bean, Diary 17, 26 September 1915, pp. 24–34. For a study of European literary evocations of the ordinary soldier as hero, see Andrew Rutherford, *The Literature of War: Five Studies of Heroic Virtues*, Macmillan, London, 1978.

35 *Age*, 1 January 1916; Bean diary, 23 December 1915 and 26 December 1915, in Fewster, *Gallipoli Correspondent*, pp. 200–1; Richard Ely, 'The First Anzac Day: Invented or Discovered?', *Journal of Australian Studies*, 17, November 1985, p. 55.

36 Bean, Diary 17, 26 September 1915, pp. 24–34; *Argus*, 22 April 1916. See also Stuart Sillars, *Art and Survival in First World War Britain*, Macmillan, Basingstoke, 1987.

37 *Bendigo Advertiser*, 19 October 1916, Bean Papers, AWM, 3DRL 6673/234.

38 McCarthy, *Gallipoli to the Somme*, pp. 351–2; H. McCann, 'When It's All Over', in C. E. W. Bean (ed.), *The Anzac Book*, Cassell & Co., London, 1916, p. 151; Thomas Loutit to Bean, 23 May 1919, quoting his son's letter of 22 June 1915, Bean Papers, AWM, 3DRL 6673/10.

39 D. A. Kent, '*The Anzac Book* and the Anzac Legend: C. E. W. Bean as Editor and Image Maker', *Historical Studies*, 21, 84, April 1985, p. 378; Winter, '*The Anzac Book*: A Reappraisal', p. 58.

40 Bean, *The Anzac Book*, p, xiv; Kent, '*The Anzac Book* and the Anzac Legend', pp. 380–90; cf. Winter '*The Anzac Book*: A Reappraisal', p. 61. See also Stanley, 'Gallipoli and Pozières', pp. 281–7, Kent's influential article has recently been the subject of severe criticism by John Barrett ('No Straw Man: C. E. W. Bean and Some Critics', *Australian Historical Studies*, 23, 90, April 1988) and Denis Winter ('*The Anzac Book*: A Reappraisal'). These critics have made some necessary empirical corrections to Kent's article, but have not provided any evidence to rebut his main arguments.

41 Signaller McCann, 'Killed in Action', in *The Anzac Book*, p. 105; Kent, '*The Anzac Book* and the Anzac Legend', pp. 380 and 383. See also Bean's own poem 'Non Nobis' in *The Anzac Book*, p. 11.

42 Kent, '*The Anzac Book* and the Anzac Legend', pp. 380 and 390.

3 Memories of war

1 Bird 1, pp. 1–2.
2 Percy Bird, Interview 2, 7 April 1987, pp. 7–8.
3 Bird 1, pp. 8–9.
4 Bird 1, p. 18.
5 Bird 1, p. 31.
6 A. W. Keown, *Forward with the Fifth*, Melbourne, Speciality Press, 1921, p. 188.
7 ibid., p. 177; Bird 1, p. 15.
8 Bird 2, pp. 6 and 24.
9 Bird 1, p. 21.
10 Bird 2, pp. 13 and 5; Bird 1, p. 13.
11 Bird 1, p. 12, and see pp. 9, 13, 15 and 20; and Bird 2, pp. 3–5.
12 Bird 1, p. 10; Bird 2, p. 22.

13   Bird 1, pp. 16 and 20; Bird 2, pp. 21–2; and see Keown, *Forward with the Fifth,* p. 22; and Bean, *Vol. vi,* p. 6.

14   Bird 1, pp. 10, 17 and 18.

15   See, for example, the interviews with Jack Glew and Jack Flannery.

16   Bird 1, p. 11; Bird 2, p. 26; Bird 1, pp. 15–16.

17   Bird 2, p. 26. For other versions of the story about the French women see Adam-Smith, *The Anzacs,* p. 307; Bean, *Anzac to Amiens,* p. 410.

18   Langham 1, pp. 1–3 and 28; Langham 2, pp. 1 and 33.

19   Langham 1, p. 6.

20   ibid., p. 7.

21   ibid., p. 9.

22   ibid., p. 7; Langham 2, p. 3.

23   Langham 2, pp. 3–4.

24   Langham 2, p. 4; Langham 1, p. 19.

25   ibid., p. 26.

26   Langham 1, p. 20.

27   ibid., 1, pp. 14 and 19.

28   ibid., p. 17. For the Jacka legend ('That's the sort of fellow Captain Jacka was') see also Farrall 1, p. 29; Farrall 2, p. 5; and Flannery, p. 18.

29   Langham 2, p. 4.

30   ibid., pp. 4, 12, 16 and 21; see Elaine Showalter, 'Rivers and Sassoon: The Inscription of Male Gender Anxieties', in *Behind the Lines.*

31   Langham 1, pp. 12–13; Langham 2, p. 15.

32   See Langham 1, pp. 14–19; Langham 2, p. 11.

33   Langham 2, p. 15. On the shared language of coping, see Fussell, *The Great War and Modern Memory,* pp. 114–35.

34   Langham 2, p. 21.

35   ibid., p. 20.

36   ibid., p. 18. For similar conversion experiences, see Morton 1, p. 18; Farrall 1, p. 22; cf. Bird, The 5th Battalion, 1916 and 1917, France. Unpublished manuscript, 1983, p. 3. The conversion effect of a meeting with the enemy was also a common theme in war literature, such as Barbusse's novel *Under Fire* (Dent, London, 1926, pp. 154–5).

37   Langham 2, pp. 20–1.

38   Langham 1, p. 10; Langham 2, pp. 8 and 19–20.

39   Langham 1, p. 11.

40   Langham 2, pp. 7, 9 and 13.

41   Bill Langham's remembering of Australian behaviour out of the line, and of the relations between officers and men in the AIF, also shows how experiences that matched the Anzac legend became highlighted in his memory and generalised in terms of the legend. See Langham 1, pp. 10, 13 and 17; Langham 2, pp. 10–13.

42   Farrall 1, pp. 1 and 8.

43   ibid., p. 4.

44   ibid., p. 59.

45   See Farrall 2, pp. 37–8.

46   Farrall 1, p. 11.

47   Farrall 2, p. 1.

48   Farrall 1, p. 23.

49   ibid., pp. 12–13.

50   Farrall 2, p. 7.
51   Farrall 1, p. 14.
52   ibid., pp. 20–1.
53   ibid., pp. 17 and 26.
54   ibid., p. 22.
55   ibid., p. 28.
56   ibid., pp. 25–26 and 62.
57   Farrall 2, p. 26b.
58   Farrall 1, p. 29.
59   Farrall 2, p. 12; Farrall 1, p. 18.
60   Farrall 1, p. 29; Farrall 2, pp. 5 and 11.
61   Farrall 2, p. 42.
62   ibid., pp. 38, 42 and 69–70.

4   The return of the soldiers

1   Bean, *Anzac to Amiens*, pp. 529–31. On repatriation see Ian Turner, '1914–19', in
     A New History of Australia, F. K. Crowley (ed.), Heinemann, Melbourne, 1974,
     p. 354; Tony Gough, 'The Repatriation of the First Australian Imperial Force',
     Queensland Historical Review, vii, 1, 1978, pp. 58–69; Marilyn Lake, *The Limits of
     Hope: Soldier Settlement in Victoria 1915–38*, Oxford University Press, Melbourne,
     1987, pp. 195–6.

2   McNair, p. 14.

3   Gough, 'The Repatriation of the First AIF', p. 62; Leslie Parker [Angela
     Thirkell], *Trooper to the Southern Cross*, Faber & Faber, London, 1934, pp. 228–9.

4   McNair, pp. 14–15.

5   Glew, p. 9; D'Altera, p. 20; and Percy Fogarty, Interview, 3 June 1983, p. 13.

6   Flannery, p. 36. See also Bird 1, p. 13; Langham 1, p. 23; McNair, pp. 21–2; and
     Glew, p. 9.

7   Flannery, p. 36; Gough 'The Repatriation of the First AIF', p. 60; and see Farrall
     1, p. 23. Other veterans fought against attempts to take pensions from them:
     Stabb, p. 22; D'Altera, p. 13; Fogarty, p. 13.

8   D'Altera, p. 22. Alf Stabb, Charles Bowden, Fred Farrall, and Albie Linton also
     used their gratuity to buy a house. Ted McKenzie, Percy Fogarty and Fred Farrall
     bought War Service Homes.

9   Norris, p. 21; Langham 1, p. 24; Flannery, p. 35. And see Gough, 'The
     Repatriation of the First AIF', pp. 60–1.

10   Bridgeman, p. 16; Cuddeford, p. 22; Bird 1, p. 27; Langham 1, pp. 27–8; Farrall
     1, p. 46; D'Altera, p. 29.

11   Morton 2, p, 27; Williams, Interview 1, 10 April 1983, p. 25; Farrall 1, pp. 18
     and 33–4. See Anthony Ellis, The Impact of War and Peace on Australian
     Soldiers 1914–1920, BA thesis, Murdoch University, 1979, p. 28; Butler, *The
     Australian Army Medical Services in the War of 1914–1918*, pp. 142–3.

12   Glew, pp. 14–15; Farrall 1, p. 54.

13   D'Altera, pp. 19 and 28; Stabb, p. 25.

14   Glew, p. 15; McGillivray, p. 16.

15   Flannery, pp. 25, 34, 39 and 53; McGillivray, p. 29.

16   D'Altera, p. 55; Cuddeford, p. 24; Morton 1, p. 26; Bowden, p. 40. See Judith A. Allen, *Sex and Secrets: Crimes Involving Australian Women Since 1880,* Oxford University Press, Melbourne, 1990, p. 131.

17   Ian Turner, '1914–19', p. 354; *Brunswick and Coburg Leader,* 1919, passim; Margaret Glyde, uptaped interview, 12 June 1982.

18   Allen, *Sex and Secrets,* p. 131. For debates about the impact of returned servicemen on the ongoing Australian battle between feminism and masculinism, see: Marilyn Lake, 'The Politics of Respectability: Identifying the Masculinist Context', *Historical Studies,* 22, 86, 1986, pp. 116–31; Judith A. Allen, 'Mundane Men: Historians, Masculinity and Masculinism', *Historical Studies,* 22, 89, 1987, pp. 617–28.

19   D'Altera, p. 17.

20   Bowden, p. 29; David, p. 37.

21   Alf Stabb, Percy Bird, E. L. Cuddeford and James McNair returned to protected employment.

22   Morton 1, p. 28; Fogarty, p. 11; D'Altera, p. 19; Farrall 1, p 44; Norris, p. 21; Glew, pp. 14–16.

23   Langham 1, p. 21; Farrall 1, p. 19.

24   Langham 1, p. 21; McGillivray, p. 14. See Lake, *The Limits of Hope.*

25   D'Altera, p. 18; Morton 1, p. 28.

26   Langham 1, p. 27. See Bird 1, p. 26.

27   D'Altera, pp. 14–19; Fogarty, p. 15, Contrast the experience of E. L. Cuddeford, p. 22.

28   See Stabb, p. 24; Cuddeford, p. 22.

29   For the role of ex-servicemen in industrial and civil violence see: Raymond Evans, '"Some Furious Outbursts of Riot": Returned Soldiers and Queensland's "Red Flag" Disturbances, 1918–1919', *War and Society,* 13, 2, September 1985, pp. 75–98; David Hood, 'Adelaide's "First Taste of Bolshevism": Returned Soldiers and the 1918 Peace Day Riots', *Journal of the Historical Society of South Australia,* 15, 1987, pp. 42–53; Anthony Perrottet, 'Politics of the Anzac Myth', *National Times,* 25 April 1986, pp. 22–3.

30   Terry King, On the Definition of 'Digger', unpublished paper, La Trobe University, 1983, p. 13.

31   *Herald,* 22 July 1919; *Argus,* 22 July 1919; *Weekly Times,* 26 July 1919; *Brunswick and Coburg Leader,* 25 July 1919, 1 August 1919.

5 The battle for the Anzac legend

1   Ian Turner, '1914–19', p. 318.

2   J. Popple, A Wary Welcome: The Political Reaction to the Demobilisation of the AIF, paper presented at the Australian War Memorial Conference, 1987, p. 1; Ian Turner, '1914–19', p. 345; McKernan, *The Australian People and the Great War,* p. 12.

3   Evans, 'Some Furious Outbursts of Riot', p. 82.

4   'One Big Union Manifesto to Returned Soldiers', 16 November 1918, quoted in King, On the Definition of 'Digger', p. 8; *Labor Call,* 20 February 1919, p. 3.

5   Quoted in Popple, A Wary Welcome, p. 12. See also: Evans, 'Some Furious Outbursts of Riot', pp. 86–90; Anthony Perrottet, 'Politics of the Anzac Myth',

pp. 22–3; King, On the Definition of 'Digger', p. 6. For an account of Australia's postwar secret armies, see Michael Cathcart, *Defending the National Tuckshop*, McPhee Gribble/Penguin, Melbourne, 1988.

6   *Brisbane Courier*, 29 March 1919, quoted in Evans, 'Some Furious Outbursts of Riot', p. 89. Terry King introduced me to the concept 'genuine digger', which he outlines in On the Definition of 'Digger' and in 'The Tarring and Feathering of J. K. McDougall: "Dirty Tricks" in the 1919 Federal Election', *Labour History*, 45, November 1983, pp. 54–67.

7   See G. L. Kristianson, *The Politics of Patriotism: The Pressure Group Activities of the Returned Servicemen's League*, Australian National University Press, Canberra, 1966, pp. 1–25; Marilyn Lake, 'The Power of Anzac', in *Australia: Two Centuries of War and Peace*, M. McKernan & M. Browne (eds), Australian War Memorial and Allen & Unwin, Canberra, 1988, pp. 200–2.

8   Terry King, W. M. Hughes and M. P. Pimental: Soolers and Soldiers, 1918–1919, unpublished paper, La Trobe University, 1984.

9   *Labor Call*, 15 May 1919, p. 10; and 26 June 1919, p. 2; Lake, 'The Power of Anzac', pp. 202–4.

10   Report of deputation to Acting Prime Minister Watt, 5 June 1918, General Files 1918, RSL papers, Ms 6609, National Library of Australia. This report is cited in Lake, 'The Power of Anzac', p. 205.

11   *Truth*, 21 January 1919; *Brunswick and Coburg Leader*, 25 July 1919, 1 August 1919, 5 September 1919. David Englander, 'Troops and Trade Unions, 1919', *Modern History*, 37, March 1987, p. 9.

12   Lake, *The Power of Anzac*, p. 200, quoting from a 1916 federal and State conference. For a local example, see *Brunswick and Coburg Gazette*, 5 June 1919.

13   Kristianson, *The Politics of Patriotism*, p. 14.

14   Morton 1, p. 23; McGillivray, p. 16.

15   Farrall 1, p. 17; D'Altera, p. 12.

16   Bowden, p. 37; Stabb, p. 27; Harold Blake, Interview, 17 May 1982, p. 30.

17   Morton 1, pp. 23–4; Norris, pp. 13–14; Farrall 1, 17–28; Les Barnes, Interview, 13 May 1982, pp. 5–6.

18   McGillivray, p. 17; Stabb, p. 28.

19   Langham 1, p. 35; D'Altera, p. 39. See also Gammage, *The Broken Years*, pp. 210–11.

20   Flannery, p. 48; H. McQueen, 'The Social Character of the New Guard', *Arena*, 40, 1975, p. 84.

21   Morton 1, p. 26.

22   Benedict Anderson, *Imagined Communities: Reflections on the Origin and Spread of Nationalism*, Verso, London, 1983; George Mosse, *The Nationalisation of the Masses*, Howard Fertig, New York, 1975; K. S. Inglis, 'Monuments and Ceremonies as Evidence for Historians', *ANZAAS Congress Papers* 834/26, September 1977. For a feminist analysis, see Carmel Shute, 'Heroines and Heroes: Sexual Mythology in Australia 1914–1918', *Hecate*, 1, 1, 1975, pp. 7–22.

23   See David Cannadine, 'War and Death, Grief and Mourning in Modern Britain', in *Mirrors of Mortality: Studies in the Social History of Death*, Joachim Whaley (ed.), Europa, London, 1981.

24   For histories of war memorials see Michael McKernan, *Here is Their Spirit: A History of the Australian War Memorial, 1917–1990*, University of Queensland Press, St Lucia, 1991; and references in Thomson, The Great War and Australian Memory, p. 170.

25 Queensland Anzac Day Commemoration Committee Minutes, 18 August 1920 and 4 August 1932, quoted in Mansfield, World War I, Foundation Myths and the Anzac tradition: A Reappraisal, paper presented at the Australian War Memorial Military Conference, 1982, pp. 9 and 14.

26 O'Neill, *Bill Harney's War*, pp. 34 and 54.

27 *Brunswick and Coburg Leader*, 21 March 1919; 23 May 1919; 4 May 1923; and see J. C. Sullivan, The Genesis of Anzac Day: Victoria and New South Wales, 1916–1926, BA thesis, University of Melbourne, 1964.

28 Quoted in Peter Sekuless and Jacqueline Rees, *Lest We Forget*, Rigby, Dee Why West, 1986, pp. 47–8.

29 Inscriptions recorded by the author at Gallipoli and in France. See McMullin, *Will Dyson*, p. 247; Inglis, 'A Sacred Place: The Making of the Australian War Memorial', *War and Society*, 3, 2, September 1985, p. 109; and E. Holt, 'Anzac Reunion', in *Best Australian One-act Plays*, W. Moore & T. J. Moore, Angus & Robertson, Sydney, 1937.

30 George Mosse, 'Two World Wars and the Myth of the War Experience', *Journal of Contemporary History*, 21, 1986, p. 494; Gerster, *Big-noting*, pp. 119–20.

31 *Brunswick and Coburg Gazette*, 2 May 1930 and 24 April 1931.

32 *Argus*, 26 April 1926.

33 Maurice French, 'The Ambiguity of Empire Day in New South Wales 1901–1921', *Australian Journal of Politics and History*, 24, April 1978, pp. 61–74.

34 Inglis, 'Monuments and Ceremonies', pp. 19–22.

35 See Mansfield, World War I, Foundation Myths and the Anzac Tradition, p. 13.

36 Inglis, 'Men, Women and War Memorials: Anzac Australia', *Daedalus*, 116, 4, Fall, 1987, p. 54, quoting from *Reveille*, 30 April 1928.

37 Queensland ADCC Minutes 9 October 1922, quoted in Mansfield, World War I, Foundation Myths and the Anzac Tradition, pp. 11 and 15; *Argus*, 26 April 1924.

38 John McQuilton, 'A Shire at War: Yackandandah, 1914–1918', *Journal of the Australian War Memorial*, 11, October 1987, pp. 11–13; Sullivan, The Genesis of Anzac Day, p. 37, quoting the official organ of the Melbourne RSSILA, *Realm*, 30 April 1923.

39 Sullivan, The Genesis of Anzac Day, p. 49.

40 *Victorian Parliamentary Debates*, Vol. 169, 23 September 1925, pp. 1271–2; *Argus*, 14 October 1925.

41 Mr J. T. Moroney in the *Daily Telegraph*, 22 April 1922. See Sullivan, The Genesis of Anzac Day, p. 26.

42 Loughlin MP, quoted by Sullivan, The Genesis of Anzac Day, p. 54.; *The Worker*, 28 April 1926.

43 *Argus*, 27 April 1926.

44 Ward, *The Australian Legend*, p. 233; Philip Kitley, 'Anzac Day Ritual', *Journal of Australian Studies*, 4, June 1979, p. 69.

45 Sullivan, The Genesis of Anzac Day, pp. 55–7.

46 Bean, *Two Men 1 Knew*, p. 226. See also the *Victorian Parliamentary Debates*, Volume 169, September 23 1925, pp. 1272–4.

47 Inglis, 'A Sacred Place', p. 104; Inglis, 'Memorials of the Great War', *Australian Cultural History*, 6, 1987, p. 12; Stephanie Brown, Anzac Rituals in Melbourne, 1916–1933: Contradictions and Resolution within Hegemonic Process, BA thesis, University of Melbourne, 1982, p. 39.

48 *Labour Daily*, 25 April 1927.

49   See Farrall 1, p. 63; D'Altera, p. 16; Morton 1, p. 19; Len Fox, *The Truth about Anzac*, Victorian Council Against War and Fascism, Melbourne, 1936; Holt, 'Anzac Reunion', pp. 214–15.

50   Williams 2, p. 31; Morton 1, pp. 40–1 and Morton 2, pp. 22–3; Norris, pp. 28–9.

51   Bridgeman, p. 25; Bowden, p. 38.

52   D'Altera, p. 53.

53   Lee Sackett, 'Marching into the Past: Anzac Day Celebrations in Adelaide', *Journal of Australian Studies*, 17, November 1985, pp. 18–30.

54   C. E. W. Bean, *Gallipoli Mission*, Australian War Memorial, Canberra, 1948, pp. 110–11.

55   C. E. W. Bean, 'Techniques of a Contemporary War Historian', *Historical Studies*, 2, 6, November 1942, pp. 78–9; Bean, 'The Writing of the Australian Official History of the Great War', pp. 85–112; Bean, 'Our War History'.

56   Bean, 'The Writing of the Australian Official History of the Great War', p. 92.

57   Bean, 'Our War History', p. 2.

58   Bean, 'Techniques of a Contemporary War Historian', p. 79; Bean, 'Our War History', p. 2. For such criticisms see S. Encel, 'The Study of Militarism in Australia', *Australian and New Zealand Journal of Sociology*, 3, 1, April 1967, pp. 2–18; Grey, *A Military History of Australia*, pp. 1–7. For comparisons with other military historians, see Keegan, *The Face of Battle*, pp. 13–72.

59   Bean, 'Techniques of a Contemporary War Historian', p. 66.

60   ibid., p. 72; Bean, 'The Writing of the Australian Official History of the Great War', p. 103.

61   Bean to Liddell Hart, 30 June 1941, Liddell Hart Papers 4/37, Liddell Hart Centre for Military Archives, King's College Library, University of London.

62   C. E. W. Bean, 'Memorandum: Censorship for Libel', 9 October 1919, quoted by Stephen Charles Ellis, 'The Censorship of the Official Naval History of Australia in the Great War', *Historical Studies*, 20, April 1983, p. 367. See also Bean, 'The Writing of the Australian Official History of the Great War', p. 86.

63   For the Sub-Committee's work see 'Report on the Work of the Historical Section', 15 February 1928, Cabinet Papers, CAB 16/52; Sub-Committee minutes, 9 March 1928, CAB 16/53, Public Records Office, Kew, London. For Foreign Office intervention see Aspinall-Oglander Papers, OG/111, Country Records Office, Isle of Wight. For correspondence between Bean and the British Historical Section see Bean Papers, AWM, 3DRL 7953/27–34. See also Peter Pederson, 'Introduction', Bean, *Vol. iii*, University of Queensland Press, St Lucia, 1982; and Denis Winter, *Haig's Command: A Reassessment*, Viking/Penguin, London, 1991.

64   See Bean, *Vol. i*, pp. 258–61, 310, 338, 425, 453, 462 and 492–6; *Vol. ii*, pp. 427–8, 658, 660 and 682; *Vol. iii*, pp. 599 and 940–1; *Vol. v*, pp. 27–30; *Vol. vi*, pp. 875–6, 933; Bean, *Anzac to Amiens*, p. 287. Edmonds to Bean, 7 February 1928, Bean Papers, AWM, 3DRL 7953/34.

65   Bean to Gellibrand, c. March 1929 and 19 May 1929, Gellibrand Papers, AWM, 3DRL 6405/14; Bean to Captain Falls, 6 July 1936, and Bean to Edmonds, 15 November 1938, Bean Papers, AWM, 3DRL 7953/30; Bean to Edmonds, 17 June 1931, Bean Papers, AWM, 3DRL 7953/27. See also Ellis, 'The Censorship of the Official Naval History of Australia in the Great War'.

66   Bean, 'Techniques of a Contemporary War Historian', p. 79; Bean, *vol. i*, p. x; Bean, *In Your Hands, Australians*, Cassell & Co., London, 1918, p. 96.

67   Bean, *Vol. vi*, p. 1078; parks and playgrounds movement quoted from Ross McMullin, 'C. E. W. Bean: A Man Who Should Be Remembered', *Canberra*

Times, 17 November 1979, p. 13; Bean to Gellibrand, 29 June 1933, Gellibrand Papers, AWM, 3DRL 6405/14. In the final volume of the history Bean placed less emphasis on the importance of the bush, and highlighted instead the general influence of the young, frontier society.

68 *Reveille*, 1 March 1936, pp. 8–9, quoted in Gerster, *Big-noting*, p. 128; Humphrey McQueen, *From Gallipoli to Petrov*, Allen & Unwin, Sydney, 1984, p. 51; Bean, *Vol. iii*, pp. 518–19; but contrast p. 599.

69 'Supermen' quote from a letter Bean wrote to Gavin Long more than a decade after the war, McCarthy, *Gallipoli to the Somme*, p. 282; Bean, *Vol. i*, pp. 606–7; Bean to Gellibrand, 29 June 1933, Gellibrand Papers, AWM, 3DRL 6405/14.

70 Bean, *Vol. vi*, pp. 7–8, 11 and 1074–5. For the nature and influence of postwar pacifism see: B. Bessant, 'Empire Day, Anzac Day, the Flag Ceremony and All That', *Historian*, 25, October 1973, pp. 36–43; Jeff Popple, A Very Temperate Reaction, BA thesis, University of New South Wales, 1980, p. 83. For Bean's own changing ideas about the causes of war, see C. E. W. Bean, *War Aims of a Plain Australian*, Angus & Robertson, Sydney, 1943, p. 143.

71 Bean, *Anzac to Amiens*, p. 142; Bean, *Vol. ii*, p. 910; Bean *Vol. vi*, pp. 401–3 and 1087–9.

72 Bean, *Anzac to Amiens*, p. 529; Bean, *Vol. i* (University of Queensland Press, 1981), Preface to the 3rd edn; and see *Times Literary Supplement*, 6 March 1922; *The Bulletin*, 17 November 1921; Alistair Thomson, 'History and "Betrayal": The Anzac Controversy', *History Today*, 43, January 1993, pp. 8–11.

73 Bean, *Anzac to Amiens*, p. 538; Bean, *Vol. vi*, p. 1085. See also C. E. W. Bean, 'Sidelights of the War on Australian Character', *Royal Australian Historical Society, Journal and Proceedings*, xiii, iv, 1927, pp. 211–17; Gerster, *Big-noting*, pp. 134–9; White, *Inventing Australia*, p. 134.

74 See, among the Bean Papers at the AWM: Historical Notes, 3DRL 8042/2–7; Extract Books, 3DRL 1722/2; Official History Manuscripts, AWM 44. See letters from Edmonds to Bean, 1927–28, 3DRL 7953/34, to see how the manuscript for *Vol. iii* was virtually unaffected by Edmonds' requests for amendments.

75 Bean to Gellibrand, 5 September 1930, 29 June 1933, Gellibrand Papers, AWM, 3DRL 6405/14.

76 Bean, *Vol. i*, pp. 462–3.

77 Bean Papers, Diary 4, 25 April 1915, p. 24 and Diary 5, 25 April 1915, pp. 33–4, AWM, 3DRL 606.

78 Bean's diary of 26 September 1915, quoted in Fewster, *Gallipoli Correspondent*, p. 157; Bean, *Vol. i*, pp. 297–8, 425, 453, 462 and 492–3; Bean, *Vol. ii*, p. 658; Bean, *Vol. iii*, pp. 940–1; Bean, *Vol. v*, pp. 27–30; Bean, *Anzac to Amiens*, p. 287. For examples of how Bean also defined away the AIF mutinies as the responsibility of a small number of 'bad soldiers', see Bean, *Vol. vi*, pp. 875–6 and 933; Bean, *Anzac to Amiens*, p. 487.

79 'Report on the Work of the Historical Section', 21 November 1924 and 22 February 1933, Memoranda 12 and 33, Sub-Committee for the Control of the Official History, CAB 16/52, Public Records Office, Kew, London. See also James Edmonds, 'Memoirs', pp. 675–77, Edmonds Papers 3/16, Liddell Hart Centre for Military Archives, King's College Library, University of London; Bean to Gellibrand, 4 July 1943, Gellibrand Papers, AWM, 3DRL 6405/14.

80 Bean to Treloar, 28 December 1923, Bean Papers, AWM, 3DRL 6673/249.

81 To trace these marketing campaigns, see Bean Papers, AWM, 3DRL 6673/249–253. The quotation citations are as follows: Bean to the Secretary, Department of Defence, 17 January 1921, 3DRL 6673/249; Treloar memo, 11 September 1928, 3DRL 6673/251; Bean to Treloar, 10 August 1933; and Treloar to Bean, 6 January 1934, 3DRL 6673/252.

82 Bean, 'Our War History', p. 2. Quotes from Sydney *Sun* (quoting the London *Observer)* 13 February 1922, and *Westralian Worker,* 6 January 1922. See generally, Bean Papers, AWM, 3DRL 8043, 'Reviews of Official History'.

83 Gavin Long, 'The Australian War History Tradition', *Historical Studies,* 6, 23 November 1954, p. 259; A. P. Rowe, 'Anzac', *The Australian Quarterly,* 29, 1, March 1957, p. 73; McKernan, *Here Is Their Spirit,* passim.

84 Heyes to Bean, 27 May 1938 and 18 March 1938, 3DRL 8040/2. For other such letters, see generally Bean Papers, AWM, 3DRL 6673/352 and 3DRL 8040.

85 Leane to Bean, 11 July 1933; Gieske to Bean, 16 July 1933; Nicholson to Bean, n.d.; Gellibrand to Bean, 2 June 1933, all in Bean Papers, AWM, 3DRL 8040.

86 McGillivray, p. 20; Cuddeford, p. 27; D'Altera, p. 31; Farrall 1, p. 29; Langham 1, p. 39. Bill Williams (Interview 2, p. 36) was the only one who expressed strong reservations about the accuracy of Bean's history.

87 The history nourished Stan D'Altera's 'mad Aussie pride' (p. 16), and was used by Fred Farrall to show social mobility in the AIF (Interview 1, p. 29).

6   Talk and taboo in postwar memories

1 Bird 1, pp. 23–4 and 27.
2 ibid., p. 26.
3 Bird p. 26; Bird 2, p. 22.
4 Bird 1, pp. 19–20 and 28–9.
5 Bird 2, pp. 15 and 18; Bird 1, p. 30.
6 ibid., p. 15.
7 Bird 2, pp. 11–12; Keown, *Forward with the Fifth,* p. 6.
8 Bird 2, pp. 9 and 23; cf. Keown, *Forward with the Fifth,* pp. 182 and 176.
9 Langham 2, p. 33; Langham 1, p. 23.
10 Langham 1, pp. 20–1 and 29.
11 Langham 1, pp. 41 and 30; Langham 2, p. 33.
12 Langham 1, pp. 27–8 and 31.
13 Langham 2, p. 20.
14 Langham 1, pp. 31–2 and 37–9; Langham 2, p. 21. For funny stories that were told at smoke nights: Langham 1, pp. 39–40; Langham 2, pp. 14–15.
15 Langham 1, p. 35.
16 Langham 1, p. 23; Langham 2, pp. 23–4.
17 Farrall 2, p. 23.
18 Farrall 1, pp. 32, 33–4, 54.
19 ibid., p. 44.
20 ibid., p. 45.
21 Farrall 1, p. 46; Farrall 2, pp. 24–5.
22 Farrall 2, pp. 28–9.
23 Farrall 1, p. 27.
24 Farrall 2, p. 28.
25 ibid., p. 26a.

26 See, O'Neill, *Bill Harney's War*, p. 54; Farrall 1, p. 49.
27 Farrall 1, p. 49.
28 ibid., pp. 31 and 60.
29 Farrall 1, p. 63; Farrall 2, p. 26a.
30 Farrall 2, p. 24.
31 Farrall 1, pp. 40 and 48.
32 ibid., Précis notes, p. 1. For a similar experience see Sid Norris (Interview, pp. 29–30), but contrast Ern Morton who tried to combine labour and digger politics (Interview 1, p. 32).
33 The historian Janet McCalman grew up as a neighbour of Fred and remembers this tension in his identity.

7 Old diggers

1 Doug Guthrie, Interview, 8 May 1983, p. 27; Bridgeman, p. 22; McGillivray, pp. 18–23.
2 McKenzie, p. 33; Fogarty, p. 23; Bowden, p. 39.
3 D'Altera, pp. 51–2; Morton 1, p. 42; Langham 1, p. 30.
4 See Peter Coleman, *Ageing and Reminiscence Processes*, Wiley, London, 1986; Joanna Bornat, 'Oral History as a Social Movement: Reminiscence and Older People', *Oral History*, 17, 2, Autumn 1989, pp. 16–25.
5 Bird 1, pp. 7–8; Langham 2, p. 35. For this analysis of 'nostalgia', I am indebted to Graham Dawson, Nostalgia, Conservatism and the Politics of Memory, unpublished manuscript, 1984; and see Coleman, *Ageing and Reminiscence Processes*.
6 Farrall 1, p. 52. See Joanna Bornat (ed.), *Reminiscence Reviewed: Achievements, Evaluations, Perspectives*, Oxford University Press, Oxford, 1993.
7 Morton 2, p. 30.
8 Bowden, p. 39.
9 *Club News* (Official Newsletter of the Yarraville Club), February 1985, *Western Times*, 19 January 1983 and 23 January 1985; *Labor Star*, May 1986; *Maryborough Advertiser*, 23 April 1986.
10 Fogarty, p. 26; McKenzie, p. 34.
11 Stabb, pp. 29–30.

8 The Anzac revival (1939–1990)

1 Bean, *War Aims of a Plain Australian*, pp. 162–3; George Johnston, 'ANZAC ... a myth for all mankind', *Walkabout*, 31, April 1965, p. 15; Gerster, *Big-noting*, pp. 180–90; Ross, *The Myth of the Digger*, p. 101.
2 See Peter Stanley, 'Reflections on Bean's Last Paragraph', *Sabretache*, xxiv, 3, July / September, 1983, pp. 4–11; R. O'Neill, 'Soldiers and Historians: Trends in Military Historiography in the Nineteenth and Twentieth Centuries', *Journal of the Royal Historical Society*, 56, 1, March 1970, pp. 36–47; M. McKernan, 'Writing about War', in *Australia: Two Centuries of War and Peace*, M. McKernan & M. Browne (eds), AWM and Allen & Unwin, Canberra, 1988.
3 Alan Seymour, *The One Day of the Year*, Penguin, 1963, p. 40; Ric Throssell, *For Valour*, Currency Press, Sydney, 1976.
4 Alex Carey, 'What Australian Troops Are Doing in Vietnam: The New Anzac Legend', *Outlook*, 14, February 1970, p. 8; Lex McAuley, *The Battle of Long Tan*

— *The Legend of Anzac Upheld*, Hutchinson, Melbourne, 1986. See Gerster, *Bignoting*, pp. 237–57.

5   Johnston, 'ANZAC', p. 13; Peter Coleman, 'Death and the Australian Legend', *The Bulletin*, 27 April 1963, pp. 18–19; Jack Woodward, 'Anzac and Australian Patriotism', *Advance Australia*, March 1973, pp. 24–5.

6   Gerard Henderson, 'The Anzac Legend after Gallipoli', *Quadrant*, July 1982, p. 62.

7   K. S. Inglis, 'The Anzac Tradition', *Meanjin Quarterly*, 24, 1, 1965, pp. 25–44; Geoffrey Serle, 'The Digger Tradition and Australian Nationalism', *Meanjin Quarterly*, 24, 2, 1965, pp. 149–58; Michael Roe, 'Comments on the Digger Tradition', *Meanjin Quarterly*, 24, 4, 1965, pp. 357–8; Noel McLachlan, 'Nationalism and the Divisive Digger', *Meanjin Quarterly*, 27, 3, 1968, pp. 302–8. See my bibliography for other, relevant publications by Inglis.

8   'Our Men, Other People's Wars', Time-Life Books publicity brochure, c. 1992. See bibliography for Lloyd Robson's valuable contributions to Australian military historiography.

9   McKernan, 'Writing about War', p. 13.

10 *Age*, 7 August 1981, 22 August 1981, and 1 December 1981.

11 Alan Attwood & Mark Dando, 'New Generation Can Bear to Look at War', *Age*, 10 October 1981.

12 See John Vader, *Anzac*, New English Library, London, 1972, p. 44; Adam-Smith, *The Anzacs*, p. 358; Joynt, *Saving the Channel Ports*, acknowledgements; Gammage, *The Broken Years*, p. xi.

13 Adam-Smith, *The Anzacs*, p. 352.

14 Henderson, 'The Anzac Legend after Gallipoli', pp. 62–4; John Robertson, *Anzac and Empire: The Tragedy and Glory of Gallipoli*, Hamby Australia, Port Melbourne, 1990, pp. 264–7; John Carroll, 'C. E. W. Bean and 1988', *Quadrant*, 32, 6, June 1988, pp. 47–9.

15 Adam-Smith, *The Anzacs*, p. vii.

16 Gammage, *The Broken Years*, pp. 58, 101, 198 and 237.

17 Tony Gough, 'The First Australian Imperial Force: C. E. W. Bean's Coloured Authenticity', *World Review*, 16, September 1977, p. 48.

18 'Anzacs: The Background', *Age Green Guide*, 24 October 1985, p. 1; John Cribben, *The Making of 'Anzacs'*, Collins/Fontana, Sydney, 1985, p. 49.

19 'Anzacs: The Background', p. 1.

20 James Wieland, 'The Romancing of Anzac', *Overland*, 105, 1986, p. 150.

21 Newspapers and magazines cited for the week from 24 April to 30 April 1987 include: *Herald, Canberra Times, Age, Australian, Sun, Times on Sunday* and *Bulletin.*

22 Ken Inglis, 'ANZAC and the Australian Military Tradition', *Current Affairs Bulletin*, 64, 11, April 1988, p. 15.

9   Living with the legend

1   Bird 2, p. 24.

2   ibid., p. 2.

3   ibid., pp. 2–6.

4   ibid., p. 25.

5   ibid., p. 16.

6   Percy Bird, untaped conversation, 7 April 1987.

7   Bird 2, pp. 7 and 9.

8   Langham 2, p. 30.

9   ibid., pp. 21–2 and 32.

10   Farrall 1, p. 52.

11   Farrall 2, p. 41. He praises Lloyd Robson's histories, Patsy Adam-Smith's *The Anzacs* and Lynn McDonald's books about the British on the Western Front.

12   ibid., p. 28.

13   Fred Farrall, Précis of untaped discussion, Anzac Day 1985.

14   Farrall 2, p. 35.

15   ibid., p. 35.

16   Farrall, Précis of untaped discussion, Anzac Day 1985.

17   *IPA Review,* 42, December/February 1988/89, p. 53.

18   Barrett, 'No Straw Man', p. 108. For my response to Barrett, see '"Steadfast Until Death"? C. E. W. Bean and the Representation of Australian Military Manhood', *Australian Historical Studies,* 23, 93, October 1989, pp. 462–78.

19   Robertson, *Anzac and Empire,* pp. 259–67.

20   To be fair, the historians who have written the introductions to the University of Queensland Press edition of Bean's history, have often been particularly sensitive to issues about Bean's Anzac legend-making. See, in particular, Peter Pederson (*Vol. iii*), Bill Gammage (*Vol. iv*) and Geoffrey Serle (*Vol. vi*).

21   Lex McAuley, *The Battle of Long Tan.* See also Patrick Hagopian, 'Oral Narratives: Secondary Revision and the Memory of the Vietnam War', *History Workshop Journal,* 32, Autumn 1991, pp. 134–50.

22   Terry Burstall, *A Soldier Returns,* University of Queensland Press, St Lucia, 1990, p. 172.

10   Searching for Hector Thomson

1   Alistair Thomson, 'Passing Shots at the Anzac Legend', in Verity Burgmann and Jenny Lee (eds), *A Most Valuable Acquisition: A People's History of Australia Since 1788*, McPhee Gribble/Penguin, Melbourne, 1988, pp. 190–204. The unpublished manuscript of 'Forgotten Anzacs' is in the Australian War Memorial, Manuscript 1180.

2   David Thomson, letter to Alistair Thomson, 27 September 1986 (in the author's possession).

3   Marina Larsson, *Shattered Anzacs: Living with the Scars of War*, University of New South Wales Press, Sydney, 2009; Damien Hadfield, 'The Evolution of Combat Stress: New Challenges for a New Generation', in Martin Crotty and Marina Larsson (eds), *Anzac Legacies: Australians and the Aftermath of War*, Australian Scholarly Publishing, North Melbourne, 2010, pp. 233–46; John Cantwell, *Exit Wounds: One Australian's War on Terror*, Melbourne University Publishing, Melbourne, 2012.

4   Larsson, *Shattered Anzacs*; Stephen Garton, *The Cost of War: Australians Return*, Oxford University Press Australia, Oxford and New York, 1996; Peter Stanley, *Men of Mont St Quentin: Between Victory and Death*, Scribe, Melbourne, 2009. For details of the National Archives of Australia (NAA) series B73, see, http://recordsearch.naa.gov.au/SearchNRetrieve/Interface/DetailsReports/SeriesDetail.aspx?series_no=B73, accessed 30 April 2013.

5   Bart Ziino, '"A Lasting Gift to his Descendants": Family Memory and the Great War in Australia', *History and Memory*, 22, 2, 2010, pp. 125–46.

6   Mr Louis D. Witts, letter to the Deputy Commissioner for Repatriation, 7 August 1929, Hector Thomson Repatriation Medical file, M58164, Series B73, NAA. All file references in this chapter are from this M58164 file, unless otherwise noted. The poem, 'Lament of the Ladies of the Lake' (referring to Lake Wellington near the Thomson farms) is in the Thomson family possession. The author signed him or herself off as 'R. Kipling'.

7   Hector Thomson, First AIF Personnel Dossier, 1914–1920, Series B2455, NAA.

8   Casualty Form, Thomson, First AIF Personnel Dossier; Record of Evidence Form, 24 July 1929, M58164; Mrs Johnston Thomson, letter to Major Lean, 1 September 1917, Thomson, First AIF Personnel Dossier.

9   Hector Thomson, Record of Evidence Form, 24 July 1929.

10   Medical Report, 21 December 1918; Hector Thomson, Record of Evidence Form, 24 July 1929; Dr Hagenauer, extract from a letter dated 14 May 1919, Evidence File, 1929; Dr Campbell, extract from a letter dated 27 November 1919, Evidence File, 1929; Dr Campbell, letter to the Deputy Commissioner, 13 November 1929.

11   David Thomson, letter to Alistair Thomson, 27 September 1986 (author's possession); N. Thomson (Mrs H.G.L. Thomson), letter to Deputy Commissioner, 19 September 1929. It is not clear if the property was named after the local district of Bungeleen or the nineteenth-century Aboriginal elder of that name (also called 'Bunjaleene' or 'Bungalene'), whose story is recounted in Don Watson, *Caledonia Australis: Scottish Highlanders on the Frontier of Australia*, Collins, Sydney, 1984, pp. 175–8.

12   Hector Thomson, Record of Evidence Form, 24 July 1929; Dr Campbell, letter to Deputy Commissioner, 13 November 1929; Mr Witts, letter to Deputy Commissioner, 7 August 1929. My father recalls his parents hosting Mr Witts and his wife at a dinner party at Bungeleen: David Thomson, 'The Thomson Story: Childhood Memories, 1924–1941', 1999, Papers of the Thomson Family, MS8600, National Library of Australia (NLA).

13   Dr Sidney Sewell, letter to Deputy Commissioner, 19 August 1929; see http://en.wikipedia.org/wiki/Encephalitis_lethargica, accessed 16 March 2013.

14   Hector Thomson, Record of Evidence Form, 24 July 1929; Dr Campbell, letter to Deputy Commissioner, 13 November 1929. See John V. Hurley, 'Sewell, Sir Sidney Valentine (1880–1949)', *Australian Dictionary of Biography*, National Centre of Biography, Australian National University, http://adb.anu.edu.au/biography/sewell-sir-sidney-valentine-8388/text14727, accessed 30 April 2013.

15   Nell Thomson, letters to Deputy Commissioner, 15 May 1929 and 19 September 1929.

16   Larsson, *Shattered Anzacs*; Marilyn Lake, *The Limits of Hope: Soldier Settlement in Victoria, 1915–38*, Oxford University Press, Melbourne, 1975.

17   Dr Scott, Medical Report, 8 November 1929; Dr Godfrey, Minute Paper, 22 November 1929; Deputy Commissioner, letter to Mr H. Thomson, 14 January 1930. For Godfrey, see Joy Damousi, *Freud in Australia: A Cultural History of Psychoanalysis in Australia*, University of New South Wales Press, Sydney, 2005, pp. 46–7.

18   Dr Campbell, letter to Deputy Commissioner, 5 February 1931; Dr Godfrey, Medical Notes, 6 February 1931.

19   Dr Dane, Medical Report, 20 March 1931. Thyroid treatment was a common response to mental illness in this period. On Dane, see Damousi, *Freud in Australia*, pp. 50–1.

20   Dr Dane, Medical Report, 31 March 1931, H58164, B73, NAA; Nell Thomson, letter to Deputy Commissioner, 1 July 1931.

21   Dr Garrett, Case Sheet, Repatriation General Hospital, Caulfield, 14 November 1931, H58164, B73, NAA; Deputy Commissioner, letter to Medical Superintendent, Royal Park Receiving House, 15 November 1931; Medical Superintendent, Royal Park Receiving House, letter to Deputy Commissioner, 19 November 1931.

22   Dr Bawm, Royal Park, Repatriation Commission Minute Paper, 24 December 1931; Dr Garrett, Repatriation Commission Minute Paper, 29 December 1931.

23   David Thomson, interview by Alistair Thomson, 4–8 August 1985, Australian Parliament's Oral History Project, TRC4900/35, NLA. David's younger brother Colin died in 2011. He found it even more difficult than my father to recall his childhood at Bungeleen and rarely talked about those times.

24   David Thomson, letter to Alistair Thomson, 27 September 1986; David Thomson interview, 1985. The Repatriation files show what Hector would not have known: that Dr Campbell had been a tardy expert witness for his pension claims, and that Nell had spent months chasing Campbell to complete a report for the Repat. See Nell Thomson, letter to Deputy Commissioner, 19 September 1929.

25   David Thomson interview, 1985.

26   David Thomson interview, 1985; David Thomson, letter to Alistair Thomson, 27 September 1986.

27   Johnny Bell, 'Needing a Woman's Hand: Child Protection and the Problem of Lone Fathers', *History Australia*, 9, 2, 2012, pp. 90–110.

28   David Thomson interview, 1985.

29   Dr Withington, Medical Report, 3 August 1933; Dr Macdonald, Medical Reports, 10 October 1934 and 15 December 1936; Dr Beveridge, Medical Report, 24 December 1936; Dr A. McKay, Medical Report, 24 December 1936. In 1937 the weekly Basic Wage for a man, dependent wife and two children was between £3.9.6 and £3.18.0 (depending on the state): Commonwealth Bureau of Census and Statistics, *Official Year Book of the Commonwealth of Australia*, 30, 1937, p. 578.

30   Dr Tyrer, Medical Report, 4 February 1937; Hector Thomson, letter to Deputy Commissioner, 25 August 1938; Dr Macdonald, Medical Report, 8 September 1938; Deputy Commissioner, letter to Dr Campbell, 8 November 1938; Dr Macdonald, Medical Report, 9 October 1939.

31   On reading a draft of this chapter, Frank Bowden (Professor of Medicine at the Australian National University and Senior Staff Specialist, Infectious Diseases, ACT Health, email to the author, 13 April 2013) suggested that Hector would now probably be diagnosed with chronic fatigue syndrome and depression — neither of which has a straightforward diagnostic test for an underlying pathological cause — and he would be given a trial of anti-depressants.

32   Dr Stephenson, Medical Case Sheet, 30 June 1948; Dr Barrett, Medical Report, 4 July 1948.

33   David Thomson interview, 1985; Hector Thomson, Medical History Sheet, 24 May 1941, Second AIF Service Records, VX56596, Series B883, NAA.

34   Dr Maxwell, Medical Report, 1 May 1947; Dr Borland, Medical Report, 10 November 1955, H58164, B73, NAA; Mrs Robertson, Social Work Report, 27 July 1956; Medical Report, 11 June 1957, H58164, B73, NAA.

35   David Thomson, letter to Alistair Thomson, 6 February 1992 (in author's possession).

36   See Ziino, "'A Lasting Gift to his Descendants'". Note that Peter Stanley suggests that we too readily overestimate this family connection with Australians at war, arguing that in the First War more than half the men of eligible age did not volunteer to enlist, and that post- World War II migrants have no direct family connection. See Peter Stanley, 'Monumental mistake? Is war the most important thing in Australian history?', in Craig Stockings (ed.), *Anzac's Dirty Dozen: 12 Myths of Australian Military History*, University of New South Wales Press, Sydney, 2012, p. 266.

11   Repat war stories

1   Dr Godfrey, Medical Report, 24 July 1939, Fred Farrall Repatriation Medical file, M101649, Series B73, NAA.

2   Farrall 1, p. 45.

3   Larsson, *Shattered Anzacs*, pp. 18 and 104; Lake, *Limits of Hope*.

4   Garton, *The Cost of War*, pp. 83–4, 92 and 74; Clem Lloyd and Jacqui Rees, *The Last Shilling: A History of Repatriation in Australia*, Melbourne University Press, Carlton, 1994, p. 419.

5   Garton, *The Cost of War*, p. 102; Larsson, *Shattered Anzacs*, p. 95; Kate Blackmore, *The Dark Pocket of Time: War, Medicine and the Australian State, 1914–1935*, Lythrum Press, Adelaide, 2008, p. 115. Larsson cites the 1920 Royal Commission on the Basic Wage which recommended £5.16.0 per week for a family of four to live in a 'reasonable standard of comfort'. By comparison in that year the 100 per cent war disability pension was set at £4.2.6 per week for a comparable household. This new research questions my assertion in the first edition (chapter 4) that Australian war pensions were more generous than those of most other combatant nations.

6   Larsson, *Shattered Anzacs*, p. 95.

7   See Blackmore, *The Dark Pocket of Time*.

8   Larsson, *Shattered Anzacs*, p. 162.

9   Blackmore, *The Dark Pocket of Time*, pp. 147 and 156.

10   Garton, *The Cost of War*, pp. 112–14; Lloyd and Rees, *The Last Shilling*, pp. 227–276 and 318.

11   Focussing on the fat Repat files might cause historians to overstate veterans' postwar difficulties, because problem cases generated the most claims and correspondence and because ex-servicemen sometimes exaggerated their ill-health. My original interview sample was selective in the opposite direction, shaped by the fact of survival and the desire to talk about the war. Though the men I interviewed in the 1980s were sometimes battered by both war and peace, they were still in reasonable mental and physical health 70 years after the war. The worst affected did not live so long.

12   Medical History Sheet, 21 December 1967, Percy Bird Repatriation Medical file, M28666, Series B73, NAA (hereafter M28666). The file is closed after 1981, and Percy's Hospital file H28666 (which covers the years after 1981) is also closed.

13   Bird, 'The 5th Battalion'; Bird 1, p. 14; Medical Report, 11 September 1917, M28666; Medical Board Report, 14 September 1917, M28666; Medical Report, 11 September 1917, M28666. With antibiotic treatments for tuberculosis since the 1950s scrofula is now rare, though it made a comeback amongst AIDS

patients in the 1980s: http://en.wikipedia.org/wiki/Tuberculous_cervical_
lymphadenitis, accessed 13 January 2013.

14 Medical Report 11 September 1917, M28666; Larsson, *Shattered Anzacs*, p. 183;
    E. Jones *et al*, 'Psychological effects of chemical weapons: a follow-up study of
    First World War veterans', *Psychological Medicine*, 38, 2008, pp. 1419–26.

15 Bird 1, p. 21.

16 Claim for Medical Treatment and War Pension, 2 December 1967, M28666;
    War Pension Statement, 7 February 1918, Percy Bird, First AIF Personnel
    Dossier, Series B2455, NAA.

17 Bird, letter to Deputy Director of Repatriation, 16 November 1967, M28666;
    Claim for Medical Treatment and War Pension, 2 December 1967, M28666;
    Medical Report, 23 February 1968, M28666; Bird, letter to Deputy Director of
    Repatriation, 26 May 1975, M28666.

18 Bill Langham, Pension Application, 18 July 1933, Bill Langham Repatriation
    Medical file, M53178, Series B73, NAA (hereafter M53178); Langham,
    Interview 1, p 41. After the interview in 1987, Bill and his wife said they had just
    celebrated their 50th wedding anniversary, and I did not do the maths to realise
    that they married in 1937 and it must have been Bill's second marriage. Thanks
    to Bill's great-niece Margaret Paulsen for updating the family history.

19 Medical Case Sheet, 5 October 1918, M53178; Medical History Form, 28
    October 1918, M53178. The M53178 file is closed from 1984, as is Bill's Hospital
    file H53178.

20 Medical Report, 31 March 1919, M53178; Langham 2, p. 33 and Langham 1, p. 23.

21 Langham, letter to Deputy Commissioner, 8 September 1927, M53178; Dr
    Craig, 25 August 1927, M53178; Dr O'Brien, 7 September 1927, M53178;
    Repatriation Board Decision, 5 October 1927, M53178.

22 Repatriation Board Notes, 17 May 1933, M53178; Langham 1, p. 31.

23 Yarraville RSL President, letter to the Deputy Commissioner, 7 April 1975,
    M53178; Pension Application, 10 April 1975, M53178.

24 Medical Report, 21 April 1980, M53178.

25 Kyla Cassells, 'Anzac Day is a celebration of war', *Socialist Alternative*,
    23 April 2010, at http://www.sa.org.au/index.php?option=com_
    k2&view=item&id=4716:anzac-day-is-a-celebration-of-war&Itemid=393,
    accessed 9 April 2013. My reprinted articles about Fred Farrall include:
    'The Return of a Soldier', *Meanjin*, 47, 4, 1988, pp. 709–716, reprinted in
    Penny Russell and Richard White (eds), *Memories and Dreams: Reflections on
    Twentieth Century Australia, Pastiche II*, Allen & Unwin, Sydney, 1997, pp.
    60–76; and 'Anzac Memories: Putting Popular Memory Theory into Practice
    in Australia', *Oral History*, Spring 1990, 18, 1, pp. 25–31, reprinted in A.
    Green and M. Troup (eds), *The Houses of History: A Critical Reader in Twentieth
    Century History and Theory*, Manchester University Press, 1999, pp. 239–252,
    and in Robert Perks and Alistair Thomson (eds) *The Oral History Reader*,
    Routledge, London and New York, 2006, pp. 244–54. Most years I receive a
    Copyright Licencing Agency fee for the use of the 'The Return of a Soldier'
    article in course readers.

26 Farrall, letters to 'My Dear Mother', 20 December 1917; to 'My Dear Laura', 16
    October 1917; and to 'Dear Sam', 10 October 1917: Box Miscellaneous 2/1/1/ –
    2/3/2, Series 1987.0418, Farrall Papers, University of Melbourne Archives. On
    soldiers' wartime letters, see Joy Damousi, *The Labour of Loss: Mourning, Memory
    and Wartime Bereavement in Australia*, Cambridge University Press, Cambridge,

1999, pp. 9–25; Michael Roper, *The Secret Battle: Emotional Survival in the Great War*, Manchester University Press, Manchester and New York, 2009, pp. 58–63; Michael Roper, 'Re-remembering the Soldier Heroes: The Psychic and Social Construction of Memory in Personal Narratives of the Great War', *History Workshop Journal*, 50, Autumn 2000, pp. 181–204; and Alistair Thomson, 'Anzac Stories: Using Personal Testimony in War History', *War and Society*, 25, 2, 2006, pp. 1–21.

27   E.F. Hill, Bryan Kelleher, Alan Miller and Ralph Gibson, with Fred Farrall, *Celebration of Fred Farrall's 90th Birthday*, Australian Society for the Study of Labour History, Melbourne, 1987, p. 20.

28   Farrall, Claim Form, 15 May 1939, Fred Farrall Repatriation Medical file, M101649, Series B73, NAA (hereafter M101649); Evidence File,1939, M101649. The M101649 file is closed from 1984, as is Fred's Hospital file H101649. The 1939 Evidence file brought together most of Fred's available wartime and interwar records. Between 1920 and 1938 the files refer to Fred's dealing with Repatriation officials in Sydney; from 1939 they cover his dealing with the Repat in Melbourne.

29   Lois Farrall, *The File on Fred: A Biography of Fred Farrall*, High Leigh Publishing, Carrum, 1992, pp. 90–98; Medical Case Sheet, 22 January 1917, Veterans Affairs Tribunal Box, Series 1987.0418, Farrall Papers; Medical Case Sheet, 15 July 1917, M101649; Farrall, letter to 'My Dear Laura', 20 December 1917; Farrall, letter to the Minister for Repatriation, 5 February 1968, M101649.

30   Medical Report, 22 October 1917, Evidence File, 1939, M101649.

31   Case Note, 15 January 1920, Evidence File, 1939, M101649; Dr Rutledge, 26 October 1920, M101649; Dr Graham, 6 November 1926, M101649; Farrall, Appeal, 26 March 53, M101649.

32   Farrall, Claim, 15 May 1939, M101649; Dr Freedman, 23 September 1949, M101649; Michael Roe, 'Arthur, Richard (1865–1932)', *Australian Dictionary of Biography*, National Centre of Biography, Australian National University, http://adb.anu.edu.au/biography/arthur-richard-5061/text8437, accessed 10 January 2013. On Arthur, see also Damousi, *Freud in Australia*, p. 26.

33   Meadowbank Manufacturing Company, Reports, 23 May 1939 and 5 November 1926, M101649; H.J. Da Silva, Report, 4 November 1926, M101649.

34   Farrall 2, pp. 24–5 (the use of the pronoun 'You' in the interview extract suggests that Fred may have been generalising about a condition he only came to understand in later years); Farrall, Appeals, 23 October 1926 and 30 October 1926, M101649; Dr Graham, 6 November 1926, M101649.

35   Doctors Willis and Smith, Medical Report, 16 December 1926, M101649.

36   Farrall, letter to the Deputy Commissioner, 12 March 1927, M101649.

37   Dr Willcocks, 31 January 1927, M101649; Dr Allen, 14 July 1928, M101649; Dr Francis, 10 February 1927, M101649; Dr Parkinson, 1 March 1927, M101649; Dr Arthur, Insurance Report, 3 March 1927, in the Evidence File, 1939, M101649. It is not clear if this is the same Dr Arthur who may have treated Fred for nerves in 1920.

38   Larsson, *Shattered Anzacs*, pp. 206–33; Lloyd and Rees, *The Last Shilling*, pp. 144 and 251–2.

39   Dr Willcocks, 31 January 1927, M101649; Dr Allen, 14 July 1928, M101649; Dr Parkinson, 1 March 1927, M101649; Clinical assessment, 17 January 1939, M101649; Dr Smith, 22 March 1927, M101649; Dr Minty, 25 January 1927, M101649; Out Patient Notes, 13 October 1931 and 11 January 1937, M101649. Fred's political development is described in our 1983 interview, but is also detailed in the intelligence report produced by Commonwealth intelligence

officers in response to a request from the Public Service Board in 1949, when Fred was a Communist public servant in Melbourne: Deputy Director, Australian Security Intelligence Organization, Central Office, Canberra, Secret Report to Secretary of Public Service Board, 23 August 1949, Control Symbol C89597, Series A367, NAA, at http://recordsearch.naa.gov.au/scripts/Imagine/ asp, accessed 10 January 2013. The report lists Fred's activities from 1931, when he was arrested carrying a 'loaded hose pipe' in his pocket during an anti-eviction rally outside a court.

40   Dr Minty , 25 January 1927, M101649; Dr Willcocks, 10 February 1927, M101649; Dr Smith, 22 March 1927, M101649.

41   Manager, State Labour Exchanges, Sydney, letter to Deputy Commissioner, 22 June 1939, M101649; Farrall, Sustenance Application, 14 February 1939, Evidence File, 1939, M101649; State Secretary, NSW RS&SILA, letter to Deputy Commissioner, 23 August 1935, M101649; Minister Rt Hon W.N. Hughes, letter to Deputy Commissioner, 9 July 1937, M101649. The League made a separate, confidential enquiry about the nature of Farrall's accepted and rejected disabilities, perhaps wary of Fred's politics and suspicious of malingering: State Secretary, NSW RS&SILA, letter to Deputy Commissioner, 9 December 1936, M101649.

42   Farrall, letter to Base Records, Canberra, 14 October 1938, Farrall, First AIF Personnel Dossier, Series B2455, NAA; Farrall, Claim Form, 17 January 1939, M101649; Michael Cannon, 'Slater, William (Bill) (1890–1960)', *Australian Dictionary of Biography*, National Centre of Biography, Australian National University, http://adb.anu.edu.au/biography/slater-william-bill-11709/text20929, accessed 25 January 2013.

43   Farrall, Claim Forms, 27 October 1938, 16 November 1938 and 17 January 1939, M101649; Dr Constable, 16 November 1938, M101649; Clinical Assessment 17 January 1939, M101649.

44   Farrall, Claim, 9 May 1939, M101649; Dr Klug, 11 May 1939, M101649; Dr Minty, 25 January 1927, M101649.

45   Dr Arthur, Medical Report for Mutual Life and Citizen's Assurance Co., 3 March 1927, Evidence File, 1939, M101649; Australasian Temperance and General Mutual Life Assurance Society, letter to Deputy Commissioner, 29 May 1939, Evidence File, 1939, M101649.

46   Dr Godfrey, 24 July 1939, M101649; Dr Crowe, 15 August 1939, M101649; State Board, 28 August 1939, M101649. Fred's appeal against the 1939 decision was rejected in August 1942.

47   This synthesis draws on the studies of shell shock in Larsson, *Shattered Anzacs*, pp. 151 and 149–77; Garton, *The Cost of War*, pp. 158 and 143–75; and Blackmore, *The Dark Pocket of Time*, pp. 173–80. See also Michael Tyquin, *Madness and the Military: Australia's Experience of the Great War*, Australian Military History Publications, Canberra, 2006.

48   Farrall, undated notes (circa 1978), Veterans Affairs Tribunal Box, Series 1987.0418, Farrall Papers; Farrall 1, p. 37; Case Sheet, Repatriation General Hospital Heidelberg (RGHH), 27 July 1944, H101649; Case Sheet, 6 July 1950, RGHH, H101649. See Sebastian Gurciullo, 'Ellery, Reginald Spencer (Reg) (1897–1955)', *Australian Dictionary of Biography*, National Centre of Biography, Australian National University, http://adb.anu.edu.au/biography/ellery-reginald-spencer-reg-10110/text17847, accessed 10 January 2013. On Ellery's 'homespun psychotherapy', see also Damousi, *Freud in Australia*, pp. 70–73.

NOTES

49   Dr Freedman, 23 September 1949, M101649. Medical convention in the 2000s
     is that duodenal ulcers are caused by bacteria and not by stress, though the link
     between ulcers and stress has had a strong popular purchase and was likely
     believed by Fred (Frank Bowden, Professor of Medicine at the Australian
     National University and Senior Staff Specialist, Infectious Diseases, ACT
     Health, email to the author, 13 April 2013).
50   Case Sheet, RGHH, 6 July 1950, H101649; Case Sheets, RGHH, 1 June
     1961 and 8 June 1961, H101649; Mr Godley, Medical Report 24 March 1983,
     H101649; Farrall, letter to 'John, Margie and Helen' (in England), 7 June 1987,
     Personal Letters Box, Series 1987.0418, Farrall Papers. Fred's correspondence
     with family and friends in the 1980s often mentioned his 'nerves', which were
     assumed to be war-caused.
51   Farrall, Pension Claim, 26 March 1953, M101649; Medical Officer Report,
     12 October 1971, M101649; Farrall, Pension Claims, 1 December 1977 and
     26 January 1981, M101649; Dr Bennett, Letter of Support for Fred Farrall, 13
     August 1982, and Repatriation Review Tribunal Notice of Decision, 20 July
     1983, Veterans Affairs Tribunal Box, Series 1987–0148, Farrall Papers.
52   Frank Crean, letter to the Minister for Repatriation, 28 March 1955, M101649;
     Farrall, letter to the Editor, *The Age*, 26 July 1961, and Deputy Commissioner
     Stephens' response to Farrall, 18 August 1961, M101649; Farrall, letter to the
     Minister for Repatriation, 5 February 1968, M101649.
53   Dr Hayes, Medical Report, 12 May 1953, M101649; Dr Morrissey, 12 October
     1955, M101649; Clinical Notes, 16 June 1961, M101649; File Note, 12 October
     1961, M101649; Drs Trinca and Summons, 26 January 1956, M101649; Dr
     Kennedy, Medical Report, 12 May 1961, M101649.
54   Lloyd and Rees, *The Last Shilling*, pp. 323–37.
55   Dr May, Report, 25 May 1950, M101649. Several publications, and new online
     and paper archives, offer a detailed picture of Fred's peacetime activism: Maureen
     Bang, 'Toorak's Pensioner Mayor', *Australian Woman's Weekly*, 3 October
     1973, pp. 4–5; Lois Farrall, *The File on Fred*; Dorothy Farrall, 'Autobiography',
     unpublished transcript of an interview by Wendy Lowenstein, transcribed by
     Fred Farrall, no date, Manuscript Collection, State Library of Victoria; Hill, et
     al, *Celebration of Fred Farrall's 90th Birthday*; Series 1983.0113 and 1987.0148,
     Farrall Papers.
56   Case Sheet, RGHH, 6 July 1950, H101649; Farrall, Claim, 1 December 1977,
     M101649; Farrall 1, p. 27.
57   Fred Farrall, 'Trade Unionism in the First A.I.F. – 1914–1918', *Recorder*, 54,
     October 1971, pp. 3–8; Medical Officer's Report, 12 October 1971, M101649;
     Dr Freed, Notes, 4 May 1972, H101649.
58   Michael Roper, 'Review of *Anzac Memories*', *Oral History*, 22, 2, 1994, p. 92.
     Jerome Bruner, 'The Narrative Construction of Reality', *Critical Inquiry*, 18,
     1, 1991, p. 16. On memory, healing and composure, see: Charlotte Linde, *Life
     Stories: The Creation of Coherence*, Oxford University Press, New York, 1993; Sean
     Field, 'Beyond "healing": trauma, oral history and regeneration', *Oral History*,
     34, 1, 2006, pp. 31–42.
59   See for example: Michael Roper, 'Re-remembering the Soldier Heroes'; Mark
     Roseman, 'Surviving Memory: Truth and Inaccuracy in Holocaust Testimony',
     The Journal of Holocaust Education, 8, 1, 1999, pp. 1–20; and Alistair Thomson,
     Moving Stories: an Intimate History of Four Women Across Two Countries,
     University of New South Wales Press, Sydney and Manchester University Press,

Manchester, 2011. Of course sometimes oral history is the main available source and we have to work with what we've got – as for example Christopher Browning does, magnificently, in his oral history of a Nazi slave labour camp: Christopher R. Browning, *Remembering Survival: Inside a Nazi Slave-Labor Camp*, W.W. Norton, New York, 2010.

60   Fred Farrall, undated notes (circa 1982), Veterans Affairs Tribunal Box, Series 1987.0418, Farrall Papers.

61   Alessandro Portelli, 'The Peculiarities of Oral History', *History Workshop Journal*, 12, Autumn 1981, pp. 96–107 (first published in Italian in 1979). See also: Alessandro Portelli, *The Death of Luigi Trastulli and Other Stories: Form and Meaning in Oral History*, State University of New York Press, Albany, 1991; Alessandro Portelli, *The Order Has been Carried Out: History, Memory, and Meaning of a Nazi Massacre in Rome*, Palgrave Macmillan, New York, 2003. For examples of other oral history books that take a similar approach, see: Penny Summerfield, *Reconstructing Women's Wartime Lives*, Manchester University Press, Manchester, 1998; Natalie Nguyen, *Memory Is Another Country: Women of the Vietnamese Diaspora*, Praeger, Santa Barbara, California, 2009. Key recent texts on oral history theory and method include: Lynn Abrams, *Oral History Theory*, Routledge, London, 2010; Donald A. Ritchie (ed.), *The Oxford Handbook of Oral History*, Oxford University Press, New York, 2011; Valerie Raleigh Yow, Recording Oral History. *A Guide for the Humanities and Social Sciences*, 2nd edition, Altamira Press, Walnut Creek, California, 2005; Thomas L. Charlton, Lois E. Myers and Rebecca Sharpless (eds), *Handbook of Oral History*, Altamira Press, Lanham, MD, 2006; Alexander Freund, and Alistair Thomson (eds), *Oral History and Photography*, Palgrave Macmillan, New York, 2011; Perks and Thomson, *The Oral History Reader*. For developments over the past 20 years in my own understandings of and approaches to oral history, see: Alistair Thomson, 'Four paradigm transformations in oral history', *Oral History Review*, 34, 1, 2007, pp. 49–70; Alistair Thomson, 'Memory and Remembering in Oral History', in Ritchie, *The Oxford Handbook of Oral History*, pp. 77–95; Thomson, *Moving Stories*.

62   On 'particular publics' see Dawson and West, '"Our Finest Hour"?', pp. 10–11. On communicative memory see: Jan Assman, 'Collective Memory and Cultural Identity', *New German Critique*, 65, 1995, pp. 125–33; Alexander Freund, 'A Canadian Family Talks about Oma's Life in Nazi Germany: Three-Generational Interviews and Communicative Memory', *Oral History Forum*, 29, 2009, pp. 1–26; Graham Smith, 'Beyond Individual / Collective Memory: Women's Transactive Memories of Food, Family and Conflict', *Oral History*, 35, 2, 2007, pp. 77–90.

63   The best oral history interviews affirm the value of the life story told whilst at the same time encouraging the narrator to elaborate and stretch their story in less well-rehearsed directions (as we saw, this was not always easy with Percy Bird or Fred Farrall). On the communicative relationship of the oral history interview, see Valerie Raleigh Yow, '"Do I Like Them Too Much?" Effects of the Oral History Interview on the Interviewer and Vice-Versa', *Oral History Review*, 24, 1, 1997, pp. 55–79; Michael Roper, 'Analysing the analysed: transference and counter-transference in the oral history encounter', *Oral History*, 31, 2, 2003, pp. 20–32; Daniel James, *Dona María's Story: Life History, Memory and Political Identity*, Duke University Press, Durham, NC, 2000; Della Pollock (ed.), *Remembering: Oral History as Performance*, Palgrave Macmillan, New York, 2005; Abrams, *Oral History Theory*, pp. 54–77.

64   For overviews about memory and oral history, see Abrams, *Oral History Theory*,
      pp. 78–105; Thomson, 'Memory and Remembering in Oral History'; Yow,
      Recording Oral History, pp. 35–67.

Postscript: Anzac postmemory

1   Graeme Davison, 'The Habit of Commemoration and the Revival of Anzac Day',
      Australian Cultural History, 23, 2003, p. 81.
2   Jay Winter, 'Shell-shock and the Cultural History of the Great War', *Journal of
      Contemporary History*, 35, 1, 2000, p. 10; Dan Todman, *The Great War: Myth and
      Memory*, Hambledon and London, London, 2005, pp. 173 and 223–24. For a
      critique of Todman, see Ziino, '"A Lasting Gift to His Descendants"', pp. 125–26.
3   Mark McKenna, 'Anzac Day: How Did it Become Australia's National Day?', in
      Marilyn Lake and Henry Reynolds, with Mark McKenna and Joy Damousi (eds),
      What's Wrong With Anzac? The Militarisation of Australian History, University
      of New South Wales Press, Sydney, 2010, p. 126 (McKenna's is the stand-out
      chapter in the book). The most comprehensive account of the Anzac resurgence
      is Ken Inglis, 'Epilogue: Towards the Centenary of Anzac', *Sacred Places: War
      Memorials in the Australian Landscape*, Melbourne University Press, Melbourne,
      3rd edition, 2008 (first published 1998), pp. 458–583.
4   Inglis, *Sacred Places*, p. xvi.
5   On 'Australia Remembers': Liz Reed, *Bigger than Gallipoli: War, History and
      Memory in Australia*, University of Western Australia Press, Crawley, 2004; and
      Inglis, *Sacred Places*, pp. 390–457.
6   Howard's speech quoted in Inglis, *Sacred Places*, p. 549.
7   Julia Gillard, speech at Anzac Cove Dawn Service, Gallipoli, 25 April
      2012, quoted in Carolyn Anne Holbrook, 'The Great War in the Australian
      Imagination Since 1915', D.Phil. thesis, University of Melbourne, 2012,
      p. 240.
8   Elizabeth Furniss, 'Timeline History and the Anzac Myth: Settler Narratives of
      Local History in a North Australian Town', *Oceania*, 71, 4, 2001, p. 279.
9   See McKenna, 'Anzac Day'.
10   Holbrook, 'The Great War in the Australian Imagination Since 1915', p. 203.
11   For Hawke's speech and a century of politicians' use and abuse of Anzac, see
      Holbrook, 'The Great War in the Australian Imagination Since 1915', pp. 203–40.
12   Howard speaking in 2003 and 2002, quoted in Matt McDonald, '"Lest
      We Forget": The Politics of Memory and Australian Military Intervention',
      International Political Sociology, 4, 2010, p. 297. On Howard and Anzac, see also:
      Holbrook, 'The Great War in the Australian Imagination Since 1915', pp. 228–
      38; McKenna, 'Anzac Day'; Inglis, *Sacred Places*, p. 550.
13   Paula Hamilton and Paul Ashton, 'At Home With the Past: Initial Findings
      From the Survey', *Australians and the Past, A special edition of Australian Cultural
      History*, 22, University of Queensland Press, St Lucia, 2003, pp. 5–30; Paul
      Ashton and Paula Hamilton, 'Connecting with History: Australians and their
      Pasts', in Paul Ashton and Hilda Kean (eds), *People and their Pasts: Public History
      Today*, Palgrave, Basingstoke, 2009, pp. 23–38.
14   Jay Winter, 'Forward to the Third Edition', Inglis, *Sacred Places*, p. iv.
15   For an account of the popular memory theory related to war remembrance,
      see T.G. Ashplant, Graham Dawson and Michael Roper, 'The politics of war
      memory and commemoration: contexts, structures and dynamics', in Timothy

Ashplant, Graham Dawson and Michael Roper (eds), *The Politics of Memory: Commemorating War*, Transaction Publishers, New Brunswick, 2004, pp. 3–85. See also: Jay Winter, *Remembering War: The Great War between History and Memory in the Twentieth Century*, Yale University Press, New Haven, 2006; Jay Winter, *Sites of Memory, Sites of Mourning: The Great War in European Cultural History*, Cambridge University Press, Cambridge, 1995; Jay Winter, and Emmanuel Sivan (eds), *War and Remembrance in the Twentieth Century*, Cambridge University Press, Cambridge, 1999; Jenny Edkins, *Trauma and the Memory of Politics*, Cambridge University Press, Cambridge, 2003.

16   Damousi, *The Labour of Loss*; Tanja Luckins, *The Gates of Memory: Australian People's Experiences and Memories of Loss in the Great War*, Curtin University Books, Fremantle, 2004.

17   Marianne Hirsch, 'The Generation of Postmemory', *Poetics Today*, 29, 1, Spring 2008, pp. 103–28; Marianne Hirsch, *The Generation of Postmemory: Writing and Visual Culture After the Holocaust*, Columbia University Press, New York, 2012.

18   Hirsch, 'The Generation of Postmemory', p. 108, citing Aby Warburg.

19   Jay Winter, 'Writing war', public lecture, State Library of Victoria, 22 August 2012.

20   See Peter Stanley, 'Monumental mistake? Is war the most important thing in Australian history?', in Craig Stockings (ed.), *Anzac's Dirty Dozen: 12 Myths of Australian Military History*, University of New South Wales Press, Sydney, 2012, pp. 260–86.

21   Dominic Bryan and Stuart Ward, 'The "Deficit of Remembrance": The Great War Revival in Australia and Ireland', in Stuart Ward and Katie Holmes (eds), *Exhuming Passions: The Pressure of the Past in Ireland and Australia*, Irish Academic Press, Dublin and Portland, Oregon, 2011, p. 169.

22   Ziino, '"A Lasting Gift to his Descendants"', p. 139. For a review of Australian family histories of the Great War, see Holbrook, 'The Great War in the Australian Imagination Since 1915', pp. 173–202.

23   Bruce Scates, *Return to Gallipoli: Walking the Battlefields of the Great War*, Cambridge, Cambridge University Press, 2006.

24   Mark McKenna and Stuart Ward, '"It was really moving, mate": The Gallipoli Pilgrimage and Sentimental Nationalism in Australia', *Australian Historical Studies*, 38, 129, 2007, p. 151; Anna Clark, *History's Children: History Wars in the Classroom*, University of New South Wales Press, Sydney, 2008, p. 62.

25   Christina Twomey, 'Trauma and the Reinvigoration of Anzac: An Argument', *History Australia*, forthcoming, December, 2013; quoting Didier Fassin and Richard Rechtman, *The Empire of Trauma: An Inquiry into the Condition of Victimhood* (2007), translated by Rachel Gomme, Princeton University Press, Princeton, 2009, p. xi.

26   Winter, 'Writing war'.

27   Peter Stanley critiques the 'turgid nationalist epics' of 'storians' like Les Carlyon and Peter Fitzsimmons, in his chapter, 'Military history: over the top', in Paul Ashton and Anna Clark (eds), *Australian History Now*, University of New South Wales Press, Sydney, forthcoming, 2013.

28   Thomson, *Moving Stories*.

29   Marilyn Lake. 'Introduction: What have you done for your country?', in Lake and Reynolds, *What's Wrong With Anzac?*, p. viii.

30   Among many others, see: Jeffrey Grey, *A Military History of Australia*, Cambridge University Press, Cambridge and Port Melbourne, 3rd edition, 2008; Craig

NOTES

Stockings, *Bardia: Myth, Reality and the Heirs of Anzac*, University of New South Wales Press, Sydney, 2009; Christopher Pugsley, 'Stories of Anzac', in Jenny Macleod (ed.), *Gallipoli: Making History*, Frank Cass, London and New York, 2004, pp. 44–58; Peter Pedersen, 'The AIF: as good as the Anzac Legend says?', *Sydney Papers*, Autumn 2007, pp. 168–177.

31   Craig Stockings, 'Myths and Australian Military History', in Stockings (ed.), *Anzac's Dirty Dozen*, p. 5; Neville Meaney, *A History of Australian Defence and Foreign Policy 1901–23: Volume 2, Australia and the World Crisis 1914–1923*, Sydney University Press, Sydney, 2009.

32   Peter Stanley, *Bad Characters: Sex, Crime, Mutiny, Murder and the Australian Imperial Force*, Pier 9, Miller's Point, 2010, cover quote.

33   Craig Stockings. 'Epilogue: Returning Zombies to their Graves', in Craig Stockings (ed.), *Zombie Myths of Australian Military History*, University of New South Wales Press, Sydney, 2010, pp. 234–38.

34   Garton, *The Cost of War*; Lloyd and Rees, *The Last Shilling*; Blackmore, *The Dark Pocket of Time*.

35   Larsson, *Shattered Anzacs*; Crotty and Larsson, *Anzac Legacies*.

36   See, among many others: Damousi, *The Labour of Loss*; Joy Damousi, *Living with the Aftermath: Trauma, Nostalgia and Grief in Post-war Australia*, Cambridge University Press, Cambridge, 2001; Holbrook, 'The Great War in the Australian Imagination Since 1915'; Inglis, *Sacred Places*; Luckins, *The Gates of Memory*; Macleod, *Gallipoli: Making History*; Reed, *Bigger than Gallipoli*; Graham Seal, *Inventing Anzac: The Digger and National Mythology*, University of Queensland Press with API Network and Curtin University of Technology, St Lucia, 2004; Scates, *Return to Gallipoli*; Bruce Scates, *A Place to Remember: A History of the Shrine of Remembrance*, Cambridge University Press, Cambridge, 2009; Alistair Thomson, 'The "Vilest Libel of the War"? Imperial Politics and the Official Histories of Gallipoli', *Australian Historical Studies*, 25, 101, 1993, pp. 628–36; Bart Ziino, *A Distant Grief: Australians, War Graves and the Great War*, University of Western Australia Press, Crawley, Western Australia, 2007.

37   For example: historian Bruce Scates at Monash University is leading several collaborative public history projects about Australians at war and in remembrance, including a major study of the history of Anzac Day (http://profiles.arts.monash.edu.au/bruce-scates/research/, accessed 4 May 2013); feminist historian Clare Wright is co-scripting a four-part television series for the First World War centenary (see http://www.screenaustralia.gov.au/news_and_events/2013/mr_130405_funding.aspx, accessed 4 May 2013); Marilyn Lake continues to debate what's wrong with Anzac in many different public contexts (see http://www.abc.net.au/radionational/programs/sundayextra/newsmaker3a-marilyn-lake/3925520, accessed 5 May 2013); and the author is among several historians advising Museum Victoria as it creates a new exhibition for the centenary of 1914–18.

38   Annette Wieviorka, *The Era of the Witness*, Cornell University Press, Ithaca, 2006.

39   Inglis, *Sacred Places*, p. 549.

40   Saul Friedlander, 'Trauma, Memory, and Transference', in Geoffrey H. Hartman (ed.), *Holocaust Remembrance: The Shapes of Memory*, Blackwell, Cambridge, Massachusetts, 1994, pp. 252–63. Thanks to my Monash University colleague Noah Shenker for this reference.

Appendix 1: Oral history and popular memory

1   Alistair Thomson, The Forgotten Anzacs, unpublished manuscript, 1986, MS1180, AWM.

2   For a summary of conservative criticisms, see Paul Thompson, *The Voice of the Past: Oral History*, Oxford University Press, Oxford, 1988, pp. 68–71, and his editorial in *Oral History*, 18, 1, Spring 1990, p. 24. For the Australian debate see the 'Oral History: Facts and Fiction' exchange in *Oral History Association of Australia Journal*, 5, 1983–84.

3   Popular Memory Group, 'Popular Memory: Theory, Politics, Method', in *Making Histories: Studies in History Writing and Politics*, Richard Johnson et al. (eds), Hutchinson, London, 1982. See also Ronald Fraser, *Blood of Spain: The Experience of Civil War 1936–1939*, Allen Lane, London, 1979; Luisa Passerini, 'Work Ideology and Consensus under Italian Fascism', *History Workshop Journal*, 8, 1979, pp. 82–108; 'Editorial — Oral History', *History Workshop Journal*, 8, Autumn 1979, pp. i–iii. For reprints of the 1977 and 1978 film debates see Tony Bennett, et al. (eds), *Popular Television and Film*, BFI Publishing / The Open University Press, London, 1981.

4   The Group's Second World War studies include: Graham Dawson & Bob West, '"Our Finest Hour"? The Popular Memory of World War Two and the Struggles over National Identity'; and Graham Dawson, 'History-Writing on The Second World War', both published in *National Fictions: World War Two in British Films and Television*. On issues of popular nationalism and the national past see Patrick Wright, *On Living in an Old Country: The National Past in Contemporary Britain*, London, Verso, 1985. Part A of the unpublished draft of a Popular Memory book (loaned by Richard Johnson) discusses memory work, 'inner stories' and the narrative form.

5   *Oral History*, 17, 2, Autumn 1989, p. 2. For criticism of the Popular Memory Group, see Trevor Lummis, *Listening to History*, Hutchinson, London, 1987, pp. 117–40. For new approaches see the 'Popular Memory' issue of *Oral History*, 18, 1, Spring 1990; the revised discussion of memory and subjectivity in Thompson, *The Voice of the Past*, 1988, pp. 150–65; the debates in the *International Journal of Oral History*, 6, February 1985; and the international anthology edited by Raphael Samuel & Paul Thompson, *The Myths We Live By*, Routledge, London, 1990. For similar developments in the United States see David Thelen, 'Memory and American History', *Journal of American History*, 75, 4, March 1989, pp. 1117–29; and Michael Frisch, *A Shared Authority: Essays on the Craft and Meaning of Oral and Public History*, State University of New York, Albany, 1990. For Australia see John Murphy, 'The Voice of Memory: History, Autobiography and Oral Memory', *Historical Studies*, 22, 87, October 1986, pp. 157–75. For ways in which psychology has also taken on these ideas, see David Middleton & Derek Edwards (eds), *Collective Remembering*, Sage, London, 1990.

6   A. Thomson, 'Shrapnel and Smiles: Memories of a Strange and Bloody Youth', *Shades*, 1, 1983, pp. 11–13; and *Shades*, 2, 1983, pp. 11–12.

7   Bridgeman, p. 27.

8   On ethical issues see Sherna Berger Gluck & Daphne Patai (eds), *Women's Words: The Feminist Practice of Oral History*, Routledge, New York, 1991; Frisch, *A Shared Authority*.

SELECT BIBLIOGRAPHY

Interviews

I recorded most of the following interviews for an oral history project funded by the Australian War Memorial. Both tapes and transcripts are lodged with the library of the Australian War Memorial and are available for public use. (NT) indicates that no tape or transcript is available. Transcripts are available online via the AWM collection, and the audio of several interviews (including Bird, Farrall and Langham) can be accessed online.

Barnes, Les. 13 May 1982. (NT)
Bird, Percy. Interview 1, 8 September 1983, 72 pages of transcript;
    Interview 2, 7 April 1987, 43pp.
Blake, Harold. 17 May 1982, 42pp.
Bowden, Charles. 2 July 1983, 40pp.
Bridgeman, Bill. 1 May 1983, 38pp.
Cuddeford, E. L. 6 September 1983, 27pp.
D'Altera, Stan. 8 May 1983, 56pp.
David, Leslie. 27 May 1983, 37pp.
Farrall, Fred. Interview 1, 7 July 1983 and 14 July 1983, 72pp;
    Interview 2, 2 April 1987, 36pp.
Flannery, Jack. 1 May 1983, 59pp.
Fogarty, Percy J. 3 June 1983, 26pp.
Glew, Jack T. 6 September 1983, 20pp.
Glyde, Margaret (pseudonym). 12 June 1982. (NT)
Guthrie, Doug. 8 May 1983, 34pp.
Langham, Bill. Interview 1, 29 May 1983, 42pp;
    Interview 2, 30 March 1987, 36pp.
Linton, Albert C. 12 June 1983, 28pp.
McGillivray, A. J. 12 September 1983, 28pp.
McKenzie, Ted. J. 29 April 1983, 39pp.
McNair, James. 27 May 1982, 29pp.
Morton, Ern. Interview 1, 21 June 1985, 42pp;
    Interview 2, 26 March 1987, 42pp.
Norris, Sid. 13 September 1983, 31pp.
Stabb, Alf. 7 September 1983, 31pp.
Williams, Bill (pseudonym). Interview 1, 10 April 1983, 54pp;
    Interview 2, 1 April 1987, 42pp.

Newspapers

Age 1914–32, 1980–93
Argus 1915–18
Australian 1981–93
Brunswick and Coburg Gazette 1928–32
Brunswick and Coburg Leader 1915–28

Bulletin 1981–93
Herald 1981–87
Smith's Weekly 1930–32
Sun 1981–93

Archival sources

Australian Film Institute

Cinedossier 1977–87
Clippings Files for the following films: *Anzacs, Break of Day, A Fortunate Life, Gallipoli, Kangaroo, The Lighthorsemen, The Monocled Mutineer, 1915*.

Australian War Memorial, Canberra

Bean Papers
Bean, C. E. W. The Beginnings of the Australian War Memorial. Unpublished typescript, 3DRL 6673/619.
Correspondence and news cuttings concerning British Official Histories, 1926–39, 3DRL 7953/27–30.
Correspondence with General Edmonds relating to the Australian Official History, 1927–35, 3DRL 7953/34.
Correspondence concerning the Official History, 3DRL 7953/35–53; 6673/10; 6673/352.
Correspondence with Edmonds, 1931–55, 3DRL 6673/201.
Diaries and Notebooks, 1914–58, 3DRL 606.
Extract Books, The Landing, 3DRL 1722/2.
Historical Notes, Gallipoli, 3DRL 8042/4–9.
Letters respecting the Official History, 3DRL 8040.
Official History Manuscripts, AWM 44.
Official History Sales, 3DRL 6673/249–253.
Press References to Bean, 1914–18, 3DRL 6673/234.
Reviews of Official History, 3DRL 8043.

Gellibrand Papers
Various diaries, notes and correspondence, 3DRL 1473/8 and 36; 3DRL 6405/14; 3DRL 1473/36.

Manuscripts
Alistair Thomson, 'The Forgotten Anzacs', MS 1180.

Private records 1914–18
I consulted the following diaries, letters or memoirs about Gallipoli by members of the AIF.

Alywn, C.
Cowtan, M.
Dix, C.
Donkin, R. L.
Grant, R. A.
Hagan, J. S.
Hampton, R. L.

Hogue, O.
Mackay, K. S.
McKern, H. T.
McLennan, J. H.
Powell, T. H. N.
Richards, E. J.
Roberts, C. R.
Rowe, F. M.
Smith, A. H.
Steele, E.
Stewart, J. C.
Tieggs, A.H.

County Records Office, Isle of Wight

Aspinall-Oglander Papers
Papers dealing with Gallipoli history, OG 111–16.

Liddell Hart Centre for Military Archives, King's College Library, University of London

Edmonds Papers
Correspondence, II.
Memoirs, III/16

Liddell Hart Papers
Correspondence with Aspinall-Oglander, 1930–58, 1/23.
Official Histories of Australia, 4/37.

National Archives of Australia

Australian Security Intelligence Organization, Series A367
Report on Fred Farrall, C89597.

First AIF Personnel Dossiers, 1914–1920, Series B2455
Percy Bird.
Fred Farrall.
Bill Langham.
Hector Thomson.

Repatriation Department Papers, Series B73
Percy Bird Medical file, M28666.
Fred Farrall Hospital file, H101649.
Fred Farrall Medical file, M101649.
Bill Langham Hospital file, H53178.
Bill Langham Medical file, M53178.
Hector Hospital file, H58164.
Hector Medical file, M58164.

Second AIF Service Records, Series B883
Hector Thomson, VX56596.

National Library of Australia

Manuscript Collection
David Thomson, 'The Thomson Story: Childhood Memories, 1924–1941', 1999, Thomson Family Papers, MS8600.

Oral History and Folklore Collection
David Thomson, interview by Alistair Thomson, 4–8 August 1985, Australian Parliament's Oral History Project, TRC4900/35.

Public Records Office, Kew, London

Cabinet Papers
Committee of Imperial Defence, Committee on the Historical Section, 1922–23, CAB 27 / 182 and CAB 27 / 212.
Committee of Imperial Defence, Sub-Committee for the Control of the Official Histories, 1923–39, CAB 16 / 52–53.

State Library of Victoria

Manuscript Collection
Dorothy Farrall, 'Autobiography', transcript of an interview by Wendy Lowenstein, transcribed by Fred Farrall, no date.

University of Melbourne Archives

Farrall Papers
Series 1987.0418

Official publications
Victorian Parliamentary Debates, 1925.
Commonwealth Bureau of Census and Statistics, *Official Year Book of the Commonwealth of Australia*, 30, 1937.

Online sources
Cassells, Kyla . 'Anzac Day is a celebration of war', *Socialist Alternative*, 23 April 2010, at http://www.sa.org.au/index.php?option=com_k2&view=item&id=4716:anzac-day-is-a-celebration-of-war&Itemid=393, accessed 9 April 2013.
Cannon, Michael. 'Slater, William (Bill) (1890–1960)', *Australian Dictionary of Biography*, National Centre of Biography, Australian National University, http://adb.anu.edu.au/biography/slater-william-bill-11709/text20929, accessed 25 January 2013.
Gurciullo, Sebastian. 'Ellery, Reginald Spencer (Reg) (1897–1955)', *Australian Dictionary of Biography*, National Centre of Biography, Australian National University, http://adb.anu.edu.au/biography/ellery-reginald-spencer-reg-10110/text17847, accessed 10 January 2013.
Hurley, John V. 'Sewell, Sir Sidney Valentine (1880–1949)'. *Australian Dictionary of Biography*, National Centre of Biography, Australian National University, http://adb.anu.edu.au/biography/sewell-sir-sidney-valentine-8388/text14727, accessed 30 April 2013.
Roe, Michael. 'Arthur, Richard (1865–1932)', *Australian Dictionary of Biography*, National Centre of Biography, Australian National University, http://adb.anu.edu.au/biography/arthur-richard-5061/text8437, accessed 10 January 2013.

Wikipedia. 'Encephalitis Lethargica', http://en.wikipedia.org/wiki/Encephalitis_
lethargica, accessed 16 March 2013.
Wikipedia. 'Tuberculous Cervical Lymphadenitis', http://en.wikipedia.org/wiki/
Tuberculous_cervical_lymphadenitis, accessed 13 January 2013.

Oral history theory and method

Abrams, Lynn. *Oral History Theory*. Routledge, London, 2010.
Assman, Jan. 'Collective Memory and Cultural identity'. *New German Critique*, 65, 1995,
pp. 125–33.
Bennett, Tony, Boyd-Bowman, Susan, Mercer, Colin & Woollacott, Janet (eds). *Popular
Television and Film*. BFI Publishing/The Open University Press, London, 1981.
Bornat, Joanna. 'Oral History as a Social Movement: Reminiscence and Older People'.
Oral History, 17, 2, Autumn 1989, pp. 16–25.
——. (ed.). *Reminiscence Reviewed: Achievements, Evaluations, Perspectives*. Oxford
University Press, Oxford, 1993.
Bromley, Roger. *Lost Narratives: Popular Fictions, Politics and Recent History*. Routledge,
London, 1988.
Browning, Christopher R. *Remembering Survival: Inside a Nazi Slave-Labor Camp*. W.W.
Norton, New York, 2010.
Bruner, Jerome. 'The Narrative Construction of Reality', *Critical Inquiry*, 18, 1, 1991, pp. 1–21.
Charlton, Thomas L., Myers, Lois E. & Sharpless, Rebecca (eds). *Handbook of Oral
History*. Altamira Press, Lanham, MD, 2006.
Coleman, Peter. *Ageing and Reminiscence Processes*. Wiley, London, 1986.
Dawson, Graham. Nostalgia, Conservatism and National Identity. Paper presented at the
Patriotism History Workshop, Oxford, 11 March 1984.
——. Nostalgia, Conservatism and the Politics of Memory. Unpublished manuscript, 1984.
——. 'Oral History: A Critique of *Voice of the Past*', Centre for Contemporary Cultural
Studies, Stencilled Paper, 1982.
——. 'Playing at War: An Autobiographical Approach to Boyhood Fantasy and
Masculinity'. *Oral History*, 18, 1, Spring 1990, pp. 44–53.
Day, Alice T. 'A Review Essay: The Fortunate Life of A. B. Facey — The Life Reviewing
Process and Personal Integration'. *Oral History Association of Australia Journal*, 4, 1982,
pp. 62–8.
Douglas, Louise. *Australia 1938 Oral History Handbook*. Australia 1938 Project, Canberra,
1981.
—— and Spearitt, Peter. 'Talking History: The Use of Oral Sources'. In *New History:
Studying Australia' Today*, G. Osborne & W. F. Mandle (eds). Allen & Unwin,
Sydney, 1982, pp. 51–68.
'Editorial: Oral History'. *History Workshop Journal*, 8, Autumn 1979, pp. i–iii.
Field, Sean. 'Beyond "healing": trauma, oral history and regeneration'. *Oral History*, 34, 1,
2006, pp. 31–42.
Formations Editorial Collective. *Formations of Nation and People*. Routledge & Kegan
Paul, London, 1984.
Fraser, Ronald. *Blood of Spain: The Experience of Civil War 1936–39*. Allen Lane, London,
1979.
——. *A Sense of the Past: The Manor House, Amnersfield, 1933–1945*. Verso, London, 1984.
Freund, Alexander. 'A Canadian Family Talks about Oma's Life in Nazi Germany: Three-
Generational Interviews and Communicative Memory'. *Oral History Forum*, 29,
2009, pp. 1–26.
——, & Thomson, Alistair (eds). *Oral History and Photography*. Palgrave Macmillan, New
York, 2011.
Friedlander, Saul . 'Trauma, Memory, and Transference'. In *Holocaust Remembrance: The*

Shapes of Memory, Geoffrey H. Hartman (ed.). Blackwell, Cambridge, Massachusetts, 1994, pp. 252–63.

Frisch, Michael. *A Shared Authority. Essays on the Craft and Meaning of Oral and Public History*. State University of New York Press, Albany, 1990.

Gluck, Sherna B. & Patai, Daphne (eds). *Women's Words: The Feminist Practice of Oral History*. Routledge, New York, 1991.

Griffiths, Gareth. 'Memory Lane: Museums and the Practice of Oral History'. *Oral History*, 17, 2, Autumn 1989, pp. 49–52.

Griffiths, Tom. 'The Debate About Oral History'. *Melbourne Historical Journal*, 13, 1981, pp. 16–21.

Hagopian, Patrick. 'Oral Narratives: Secondary Revision and the Memory of the Vietnam War'. *History Workshop Journal*, 32, Autumn 1991, pp. 134–50.

Healy, Chris. 'History, History Everywhere but …'. *Making the Bicentenary — Australian Historical Studies*, 23, 91, October 1988, pp. 180–92.

Henige, David. *Oral Historiography*. Longman, London, 1982.

Hirsch, Marianne. 'The Generation of Postmemory'. *Poetics Today*, 29, 1, 2008, pp. 103–28.
——. *The Generation of Postmemory: Writing and Visual Culture After the Holocaust*. Columbia University Press, New York, 2012.

Hobsbawm, Eric & Ranger, Terence. *The Invention of Tradition*. Cambridge University Press, Cambridge, 1983.

Humphries, Stephen. *The Handbook of Oral History: Recording Life Stories*. Inter-Action Imprint, London, 1984.

Hurd, Geoff. *National Fictions: World War Two in British Film and Television*. BFI Publishing, London, 1984.

Hutton, Anne. 'Australian Historical Film: The Production of Popular Memory'. In *The First Australian History and Film Conference Papers*, Anne Hutton (ed.). Australian Film and Television School, Sydney, 1982, pp. 206–23.

International Oral History Committee. *Papers presented at the Sixth International Oral History Conference*. Oxford, 11–13 September 1987.
——. *Papers presented at the Seventh International Oral History Conference*. Essen, 31 March 1990.

James, Daniel. *Dona María's Story: Life History, Memory and Political Identity*. Duke University Press, Durham, NC, 2000.

Johnson, Richard. 'Cultural Studies and English Studies — Approaches to Nationalism, Narrative and Identity'. *Hard Times* 31, 1987, pp. 3–10.
——. McLennan, Gregor, Schwartz, Bill & Sutton, David. *Making Histories: Studies in History Writing and Politics*. Hutchinson, London, 1982.

Jung, C. G. *Memories, Dreams and Reflections*. Collins/Routledge & Kegan Paul, London, 1963.

Laclau, Ernesto. *Politics and Ideology in Marxist Theory*. New Left Books, London, 1977.

Linde, Charlotte. *Life Stories: The Creation of Coherence*. Oxford University Press, New York, 1993.

Lowenstein, Wendy. 'The Interpretation and Use of Oral History'. *Oral History Association of Australia Journal*, 1, 1978–79, pp. 67–72.

Lummis, Trevor. *Listening to History*. Hutchinson, London, 1987.

McArthur, Colin. 'Days of Hope'. *Screen*, 16, 4, Winter 1975/76.
——. *Television and History*. BFI Publishing, London, 1978.

McCabe, Colin. '"Days of Hope" — A Response to Colin McArthur'. *Screen*, 16, 4, Winter 1975/76.

McConville, Chris. 'Oral History or Popular Memory? The Power of Talk'. Paper presented at the Oral History Association of Australia Annual Conference, 1985.

Middleton, David & Edwards, Derek (eds). *Collective Remembering*. Sage, London, 1990.

Murphy, John. 'Conscripting the Past: The Bicentenary and Everyday Life'. *Making the*

Bicentenary — Australian Historical Studies, 23, 91, October 1988, pp. 45–54.

——. 'The Voice of Memory: History, Autobiography and Oral Memory'. *Historical Studies*, 22, 87, October 1986, pp. 157–75.

Nguyen, Natalie. *Memory Is Another Country: Women of the Vietnamese Diaspora*. Praeger, Santa Barbara, California, 2009.

Passerini, Luisa. 'On the Use and Abuse of Oral History'. Mimeo translated from *Storia Orale: Vita Quotidiana e Cultura Materiale delli Classe Subalterne*. Rosenberg & Sellier, Turin, 1978.

——. 'Work Ideology and Consensus Under Italian Fascism'. *History Workshop Journal*, 8, 1979, pp. 82–108.

——. (ed.) *International Yearbook of Oral History and Life Stories, Vol. I, Memory and Totalitarianism*. Oxford University Press, Oxford, 1992.

——. *Fascism in Popular Memory: The Cultural Experience of the Turin Working Class*. Cambridge University Press, Cambridge, 1987.

Perks, Robert & Thomson, Alistair (eds). *The Oral History Reader*. Routledge, London and New York, 2006.

Pollock, Della (ed.). *Remembering: Oral History as Performance*. Palgrave Macmillan, New York, 2005.

Popular Memory Group. Unpublished manuscript of Part A of the 'Popular Memory' book.

——. 'What Do We Mean By Popular Memory'. Centre for Contemporary Cultural Studies, Stencilled Paper 67, January 1982.

Portelli, Alessandro. 'The Peculiarities of Oral History'. *History Workshop Journal*, 12, Autumn 1981, pp. 96–107.

——. The *Death of Luigi Trastuli and Other Stories: Form and Meaning in Oral History*. State University of New York, Albany, 1991.

——. *The Order Has been Carried Out: History, Memory, and Meaning of a Nazi Massacre in Rome*. Palgrave Macmillan, New York, 2003.

Ritchie, Donald A. (ed.). *The Oxford Handbook of Oral History*. Oxford University Press, New York, 2011.

Roper, Michael. 'Analysing the analysed: transference and counter-transference in the oral history encounter'. *Oral History*, 31, 2, 2003, pp. 20–32.

——. 'Re-remembering the Soldier Heroes: The Psychic and Social Construction of Memory in Personal Narratives of the Great War'. *History Workshop Journal*, 50, Autumn 2000, pp. 181–204.

——. 'Review of *Anzac Memories*'. *Oral History*, 22, 2, 1994, p. 92.

Roseman, Mark . 'Surviving Memory: Truth and Inaccuracy in Holocaust Testimony'. *The Journal of Holocaust Education*, 8, 1, 1999, pp. 1–20

Samuel, Raphael (ed.). *People's History and Socialist Theory*. Routledge & Kegan Paul, London, 1981.

—— & Thompson, Paul (eds). *The Myths We Live By*. Routledge, London, 1990.

Sheridan, Dorothy. 'Ambivalent Memories: Women and the 1939–45 War in Britain'. *Oral History*, 18, 1, Spring 1990, pp. 32–40.

Smith, Graham. 'Beyond Individual / Collective Memory: Women's Transactive Memories of Food, Family and Conflict'. *Oral History*, 35, 2, 2007, pp. 77–90.

Summerfield, Penny. *Reconstructing Women's Wartime Lives*. Manchester University Press, Manchester, 1998.

Thelen, David. 'Memory and American History'. *Journal of American History*, 75, 4, March 1989, pp. 1117–29.

Thompson, Paul. 'Editorial'. *Oral History*, 17, 2, Autumn 1989, p. 2.

—— (ed.). *Our Common History: The Transformation of Europe*. Pluto Press, London, 1982.

—— *The Voice of the Past: Oral History*. Oxford University Press, Oxford, 1978 and 1988.

——, Passerini, Luisa, Bertaux-Wiame, Isabelle & Portelli, Alessandro. 'Between Social Scientist: Responses to Louisa A. Tilly'. *International Journal of Oral History*, 6,

February 1985, pp. 19–39.

Thomson, Alistair. 'Anzac Memories: Putting Popular Memory Theory into Practice in Australia'. *Oral History*, 18, 1, Spring 1990, pp. 25–31.

——. 'Anzac Stories: Using Personal Testimony in War History', *War and Society*, 25, 2, 2006, pp. 1–21.

——. 'Four paradigm transformations in oral history'. *Oral History Review*, 34, 1, 2007, pp. 49–70.

——. 'Memory and Remembering in Oral History'. In *The Oxford Handbook of Oral History*, Donald A. Ritchie (ed.). Oxford University Press, New York, 2011, pp. 77–95.

——. *Moving Stories: an Intimate History of Four Women Across Two Countries*. University of New South Wales Press, Sydney and Manchester University Press, Manchester, 2011.

Tilly, Louisa A. 'People's History and Social Science History'. *International Journal of Oral History*, 6, February 1985, pp. 5–18.

Tribe, Keith. 'History and the Production of Memories'. *Screen*, xvii, 4, 1977/78.

Vansina, Jan. 'Memory and Oral Tradition'. In *The African Past Speaks*, Joseph C. Miller (ed.). Dawson & Sons, Folkestone, 1980, pp. 262–79.

Various Authors. 'Oral History: Facts and Fiction Debate'. *Oral History Association of Australia Journal*, 5, 1982–3, passim.

Wollheim, Richard. *Freud*. Fontana, Glasgow, 1971.

Wright, Patrick. *On Living In An Old Country: The National Past in Contemporary Britain*. Verso, London, 1985.

Yow, Valerie Raleigh. '"Do I Like Them Too Much?" Effects of the Oral History Interview on the Interviewer and Vice-Versa'. *Oral History Review*, 24, 1, 1997, pp. 55–79.

——. *Recording Oral History. A Guide for the Humanities and Social Sciences*. Altamira Press, Walnut Creek, California, 2nd edition, 2005.

Other books, articles and theses

Adam-Smith, Patsy. *The Anzacs*. Thomas Nelson, Melbourne, 1978.

Allen, J. A. 'Mundane Men: Historians, Masculinity and Masculinism'. *Historical Studies*, 22, 89, 1987, pp. 617–28.

——. *Sex and Secrets: Crimes Involving Australian Women Since 1880*. Oxford University Press, Melbourne, 1990.

Altman, Denis. 'The Myth of Mateship'. *Meanjin*, 46, 2, June 1987, pp. 163–72.

Anderson, Benedict. *Imagined Communities: Reflections on the Origin and Spread of Nationalism*. Verso, London, 1983.

The Anzac Tradition: Between the Lines. Australia Post, Canberra, 1990.

Ashplant, T.G., Dawson, Graham & Roper, Michael. 'The politics of war memory and commemoration: contexts, structures and dynamics'. In Timothy Ashplant, Graham Dawson and Michael Roper (eds). *The Politics of Memory: Commemorating War*, Transaction Publishers, New Brunswick, 2004, pp. 3–85.

Ashton, Paul and Kean, Hilda (eds). *People and their Pasts: Public History Today*. Palgrave, Basingstoke, 2009.

Ashworth, A. E. *Trench Warfare, 1914–1918: The Live and Let Live System*. Macmillan, London, 1980.

Aspinall-Oglander, C. F. *Military Operations, Gallipoli, Vol. I*. Heinemann, London, 1929.

Bang, Maureen. 'Toorak's Pensioner Mayor'. *Australian Woman's Weekly*, 3 October 1973, pp. 4–5.

Barbusse, M. *Under Fire*. Dent, London, 1926.

Barrett, John. 'No Straw Man: C. E. W. Bean and Some Critics', *Australian Historical Studies*, 23, 90, April 1988, pp. 102–14.

Baxter, Archibald. *We Will Not Cease*. The Caxton Press, Christchurch, 1968 (1939).

Bazley, A. W. 'Writing the Official History of World War I at Tuggeranong'. Mimeo of address delivered to Canberra and District Historical Society, 10 April 1959.

Bean, C. E. W. *Anzac to Amiens*. Australian War Memorial, Canberra, 1946.

—— (ed.). *The Anzac Book*. Cassell & Co., London, 1916.

——. 'The Australian Forces in the War of 1914–18'. In *The Book of the Anzac Memorial, NSW*, S. Elliot Napier (ed.). Beacon Press, Sydney, 1934.

——. *Australians in Action: The Story of Gallipoli*. W. A. Gullick, Government Printer, Sydney, 1915.

——. *The Dreadnought of the Darling*. Alston Rivers, London, 1911; republished Angus & Robertson, Sydney, 1956.

——. *Flagships Three*. Alston Rivers, London, 1913.

——. *Gallipoli Mission*. Australian War Memorial, Canberra, 1948.

——. *Here, My Son. An Account of the Independent and Other Corporate Schools of Australia*. Angus & Robertson, Sydney and London, 1950.

——. *In Your Hands, Australians*. Cassell & Co., London, 1918.

——. 'League of Nations'. *Sydney Morning Herald*, 13 November 1930, p. 4.

——. *Letters from France*. Cassell & Co., London, 1917.

——. *The Official History of Australia in the War of 1914–1918, Vol. i, The Story of Anzac: The First Phase*. Angus & Robertson, Sydney, 1921.

——. *Vol. ii, The Story of Anzac: From May 4, 1915 to the Evacuation*. Angus & Robertson, Sydney, 1924.

——. *Vol. iii, The AIF in France 1916*. Angus & Robertson, Sydney, 1929.

——. *Vol. iv, The AIF in France 1917*. Angus & Robertson, Sydney, 1937.

——. *Vol. v, The AIF in France during the Main German Offensive, 1918*. Angus & Robertson, Sydney, 1937.

——. *Vol. vi, The AIF in France: May — The Armistice 1918*. Angus & Robertson, Sydney, 1942.

——. *On The Wool Track*. Alston Rivers. London, 1910; republished Charles Scribners, New York, 1947.

——. 'Our War History'. *Bulletin*, 27 May 1942, p. 2.

——. 'Sidelights of the War on Australian Character'. Royal Australian Historical Society, *Journal and Proceedings*, xiii, iv, 1927, pp. 209–23.

——. 'Techniques of a Contemporary War Historian'. *Historical Studies*, 2, 6, November 1942, pp. 65–79.

——. *Two Men I Knew*. Angus & Robertson, Sydney, 1957.

——. *War Aims of a Plain Australian*. Angus & Robertson, Sydney, 1943.

——. *What to Know in Egypt ... A Guide for Australian Soldiers*. Société Orientale de Publicité, Cairo, 1915.

——. *With the Flagship in the South*. William Brooks, Sydney, 1909.

—— 'The Writing of the Australian Official History of the Great War — Sources, Methods and Some Conclusions'. Royal Australian Historical Society, *Journal and Proceedings*, xxiv, 2, 1938, pp. 85–112.

Bell, Johnny. 'Needing a Woman's Hand: Child Protection and the Problem of Lone Fathers'. *History Australia*, 9, 2, 2012, pp. 90–110.

Bessant, B. 'Empire Day, Anzac Day, the Flag Ceremony and All That'. *Historian*, 25, October 1973, pp. 36–43.

Bird, Percy. The 5th Battalion, 1916 and 1917, France. Unpublished manuscript, 1983.

Blackmore, Kate. *The Dark Pocket of Time: War, Medicine and the Australian State, 1914–1935*. Lythrum Press, Adelaide, 2008.

Bond, Brian (ed.). *The First World War and British Military History*. Clarendon Press, London. 1991.

Broadbent, Harvey. *The Boys Who Came Home: Recollections of Gallipoli*. ABC Enterprises, Crows Nest, 1990.

Brown, Stephanie. Anzac Rituals in Melbourne, 1916–1933: Contradictions and Resolutions within Hegemonic Processes. BA thesis, University of Melbourne, 1982.

Brugger, Suzanne. *Australians and Egypt, 1914–1919*. Melbourne University Press, Carlton, Victoria, 1980.

Bryan, Dominic & Ward, Stuart. 'The "Deficit of Remembrance": The Great War Revival in Australia and Ireland'. In *Exhuming Passions: The Pressure of the Past in Ireland and Australia*, Stuart Ward and Katie Holmes (eds). Irish Academic Press, Dublin and Portland, Oregon, 2011, pp. 163–86.

Buchanan, Margaret. The Shrine of Remembrance and its Forgotten Controversies (1921–35). BA thesis, University of Melbourne, 1984.

Burgess, Pat. *Warco: Australian Reporters at War*. Heinemann, Melbourne, 1986.

Burstall, Terry. *A Soldier Returns: A Long Tan Veteran Discovers the Other Side of Vietnam*. University of Queensland Press, St Lucia, 1990.

Butler, A. G. *The Australian Army Medical Services in the War of 1914–1918, Vol. iii, Special Problems and Services*. Australian War Memorial, Canberra, 1943.

——. *The Digger: A Study in Democracy*. Angus & Robertson, Sydney and London, 1945.

Cannadine, David. 'War and Death, Grief and Mourning in Modern Britian'. In *Mirrors of Mortality; Studies in the Social History of Death*, Joachim Whaley (ed.). Europa, London, 1981.

Cantwell, John. *Exit Wounds: One Australian's War on Terror*. Melbourne University Publishing, Melbourne, 2012.

Carey, Alex. 'What Australian Troops are Doing in Vietnam: The New Anzac Legend'. *Outlook*, 14, February 1970, pp. 5–8.

Carroll, John. 'C. E. W. Bean and 1988'. *Quadrant*, 32, 6, June 1988. pp. 47–9.

—— (ed.). *Intruders in the Bush: The Australian Quest for Identity*. Oxford University Press, Melbourne, 1982.

Cathcart, Michael. *Defending the National Tuckshop: Australia's Secret Army Intrigue of 1931*. McPhee Gribble/Penguin, Melbourne, 1988.

Charlton, Peter. *Pozières: Australians on the Somme, 1916*. Methuan Haynes, North Ryde, 1980.

Clark, Anna. *History's Children: History Wars in the Classroom*. University of New South Wales Press, Sydney, 2008.

Clissold, Barry. 'Military Historian C. E. W. Bean: Another View'. *Sabretache*, 23, April–June 1982, pp. 11–14.

Coleman, Peter. 'Death and the Australian Legend'. *Bulletin*, 85, 27 April 1963, pp. 18–19.

Connell, R. W. and Irving, T. H. *Class Structure in Australian History*. Longman Cheshire, Melbourne, 1980.

Conway, Ronald. 'Last Post for Anzac'. *Bulletin*, 29 April 1980, pp. 55–8.

Cottle, Drew. 'A Comprador Countryside: Rural New South Wales, 1919–1939'. In *Capital Essays: Selected Papers from the General Studies Conference on Australian Capital History*, Drew Cottle (ed.). D. Cottle, Sydney, 1984.

Cribben, John. *The Making of 'Anzacs'*. Collins/Fontana, Sydney, 1985.

Crotty, Martin & Larsson, Marina (eds). *Anzac Legacies: Australians and the Aftermath of War*. Australian Scholarly Publishing, North Melbourne, 2010.

Crowley, F. K. *Modern Australia in Documents, Volume 1, 1901–1939*. Wren, Melbourne, 1973.

——. *Modern Australia in Documents, Volume 2, 1939–1970*. Wren, Melbourne, 1973.

—— (ed.). *A New History of Australia*. Heinemann, Melbourne, 1974.

Dallas, Gloden and Gill, Douglas. *The Unknown Army: Mutinies in the British Army in World War 1*. Verso, London, 1985.

Damousi, Joy. *Freud in Australia: A Cultural History of Psychoanalysis in Australia*. University of New South Wales Press, Sydney, 2005.

——. *The Labour of Loss: Mourning, Memory and Wartime Bereavement in Australia*. Cambridge University Press, Cambridge, 1999.

——. *Living with the Aftermath: Trauma, Nostalgia and Grief in Post-war Australia*.

Cambridge University Press, Cambridge, 2001.

Davison, F. D. *The Road to Yesterday*. Angus & Robertson, Sydney, 1965.

Davison, Graeme. 'The Habit of Commemoration and the Revival of Anzac Day'. *Australian Cultural History*, 23, 2003, pp. 73–82.

——. 'Sydney and the Bush: An Urban Context for the Australian Legend'. *Historical Studies*, 18, 71, October 1978, pp. 191–209.

Dawson, Graham. *Soldier Heroes: Britishness, Colonial Adventure and the Imagining of Masculinities*. Routledge, London, 1994.

Dennis, C. J. *Digger Smith*. Angus & Robertson, Sydney, 1918.

——. *The Moods of Ginger Mick*. Angus & Robertson, Sydney, 1916.

Denton, Kit. *Gallipoli, One Long Grave*. Time Life Books & John Ferguson, Sydney, 1986.

Dyson, W. *Australians at War*. G. Palmer & Hayward, London, 1918.

Duncan, Tim. 'History as a Kangaroo Court'. *IPA Review*, 42, 3, December–February 1988/89, pp. 52–3.

Edkins, Jenny. *Trauma and the Memory of Politics*. Cambridge University Press, Cambridge, 2003.

Edwards, P. G. 'Official History: Does It Merit Suspicion?' *Canberra Historical Journal*, new series 17, March 1986, pp. 17–21.

Ellis, Anthony. The Impact of War and Peace on Australian Soldiers 1914–1920. BA thesis, Murdoch University, 1979.

Ellis, Stephen Charles. 'The Censorship of the Official Naval History of Australia in the Great War'. *Historical Studies*, 20, April 1983, pp. 367–82.

——. C. E. W. Bean: A Study of His Life and Works. MA thesis, University of New England, 1969.

——. 'Some Aspects of Military History in Australia'. *Armidale and District Historical Society Journal*, 10, 1967, pp. 9–16.

Ely, Richard. 'The First Anzac Day: Invented or Discovered?'. *Journal of Australian Studies*, 17, November 1985, pp. 41–58.

Encel, S. 'The Study of Militarism in Australia'. *Australian and New Zealand Journal of Sociology*, 3, 1, April 1967, pp. 2–18.

Englander, David. 'Troops and Trade Unions, 1919'. *Modern History*, 37, March 1987, pp. 8–13.

Evans, Raymond. *Loyalty and Disloyalty: Social Conflict on the Queensland Home Front, 1914–1918*. Allen & Unwin, Sydney, 1987.

——. '"Some Furious Outburst of Riot": Returned Soldiers and Queensland's "Red Flag" Disturbances, 1918–1919'. *War and Society*, 13, 2, September 1985, pp. 75–98.

Farrall, Fred. 'Trade Unionism in the First A.I.F. – 1914–1918'. *Recorder*, 54, October 1971, pp. 3–8.

Farrall, Lois. *The File on Fred: A Biography of Fred Farrall*. High Leigh Publishing, Carrum, 1992.

Fassin, Didier & Rechtman, Richard. *The Empire of Trauma: An Inquiry into the Condition of Victimhood*. Translated by Rachel Gomme, Princeton University Press, Princeton, 2009.

Fewster, Kevin. 'Ellis Ashmead-Bartlett and the Making of the Anzac Legend'. *Journal of Australian Studies*, 10, June 1982, pp. 17–30.

——. Expression and Suppression: Aspects of Military Censorship in Australia during the Great War. PhD thesis, University of New South Wales, 1980.

—— (ed.). *Gallipoli Correspondent: The Frontline Diary of C. E. W. Bean*. Allen & Unwin, Sydney. 1983.

——. 'The Wazza Riots'. *Journal of the Australian War Memorial*, 4, April 1984, pp. 47–53.

Firth, Stewart & Hoorn, Jeanette. 'From Empire Day to Cracker Night'. In *Australian Popular Culture*, Peter Spearritt and David Walker (eds). Allen & Unwin, Sydney, 1979, pp. 17–38.

Fox. L. F. *The First World War — and the Second?* Victorian Council Against War and
  Fascism, Melbourne, 1935.
——. *The Truth about Anzac.* Victorian Council Against War and Fascism, Melbourne, 1936.
French, Maurice. 'The Ambiguity of Empire Day in New South Wales 1901–21'.
  Australian Journal of Politics and History, 24, April 1978, pp. 61–74.
From the Australian Front. Cassell & Co., London, 1917.
Fuller, J. G. *Troop Morale and Popular Culture in the British and Dominion Armies 1914–
  1918.* Clarendon Press, London, 1991.
Furniss, Elizabeth. 'Timeline History and the Anzac Myth: Settler Narratives of Local
  History in a North Australian Town'. *Oceania*, 71, 4, 2001, pp. 279–97.
Fussell, Paul. *The Great War and Modern Memory.* Oxford University Press, London, 1975.
Gammage, Bill. 'Australians and the Great War'. *Journal of Australian Studies*, 6, June 1980,
  pp. 26–35.
——. *The Broken Years: Australian Soldiers in the Great War.* Penguin, Ringwood, Victoria,
  1975.
——. Introduction to Bean, *The Official History of Australia in the War of 1914–1918, Vol.
  iv.* University of Queensland Press, St Lucia, 1982.
——. 'Working on *Gallipoli*'. In Anne Hutton (ed.), *The First Australian History and Film
  Conference Papers.* Australian Film and Television School, North Ryde, New South
  Wales, 1982.
—— and Williamson, David. *The Story of* Gallipoli. Penguin, Ringwood, Victoria, 1981.
Garton, Stephen. *The Cost of War: Australians Return.* Oxford University Press Australia,
  Oxford and New York, 1996.
Gerster, Robin. *Big-noting: The Heroic Theme in Australian War Writing.* Melbourne
  University Press, Melbourne, 1987.
Glenister, Richard. C. E. W. Bean: Colonial Historian: The Writing of the History of the
  AIF in France during World War I. Unpublished paper, Mitchell College, Bathurst,
  1983.
——. Desertion Without Execution: Decisions that Saved Australian Imperial Force Deser-
  ters from the Firing Squad in World War I. BA thesis, La Trobe University, 1984.
Gough, Tony. 'The First Australian Imperial Force: C. E. W. Bean's Coloured
  Authenticity'. *World Review*, 16, September 1977, pp. 40–9.
——. 'The Repatriation of the First Australian Imperial Force'. *Queensland Historical
  Review*, vii, 1, 1978, pp. 58–69.
Grey, Jeffrey. *A Military History of Australia.* Cambridge University Press, Cambridge, 1990.
——. *A Military History of Australia.* Cambridge University Press, Cambridge and Port
  Melbourne, 3rd edition, 2008.
Griffiths, Tom. 'Anzac Day'. *Overland*, 87, 1982, pp. 5–8.
Hadfield, Damian. 'The Evolution of Combat Stress: New Challenges for a New Generat-
  ion'. In Martin Crotty and Marina Larsson (eds), *Anzac Legacies: Australians and the
  Aftermath of War*, Australian Scholarly Publishing, North Melbourne, 2010, pp. 233–46
Hamilton, Paula & Ashton, Paul. 'At Home With the Past: Initial Findings From the
  Survey'. *Australians and the Past, A special edition of Australian Cultural History*, 22,
  University of Queensland Press, St Lucia, 2003, pp. 5–30.
Hastings, Anthony Paul. C. E. W. Bean — Australian War Historian. An Analysis of Bean's
  Six Volumes of the Official History. BA thesis, University of Melbourne, 1979.
——. 'Writing Military History in Australia'. *Melbourne Historical Journal*, 13, 1981, pp. 51–5.
Henderson, Gerard. 'The Anzac Legend after Gallipoli', *Quadrant*, 26 July 1982, pp. 62–4.
Hetherington, John. *The Australian Soldier.* F. H. Johnstone, Sydney, 1943.
Higham, Robin (ed.). *Official Histories: Essays and Bibliographies from around the World.*
  Kansas State University Library, Manhattan, Kansas, 1970.
Higonnet, M. R., Jenson, J., Michel, S. & Weitz, M. C. (eds). *Behind the Lines: Gender and
  the Two World Wars.* Yale University Press, New Haven, 1987.

Hill, E.F., Kelleher, B., Miller, A. & Gibson, R., with Fred Farrall. *Celebration of Fred Farrall's 90th Birthday*. Australian Society for the Study of Labour History, Melbourne, 1987.

Hirst, J. B. 'The Pioneer Legend'. *Historical Studies*, 18, 71, October 1978, pp. 316–37.

Holbrook, Carolyn Anne. The Great War in the Australian Imagination Since 1915, D.Phil. thesis, University of Melbourne, 2012.

Holmes, Katie. Between the Lines: The Letters and Diaries of First World War Australian Nurses. BA thesis. University of Melbourne, 1984.

Holt, E. 'Anzac Reunion'. In *Best Australian One-act Plays*, W. Moore & T. J. Moore (eds). Angus & Robertson, Sydney, 1937.

Hood, David. 'Adelaide's First "Taste of Bolshevism": Returned Soldiers and the 1918 Peace Day Riots'. *Journal of the Historical Society of South Australia*, 15, 1987, pp. 42–53.

Howe, Adrian. 'Anzac Mythology and the Feminist Challenge'. *Melbourne Journal of Politics*, 15, 1983–84, pp. 16–23.

Hyde, Robin. *Nor the Years Condemn*. Hurst & Blackett, London, 1938.

Inglis, K. S. 'ANZAC and the Australian Military Tradition', *Current Affairs Bulletin*, 64, 11, April 1988, p. 15.

——. 'Anzac and Christian: Two Traditions or One?'. *St Mark's Review*, 42, November 1965, pp. 3–12.

——. 'The Anzac Tradition'. *Meanjin Quarterly*, 24, 1, 1965, pp. 25–44.

——. The Australian Colonists: An Exploration of Social History 1788–1870. Melbourne University Press, Melbourne, 1974.

——. 'The Australians at Gallipoli'. *Historical Studies*, 14, 54 and 55, April and October 1970, pp. 219–30 and 361–75.

——. *C. E. W. Bean, Australian Historian*. University of Queensland Press, St Lucia, 1970.

——. Introduction to Bean, The *Official History of Australia in the War of 1914–1918, Vol. i*. University of Queensland Press, St Lucia, 1981.

——. 'Memorials of the Great War'. *Australian Cultural History*, 6, 1987, pp. 5–17.

——. 'Men, Women and War Memorials: Anzac Australia'. *Daedalus*, 116, 4, Fall, 1987, pp. 35–59.

——. 'Monuments and Ceremonies as Evidence for Historians'. *Anzaas Congress Papers*, 834/26, September 1977.

——. *The Rehearsal: Australians at War in the Sudan 1885*. Rigby, Adelaide, 1985.

——. *Sacred Places: War Memorials in the Australian Landscape*. Melbourne University Press, Melbourne, 3rd edition, 2008.

——. 'A Sacred Place: The Making of the Australian War Memorial'. *War and Society*, 3, 2, September 1985, pp. 99–126.

Johnston, George. 'ANZAC … A Myth for all Mankind'. *Walkabout*, 31, April 1965, pp. 12–16.

Jones, E., Everitt, B., Ironside, S., Palmer, I. & Wessely, S. 'Psychological effects of chemical weapons: a follow-up study of First World War veterans'. *Psychological Medicine*, 38, 2008, pp. 1419–26.

Jones, Margaret. Repatriation after the First World War. BA thesis, University of Melbourne, 1955.

Joynt, W. D. *Saving the Channel Ports*. Wren, Melbourne, 1975.

Keegan, John. *The Face of Battle*. Jonathan Cape, London, 1976.

Kemp. J. 'Anzac: The Substitute Religion'. *Nation*, 42, 23 April 1960, pp. 7–9.

Kent, D. A. 'The Anzac Book and the Anzac Legend: C. E. W. Bean as Editor and Image Maker'. *Historical Studies*, 21, 84, April 1985, pp. 376–90.

——. 'The Anzac Book: A Reply to Denis Winter'. *Journal of the Australian War Memorial*, 17, October 1990, pp. 54–5.

——. 'From the Sudan to Saigon: A Critical View of Historical Works'. *Australian Literary Studies* 12, 2, October 1985, pp. 155–65.

——. 'Troopship Literature: "A Life on the Ocean Wave", 1914–19'. *Journal of the Australian War Memorial*, 10, April 1987, pp. 3–10.

Keown, A. W. *Forward with the Fifth*. Speciality Press, Melbourne, 1921.

King, Terry. The Adoration of the Magsman. Paper presented at a postgraduate seminar, History Department, La Trobe University, June 1987.

——. On the Definition of 'Digger'. Paper presented at a postgraduate seminar. History Department, La Trobe University, June 1983.

——. 'The Tarring and Feathering of J. K. McDougall: "Dirty Tricks" in the 1919 Federal Election'. *Labour History*, 45, November 1983, pp. 54–67.

——. 'Telling the Sheep from the Goats: "Dinkum Diggers" and Others'. In *An Anzac Muster*, Judith Smart & Tony Wood (eds). Monash Publications in History, Melbourne, 1992.

——. W. M. Hughes & M. P. Pimental: Soolers and Soldiers, 1918–1919. Paper presented at a postgraduate seminar, History Department, La Trobe University, 1984.

Kitley, Philip. 'Anzac Day Ritual', *Journal of Australian Studies*, 4, June 1979, pp. 58–69.

Knightley, Phillip. *The First Casualty: The War Correspondent as Hero, Propagandist and Myth Maker From the Crimea to Vietnam*, Andre Deutsch, London, 1975.

Kristianson, G. L. *The Politics of Patriotism: The Pressure Group Activities of the Returned Servicemen's League*. Australian National University Press, Canberra, 1966.

Lake, Marilyn. *A Divided Society: Tasmania During World War I*. Melbourne University Press, Melbourne, 1975.

——. *The Limits of Hope: Soldier Settlement in Victoria 1915–38*. Oxford University Press, Melbourne, 1987.

——. 'The Politics of Respectability: Identifying the Masculinist Context'. *Historical Studies*, 22, 86, 1986, pp. 116–31.

——. 'The Power of Anzac'. In *Australia: Two Centuries of War and Peace*, M. McKernan & M. Browne (eds). Australian War Memorial and Allen & Unwin, Canberra, 1988, pp. 194–222.

—— & Reynolds, Henry, with Mark McKenna and Joy Damousi (eds). *What's Wrong With Anzac? The Militarisation of Australian History*. University of New South Wales Press, Sydney, 2010.

Larsson, Marina. *Shattered Anzacs: Living with the Scars of War*. University of New South Wales Press, Sydney, 2009.

Leed, Eric J. *No Man's Land: Combat and Identity in World War I*. Cambridge University Press, Cambridge, 1979.

Lewis, R. 'The Spirit of Anzac — Myth or Reality'. *Journal of History*, xi, 4, 1980, pp. 1–11.

Liddle, Peter. *Men of Gallipoli*. Allen Lane, London, 1976.

Lindstrom, R. G. Stress and Identity: Australian Soldiers during the First World War. MA thesis, University of Melbourne, 1985.

Lloyd, Clem & and Rees, Jacqui. *The Last Shilling: A History of Repatriation in Australia*. Melbourne University Press, Carlton, 1994.

Lohrey, Amanda. 'Australian Mythologies: *Gallipoli* — Male Innocence as a Marketable Commodity'. *Island*, 9/10, March 1982, pp. 29–34.

Long, Gavin. 'The Australian War History Tradition'. *Historical Studies*, 6, 23, November 1954, pp. 249–60.

Luckins, Tanja. *The Gates of Memory: Australian People's Experiences and Memories of Loss and in Great War*. Curtin University Books, Fremantle, 2004.

Mandle, W. F. *Going it Alone: Australia's National Identity in the Twentieth Century*. Penguin, Ringwood, Victoria, 1980 (1978).

Mansfield, Wendy. 'The Importance of Gallipoli: The Growth of an Australian Folklore'. *Queensland Historical Review*, 6, 1, 1977, pp. 41–52.

——. World War I, Foundation Myths and the Anzac Tradition: A Reappraisal. Paper presented at the Australian War Memorial Military History Conference, Canberra, 1982.

SELECT BIBLIOGRAPHY

Masefield, John. *Gallipoli*. Heinemann, London, 1916.

McAuley, Lex. *The Battle of Long Tan — The Legend of Anzac Upheld*. Hutchinson, Melbourne, 1986.

McCarthy, Dudley. *Gallipoli to the Somme: The Story of C. E. W. Bean*. John Ferguson, Sydney, 1983.

McConville, Chris. 'Rough Women, Respectable Men and Social Reform: A Response to Lake's "Masculinism"'. *Historical Studies*, 22, 1987, pp. 432–40.

McDonald, Matt. '"Lest We Forget": The Politics of Memory and Australian Military Intervention'. *International Political Sociology*, 4, 2010, pp. 287–302.

McFarlane, Susannah. The Anzac Legend Affirmed: The Representation of Australians at War in Australian Feature Films 1915–1981. BA thesis, University of Melbourne, 1986.

McKay, Judith. Lest We Forget: A Study of War Memorials in Queensland: First Report. Unpublished report to the Queensland branch of the RSL, Brisbane, 1983.

McKenna, Mark. 'Anzac Day: How Did it Become Australia's National Day?'. In *What's Wrong With Anzac? The Militarisation of Australian History*, Marilyn Lake and Henry Reynolds, with Mark McKenna and Joy Damousi (eds). University of New South Wales Press, Sydney, 2010, pp. 110–34.

——. and Ward, Stuart. '"It was really moving, mate": The Gallipoli Pilgrimage and Sentimental Nationalism in Australia'. *Australian Historical Studies*, 38, 129, 2007, pp. 141–51.

McKernan, Michael. *The Australian People and the Great War*. Collins, Sydney, 1984 (1980).

——. *Here is Their Spirit. A History of the Australian War Memorial 1917–1990*. University of Queensland Press, St. Lucia, 1991.

—— and Browne. M (eds). *Australia: Two Centuries of War and Peace*. Australian War Memorial and Allen & Unwin, Canberra, 1988.

—— and Stanley, Peter. *Anzac Day Seventy Years On*. Collins, Sydney, 1986.

McLachlan, Noel. 'Nationalism and the Divisive Digger'. *Meanjin Quarterly*, 27, 3, 1968, pp. 302–8.

Macleod, Jenny (ed.). *Gallipoli: Making History*. Frank Cass, London and New York, 2004.

McMullin, Ross. *Will Dyson: Cartoonist, Etcher and Australia's Finest War Artist*. Angus & Robertson, Sydney, 1984.

McQueen, Humphrey. *From Gallipoli to Petrov*. Allen & Unwin, Sydney, 1984.

——. *A New Britannia*. Penguin, Ringwood, Victoria, 1975.

——. 'Shoot the Bolshevik! Hang the Profiteer! Reconstructing Australian Capitalism, 1918–21'. In *Essays in the Political Economy of Australian Capitalism, vol. 2*, E. L. Wheelwright & K. Buckley (eds). Australian and New Zealand Book Company, Sydney, 1978, pp. 185–206.

——. 'The Social Character of the New Guard'. *Arena*, 40, 1975, pp. 67–86.

McQuilton, John. 'A Shire at War: Yackandandah, 1914–18'. *Journal of the Australian War Memorial*, 11, October 1987, pp. 3–15.

Meaney, Neville. *A History of Australian Defence and Foreign Policy 1901–23: Volume 2, Australia and the World Crisis 1914–1923*. Sydney University Press, Sydney, 2009.

Moore, Andrew. 'Guns Across the Yarra'. In *What Rough Beast*, Sydney Labour History Group. Allen & Unwin, Sydney, 1982.

Moorehead, Alan. *Gallipoli*. Andre Deutsch, London, 1989 (1956).

Morris, Tim. The Writings of C. E. W. Bean in France during 1916. BA thesis, University of Melbourne, 1985.

Mosse, George. *Fallen Soldiers. Reshaping the Memory of the World Wars*. Oxford University Press, New York, 1990.

——. *The Nationalisation of the Masses*. Howard Fertig, New York, 1975.

——. 'Two World Wars and the Myth of the War Experience'. *Journal of Contemporary History*, 21, 1986, pp. 491–513.

Napier, S. Elliot (ed.). *The Book of the Anzac Memorial, NSW*. Beacon Press, Sydney, 1934.

O'Neil, Currey (ed.). *Bill Harney's War*. Currey O'Neil Press, Melbourne, 1983.

O'Neill, R. 'Soldiers and Historians: Trends in Military Historiography in the Nineteenth and Twentieth Centuries'. *Journal of the Royal Australian Historical Society*, 56, Part 1, March 1970, pp. 36–47.

Parker, Leslie [Angela M. Thirkell]. *Trooper to the Southern Cross*. Faber & Faber, London, 1934.

Pascoe, Rob. *The Manufacture of Australian History*. Oxford University Press, Melbourne, 1979.

Pederson, P. A. 'The AIF: as good as the Anzac Legend says?'. *Sydney Papers*, Autumn 2007, pp. 168–177.

——. Introduction to Bean, *The Official History of Australia in the War of 1914–1918, Vol iii*, University of Queensland Press, St Lucia, 1982.

Perrottet, Anthony. 'Politics of the Anzac Myth'. *National Times*, 25 April 1986. pp. 22–3.

Phillips, Jock, Boyack, Nicholas & Malone, E. P. *The Great Adventure: New Zealand Soldiers Describe the First World War*. Allen & Unwin / Port Nicholson Press, Wellington, 1988.

Piggott, Michael. *A Guide to the Personal, Family and Official Papers of C. E. W. Bean*. Australian War Memorial, Canberra, 1983.

Popple, Jeff. A Very Temperate Reaction: A Study of the First Nine Months of the Second World War in Australia, with Particular Emphasis on Enlistment in the Second AIF. BA thesis, University of New South Wales, 1980.

——. A Wary Welcome: The Political Reaction To The Demobilisation of the AIF. Paper presented at the Australian War Memorial History Conference, July, 1987.

Pringle, Rosemary. 'Rape: The Other Side of Anzac Day'. *Refractory Girl*, 26, June 1983, pp. 31–5.

Pritchard, Katharine Susannah. *Golden Miles*. Virago, London, 1984 (1948).

Pryor, L. J. The Origins of Australia's Repatriation Policy, 1914–1920. MA thesis, University of Melbourne, 1932.

Pugsley, Christopher. 'Stories of Anzac'. In *Gallipoli: Making History*, Jenny Macleod (ed.). Frank Cass, London and New York, 2004, pp. 44–58.

Reed, Liz. *Bigger than Gallipoli: War, History and Memory in Australia*. University of Western Australia Press, Crawley, 2004.

Remarque. E. M. *All Quiet on the Western Front*. Putnam, London, 1929.

Rhodes-James, Robert. *Gallipoli*. B. T. Batsford, London, 1965.

Robertson, John. *Anzac and Empire: The Tragedy and Glory of Gallipoli*. Hamby Australia, Port Melbourne, 1990.

Robson, Lloyd. 'The Anzac Tradition'. *Journal of History for Senior Students*, 4, 2, May 1973, pp. 57–62.

——. 'Behold a Pale Horse: Australian War Studies' *Australian Historical Studies*, 23, 90, April 1988, pp. 115–26.

——. 'C. E. W. Bean: A Review Article'. *Journal of the Australian War Memorial*, 4, 1984, pp. 54—57.

——. 'The Origin and Character of the First AIF: 1914–1918. Some Statistical Evidence'. *Historical Studies*, 61, October, 1973, pp. 737–49.

Roe, Michael. 'Comments on the Digger Tradition'. *Meanjin Quarterly*, 24, 4, 1965, pp. 357–8.

Roper, Michael. 'Memories of the Depression'. *Melbourne Historical Journal*, 13, 1981, pp. 22–8.

——. *The Secret Battle: Emotional Survival in the Great War*. Manchester University Press, Manchester and New York, 2009.

—— & Tosh, John (eds.). *Manful Assertions: Masculinities in Britain since 1800*. Routledge, London, 1991.

Ross, Jane. *The Myth of the Digger*. Hale & Iremonger, Sydney, 1985.

Rowe, A. P. 'Anzac'. *The Australian Quarterly*, 29, 1, March 1957, pp. 67–73.

Rowe, Lesley. Returned Servicemen and Politics in Victoria 1915–1925. BA thesis, University of Melbourne, 1966.

Rowse, Tim. *Australian Liberalism and National Character*. Kibble Books, Melbourne, 1978.

Rutherford, Andrew. *The Literature of War: Five Studies of Heroic Virtues*. Macmillan, London, 1978.

Sackett, Lee. 'Marching into the Past: Anzac Day Celebrations in Adelaide'. *Journal of Australian Studies*, 17, November 1985, pp. 18–30.

Scates, Bruce. *A Place to Remember: A History of the Shrine of Remembrance*. Cambridge University Press, Cambridge, 2009.

——. *Return to Gallipoli: Walking the Battlefields of the Great War*. Cambridge, Cambridge University Press, 2006.

Scott, Ernest. *Official History of Australia in the War of 1914–18, Vol. xi, Australia During the War*. Angus & Robertson, Sydney, 1936.

Seal, Graham. *Inventing Anzac: The Digger and National Mythology*. University of Queensland Press with API Network and Curtin University of Technology, St Lucia, 2004.

——. 'Two Traditions: The Folklore of the Digger and the Invention of Anzac'. *Australian Folklore*, 5, 1990, pp. 37–60.

——. '"Written in the Trenches": Trench Newspapers of the First World War'. *Journal of the Australian War Memorial*, 16, April 1990, pp. 30–8.

Sekuless, Peter & Rees, Jaqueline. *Lest We Forget*. Rigby, Dee Why West, New South Wales, 1986.

Serle, Geoffrey. 'The Digger Tradition and Australian Nationalism'. *Meanjin Quarterly*, 24, 2, 1965, pp. 149–58.

——. 'Introduction' to Bean, *The Official History of Australia in the War of 1914–1918, Vol. vi*, University of Queensland Press, St. Lucia, 1983.

Seymour, Alan. *The One Day of the Year*. In *Three Australian Plays*. Penguin, Melbourne, 1963.

——. and Nile, Richard (eds). *Anzac: Meaning, Memory and Myth*. Sir Robert Menzies Centre for Australian Studies, London, 1991.

Showalter, Elaine. *The Female Malady: Women, Madness and English Culture*. Virago, London, 1987.

Shute, Carmel. 'Heroines and Heroes: Sexual Mythology in Australia 1914–1918'. *Hecate*, 1, 1, 1975, pp. 7–22.

Sillars, Stuart. *Art and Survival in First World War Britain*. Macmillan, Basingstoke, 1987.

Souter, Gavin. 'The Inadequacy of Anzac Day'. *Hemisphere*, 17, April 1973, pp. 2–7.

——. *Lion and Kangaroo: Australia 1901–1919 The Rise of a Nation*. Fontana, Sydney, 1978.

Stanley, Peter. *Bad Characters: Sex, Crime, Mutiny, Murder and the Australian Imperial Force*. Pier 9, Miller's Point, 2010.

——. 'Gallipoli and Pozières: A Legend and a Memorial — Seventieth Anniversary of the Gallipoli Landing'. *Australian Foreign Affairs Record*, 56, 4, April 1985, pp. 281–7.

——. *Men of Mont St Quentin: Between Victory and Death*. Scribe, Melbourne, 2009.

——. 'Military history: over the top'. In *Australian History Now*, Paul Ashton and Anna Clark (eds). University of New South Wales Press, Sydney, forthcoming, 2013.

——. 'Monumental mistake? Is war the most important thing in Australian history?'. In Craig Stockings (ed.). *Anzac's Dirty Dozen: 12 Myths of Australian Military History*. University of New South Wales Press, Sydney, 2012, pp. 260–86.

——. 'Reflections on Bean's Last Paragraph'. *Sabretache*, xxiv, 3, July/September 1983, pp. 4–11.

Stockings, Craig (ed.). *Anzac's Dirty Dozen: 12 Myths of Australian Military History*. University of New South Wales Press, Sydney, 2012.

——. *Bardia: Myth, Reality and the Heirs of Anzac*. University of New South Wales Press, Sydney, 2009.

——. *Zombie Myths of Australian Military History*, University of New South Wales Press, Sydney, 2010

Strong, A. *The Story of the Anzacs*. James Ingram & Son, Melbourne, 1917.

Sullivan, J. C. The Genesis of Anzac Day: Victoria and New South Wales, 1916–1926. BA thesis, University of Melbourne, 1964.

Tanner, Lindsay. Anzac Day and the Commemoration of Australia's War Dead 1916–1929. Unpublished paper, History Department, University of Melbourne, 1979.

Taussig, Mick. 'An Australian Hero'. *History Workshop Journal*, 24, Autumn 1987, pp. 111–33.

Thomson, Alistair. "Back to My Own Native Land": The Political Impact of Returned Servicemen in Brunswick, 1919–1931. Unpublished essay, University of Melbourne, 1982.

——. The Forgotten Anzacs: Radical Diggers Challenge an Australian Legend. Unpublished manuscript, 1986.

——. 'Gallipoli — A Past That We Can Live By?' *Melbourne Historical Journal*, 14, 1982, pp. 56–72.

——. 'The Great War and Australian Memory: A Study of Myth, Remembering and Oral History'. D.Phil, thesis, University of Sussex, 1990.

——. 'History and "Betrayal": The Anzac Controversy', *History Today*, 43, January 1993, pp. 8–11.

——. 'Known Unto God'. *Ormond Chronicle*, 1981, pp. 31–4.

——. 'O Valiant Hearts — How the Mighty Have Fallen'. *Farrago*, 61, 22 April 1983, pp. 10–11.

——. 'Passing Shots at the Anzac Legend'. In *A Most Valuable Acquisition: A People's History of Australia since 1788*, Verity Burgmann & Jenny Lee. McPhee Gribble/Penguin Books, Melbourne, 1988, pp. 189–204.

——. 'A Past You Can Live With: Digger Memories and the Anzac Legend', *Journal of the Australian War Memorial*, April, 1992.

——. 'Remembering Stan D'Altera'. *City of Footscray Historical Newsletter*, 1985, pp. 436–9.

——. 'The Return of the Soldier'. *Meanjin*, 47, 4, 1988.

——. 'Shirkers and Stragglers: An Anzac Imperial Controversy'. *Working Papers in Australian Studies*, 67, Australian Studies Centre, University of London, 1991.

——. 'Shrapnel and Smiles: Memories of a Strange and Bloody Youth'. *Shades*, 1, pp. 11–13, and 2, pp. 11–12, 1983.

——. '"Steadfast Until Death"? C. E. W. Bean and the Representation of Australian Military Manhood'. *Australian Historical Studies*, 23, 93, October 1989, pp. 462–78.

——. '"They Were Finished With You and You Were Finished": Memories of Western Suburbs Veterans of the Great War of 1914–1918'. *City of Footscray Historical Society Newsletter*, 1984, pp. 389–90.

——. 'The "Vilest Libel of the War"? Imperial Politics and the Official Histories of Gallipoli'. *Australian Historical Studies*, 25, 101, 1993, pp. 628–36.

Thomson, Boyd C. *Boyhood's Fancies*. Brown, Prior & Co., Melbourne, 1917.

Throssell, Ric. *For Valour*. Currency Press, Sydney, 1976 (1960).

Todman, Dan. *The Great War: Myth and Memory*. Hambledon and London, London, 2005.

Tognolini, John 'Togs'. 'Red Anzacs'. *Newswit*, 25 May 1988, p. 41.

Travers, Tim. *The Killing Ground: The British Army, the Western Front and the Emergence of Modern Warfare, 1900–1918*. Allen & Unwin, London, 1987.

Turner, Ian. '1914–19'. In *A New History of Australia*, F. K. Crowley (ed.). Heinemann, Melbourne, 1974.

Turner, L. C. F. 'Australian Historians and the Study of War, 1964–74'. In *Historical Disciplines and Culture in Australasia*, J. A. Moses (ed.). University of Queensland Press, St Lucia, 1979.

——. 'Introduction' to Bean, *The Official History of Australia in the War of 1914–1918, Vol. v*. University of Queensland Press, St. Lucia, 1983.

Twomey, Christina. 'Trauma and the Reinvigoration of Anzac: An Argument', *History Australia*, forthcoming, December, 2013.

Tyler, Deborah. 'Making Nations, Making Men: Feminists and the Anzac Tradition'. *Melbourne Historical Journal*, 16, 1984, pp. 24–33.

Tyquin, Michael. *Madness and the Military: Australia's Experience of the Great War.* Australian Military History Publications, Canberra, 2006.

Vader, John. *Anzac.* New English Library, London, 1972.

Ward, Russell. *The Australian Legend.* Oxford University Press, Melbourne, 1958.

Waters, J. C. *Crosses of Sacrifice — The Story of the Empire's Million War Dead and Australia's 60 000.* Angus & Robertson, Sydney, 1932.

Watson, Don. *Caledonia Australis: Scottish Highlanders on the Frontier of Australia.* Collins, Sydney, 1984.

Welborn, Suzanne. *Lords of Death: A People, A Place, A Legend.* Fremantle Arts Centre Press, Fremantle, 1982.

White, Richard. *Inventing Australia.* Allen & Unwin, Sydney, 1981.

——. 'Motives for Joining Up: Self-sacrifice, Self-interest and Social Class, 1914–18'. *Journal of the Australian War Memorial*, 9, October 1986, pp. 3–16.

——. 'The Soldier as Tourist: The Australian Experience of the Great War'. *War and Society*, 5, 1, May 1987, pp. 63–77.

——. 'Sun, Sand and Syphilis: Australian Soldiers and the Orient, Egypt 1914'. *Australian Cultural History*, 9, 1990, pp. 49–64.

Wieland, James. 'The Romancing of Anzac'. *Overland*, 105, 1986.

Wieviorka, Annette. *The Era of the Witness.* Cornell University Press, Ithaca, 2006.

Williams, Jeffrey. Discipline on Active Service: The 1st Brigade, First AIF. 1914–1919. B. Litt. thesis, Australian National University, 1982.

——. The First AIF Overseas. Paper presented at the Australian War Memorial History Conference, 8–12 February 1983.

Wilson, Mary. 'The Making of Melbourne's Anzac Day'. *Australian Journal of Politics and History*, 20, August 1974, pp. 197–209.

Winter, Denis. 'The Anzac Book: A Reappraisal'. *Journal of the Australian War Memorial*, 16, April 1990, pp. 58–61.

——. *Death's Men: Soldiers of the Great War.* Allen Lane, London, 1978.

——. *Haig's Command: A Reassessment.* Viking/Penguin, London, 1991.

Winter, Jay. *Remembering War: The Great War between History and Memory in the Twentieth Century.* Yale University Press, New Haven, 2006.

——. 'Shell-shock and the Cultural History of the Great War'. *Journal of Contemporary History*, 35, 1, 2000, pp. 7–11.

——. *Sites of Memory, Sites of Mourning: The Great War in European Cultural History,* Cambridge University Press, Cambridge, 1995.

——. 'Writing war', Public Lecture, State Library of Victoria, 22 August 2012.

——. & Sivan, Emmanuel (eds). *War and Remembrance in the Twentieth Century.* Cambridge University Press, Cambridge, 1999.

Wohl, R. *The Generation of 1914.* Harvard University Press, Cambridge, Massachusetts, 1979.

Woodward, Jack. 'Anzac and Australian Patriotism'. *Advance Australia*, March 1973, pp. 24–5.

Ziino, Bart. *A Distant Grief: Australians, War Graves and the Great War.* University of Western Australia Press, Crawley, Western Australia, 2007.

——. '"A Lasting Gift to his Descendants": Family Memory and the Great War in Australia'. *History and Memory*, 22, 2, 2010, pp. 125-46.

INDEX

(Note: page reference followed by '*p*' indicates photograph; page reference followed by '*f*' indicates figure)

INDEX